Industrial Energy Conservation

ENERGY, POWER, AND ENVIRONMENT

A Series of Reference Books and Textbooks

Editor

PHILIP N. POWERS

Director, Energy Engineering Center
Purdue University
West Lafayette, Indiana

Consulting Editor
Energy Management and Conservation

PROFESSOR WILBUR MEIER, JR.

Head, School of Industrial Engineering
Purdue University
West Lafayette, Indiana

Additional Volumes in Preparation

Industrial Energy Conservation

MELVIN H. CHIOGIOJI

Professorial Lecturer
George Washington University
Washington, D.C.

MARCEL DEKKER, INC.   New York and Basel

Library of Congress Cataloging in Publication Data

Chiogioji, Melvin H   [Date]
    Industrial energy conservation.

    (Energy, power, and environment; 4)
    Bibliography: p.
    Includes index.
    1. Industry--Energy conservation. I. Title.
II. Series
TJ163.3.C45   670   79-15191
ISBN 0-8247-6809-4

MARCEL DEKKER, INC.
270 Madison Avenue, New York, New York 10016

Current printing (last digit)
10 9 8 7 6 5 4 3 2 1

PRINTED IN THE UNITED STATES OF AMERICA

PREFACE

Until quite recently, our nation had been blessed with abundant domestic supplies of readily available fuels. But because of the growing constraints on the availability of these fuels, the situation is rapidly changing. Since industry currently uses approximately 40% of the total U.S. energy demand, energy conservation offers one of the best opportunities for industry to contribute to the mitigation of the overall energy dilemma and offers a positive approach for countering rising energy costs.

Although gross energy consumption by the industrial sector is increasing, energy use per unit of output has been steadily declining at a 1.6% average annual rate over the past quarter century; at the same time, industrial production has increased by more than 150%. The principal method by which energy use per unit of output was reduced was through substitution of capital-embodying technology for energy. There was, also, within manufacturing, a shift away from the energy intensive industries as a group, and toward the less energy intensive industries. The general belief is that recent sharp increases in energy prices, together with present expected interruptions in supplies of energy, will result in an acceleration in energy savings in manufacturing.

In the short run, approximately 30% of the energy used in industrial processes can be saved through the application of existing techniques that are commonly justifiable. Predicted increases in fuel prices are expected to make energy conservation measures even more attractive in the future.

This text can be used as a ready reference on energy consumption patterns and conservation measures available in typical manufacturing processes or as a college text on industrial energy conservation. The information contained in this text was obtained primarily from open literature sources, government publications and studies, and discussions with other authors and with representatives of trade associations and firms in the industrial sector. Because of the diverse nature of manufacturing processes in each industry, energy usage and savings quoted are, of necessity, averages. However, the information is believed to be accurately represented in supporting the material presented.

Much of the primary works cited and used within this text were obtained as a result of studies which had been initiated and directed by the author. Particular appreciation is expressed to the following organizations and individuals: Gordian Associates for their nine-volume study on energy usage in industry; Arthur D. Little, Inc. for their fifteen-volume study on energy conservation in industry and for their study on electric motors; National Bureau of Standards on waste heat management; KVB, Inc. on boilers; Resource Planning Associates and ThermoElectron Corporation on co-generation; Battelle Columbus Laboratories on primary metals; The Portland Cement Association and George McCord on cement, A. Kouvalis and Stephen Kaplan on pulp and paper; and Virgil Haynes on petroleum refining. Without their extensive studies and analyses, this text would not have been possible.

Finally, I would like to express appreciation to my wife and family for their patience during the countless months while I was less than husband and father. Particular appreciation must be given to my wife, Eli, who assisted me by editing the entire text and providing helpful criticism and suggestions.

Melvin H. Chiogioji

CONTENTS

FIGURES

TABLES

Industrial Energy
Conservation

Chapter 1

THE ENERGY PROBLEM

Energy in sufficient quantities is a necessity for maintaining
U.S. industrial production and for enabling a modern society to do the
things it wants to do. Such goals as meeting the standards of living
for a growing population, upholding national security, improving the
quality of life, and assisting less developed countries can only be
achieved with increasingly large amounts of energy.

Beginning in the 1930s the government adopted measures which have
generally amounted to a de facto policy of encouraging energy consumption. This is evident by the following:

- Sale of low cost and plentiful electric power by Federal agencies
  such as the Tennessee Valley Authority, Bonneville Power Authority, and, later, by the Rural Electrification Authority.

- Control of well head natural gas prices when large natural gas
  fields were developed.

- Importation of large quantities of foreign oil, particularly
  after 1970.

The results of such measures helped to induce the growth and prosperity
of this country. With few incentives to use energy efficiently or to
conserve energy, our use of energy in the past has followed the same
"throw away" philosophy we have practiced for our consumer goods.

The advent of the Arab oil embargo in October of 1973 sent shock
waves through the high energy industrial communities in the United
States, Europe, and Asia. To make matters worse, in January of 1974
the oil producing nations proceeded to double the price of crude oil.
To the energy dependent industrial nations, the dual impact of having
to pay billions more for oil could only have repercussions. The Arab
embargo was a signal, a warning of more serious problems which are sure
to come, but to date, we have yet to feel the full impact of the energy
shortage.

The root cause of the energy situation is that in recent years,
while there has been no limit on growth in energy demand, production
of energy from existing sources has come under mounting pressure. Each
year we fall farther and farther behind in meeting demand through domestic production, creating shortages which can only be met through greater
imports. Although projections of recoverable energy resources vary
considerably, experts agree that the rate of recovery will decrease,
while consumption will continue to steadily increase. As a result, a
shortfall will occur, as tabulated in Table 1-1. This shortfall can
be satisfied by importing oil and natural gas, but the solution is temporary, at best, and is counter to stated national policy.

In a two year study on energy, the Workshop on Alternative Energy
Strategy (WAES) reported the following conclusions:

> 1. The supply of oil will fail to meet increasing
> demand before the year 2000, most probably between 1985
> and 1995, even if energy prices rise 50% above current
> levels in real terms. Additional constraints on oil
> production will hasten this shortage, thereby reducing
> the time available for action on alternatives.

> 2. Demand for energy will continue to grow even if
> governments adopt vigorous policies to conserve energy.
> This growth must increasingly be satisfied by energy
> resources other than oil, which will be progressively
> reserved for uses that only oil can satisfy.

> 3. The continued growth of energy demand requires
> that energy resources be developed with the utmost
> vigor. The change from a world economy dominated by

Table 1-1

United States Energy Shortfall

	10^{15} BTU				
	1971	1975	1980	1985	2000
Domestic Supply					
Natural Gas	21.810	22.640	22.960	22.510	22.850
Petroleum	22.569	22.130	23.770	23.600	21.220
Coal	12.560	13.825	16.140	21.470	31.360
Hydro	2.833	3.570	3.990	4.320	5.950
Nuclear	0.391	2.560	6.720	11.750	49.230
Total	60.163	64.725	73.580	83.650	130.610
Domestic					
Consumption	68.728	80.265	90.075	105.300	162.450
Shortfall to be					
Satisfied by Imports	8.656	15.540	16.495	21.650	31.840
(10^6 BBL Oil)	(1,546)	(2,775)	(2,946)	(3,866)	(5,686)

Source: United States Department of the Interior, "United States Energy Through the Year 2000,"
December 1972; Also See K. C. Hoffman, M. Beller, and A. B. Doernberg, "Current BNL
Reference Energy System Projections: 'Base Case (Sept. 19, 1975),' Brookhaven National
Laboratory, September 1975."

oil must start now. The alternatives require 5 to 15
years to develop, and the need for replacement fuels
will increase rapidly as the last decade of the century
is approached.

4. Electricity from nuclear power is capable of
making an important contribution to the global energy
supply, although worldwide acceptance of it on a sufficiently large scale has yet to be established.
Fusion power will not be significant before the year
2000.

5. Coal has the potential to contribute substantially to future energy supplies. Coal reserves are
abundant, but taking advantage of them requires an active program of development by both producers and consumers.

6. Natural gas reserves are large enough to meet
projected demand provided the incentives are sufficient
to encourage the development of extensive and costly
intercontinental gas transportation systems.

7. Although the resource base of other fossil fuels such as oil sands, heavy oil, and oil shales is very large, they are likely to supply only small amounts of energy before the year 2000.

8. Other than hydroelectric power, renewable resources of energy--e.g., solar, windpower, wave power-- are unlikely to contribute significant quantities of additional energy during this century at the global level, although they could be of importance in particular areas. They are likely to become increasingly important in the 21st century.

9. Energy efficiency improvements, beyond the substantial energy conservation assumptions built in, can further reduce energy demand and narrow the prospective gaps between energy demand and supply. Policies for achieving energy conservation should continue to be key elements of all future energy strategies.

10. The critical interdependence of nations in the energy field requires an unprecendented degree of international collaboration in the future. In addition it requires the will to mobilize finance, labor, research and ingenuity with a common purpose never before attained in time of peace; and it requires it now (1).

While the technologies and their implications for ultimate relief of our energy problem are complex, the basic energy equation is simple: supply must equal or exceed demand. In the event supply is not available to meet demand, then demand must be reduced to regain equilibrium. Our nation is faced, in the short run, with precisely that requirement.

Of the two avenues open to us--increasing supply or reducing demand--only the latter one can be taken quickly. Augmenting supply takes large amounts of capital as well as long lead times. But conservation can begin almost as soon as we decide to use less energy and to use it more efficiently.

HISTORIC ENERGY USAGE

Until quite recently, this nation has been blessed with abundant domestic supplies of readily available fuels. Even today, U.S. energy costs relative to those of other commodities are less than in any other

industrialized country. Figure 1-1 provides an indication of the energy
consumption in the U.S. from 1947 to 1975. During this period, gross
energy consumption increased at an annual rate of 2.8%. Table 1-2 provides this information in tabular form. The chart also indicates that
energy usage has been accelerating in the 1960 to 1973 period, when
energy usage increased at an annual rate of 4.1%. This is relative to
energy production, which only grew at a 3.1% annual rate. As a result
of the disparity between domestic consumption and domestic production,
net energy inflows increased during 1960 to 1973 at an annual rate of
11.9%. See Fig. 1-2.

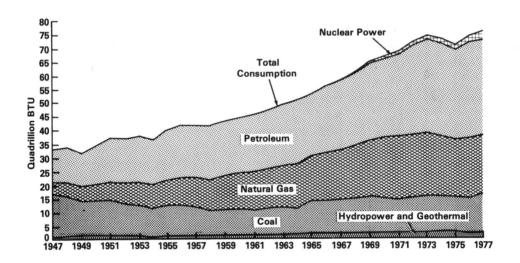

Figure 1-1. Total U.S. Energy Consumption. Source: U.S. Department of the Interior, Energy Perspectives 2 (Washington, D.C.: Government Printing Office, June 1976), p. 62.

Table 1-2
Total U.S. Energy Consumption, 1947-1976

Year	Coal[1] Trillion Btu	Coal[1] Million Short Tons	Natural Gas Trillion Btu	Natural Gas Billion ft³	Petroleum[2] Trillion Btu	Petroleum[2] Million bbl	Hydropower and Geothermal Trillion Btu	Hydropower and Geothermal Billion kWh	Nuclear Power Trillion Btu	Nuclear Power Billion kWh	Total Gross Energy Consumption (Trillion Btu)
1947	15,824	605	4,518	4,366	11,367	1,990	1,326	85.0	—	—	33,035
1948	14,897	570	5,033	4,862	12,558	2,020	1,393	88.5	—	—	33,881
1949	12,631	483	5,289	5,110	12,120	2,128	1,449	96.4	—	—	31,489
1950	12,913	494	6,150	5,942	13,489	2,375	1,440	102.7	—	—	33,992
1951	13,225	506	7,248	7,003	14,848	2,584	1,454	106.6	—	—	36,775
1952	11,868	454	7,760	7,498	15,334	2,671	1,496	112.0	—	—	36,458
1953	11,893	455	8,156	7,870	16,098	2,775	1,439	111.6	—	—	37,586
1954	10,195	390	8,548	8,259	16,132	2,849	1,388	112.0	—	—	36,263
1955	11,540	447	9,232	8,920	17,524	3,100	1,407	120.3	—	—	39,703
1956	11,752	457	9,834	9,502	18,627	3,233	1,487	129.8	—	—	41,700
1957	11,168	434	10,416	10,064	18,570	3,234	1,551	137.0	1	(3)	41,706
1958	9,849	386	10,995	10,623	19,214	3,371	1,636	147.6	2	0.2	41,696
1959	9,810	385	11,990	11,585	19,747	3,481	1,591	145.0	2	0.2	43,140
1960	10,140	398	12,699	12,269	20,067	3,611	1,657	154.0	6	0.5	44,569
1961	9,906	390	13,228	12,750	20,487	3,641	1,680	157.8	18	1.7	45,319
1962	10,189	403	14,121	13,612	21,267	3,796	1,821	172.5	24	2.3	47,422
1963	10,714	423	14,843	14,341	21,950	3,925	1,767	168.6	34	3.2	49,308
1964	11,264	446	15,648	15,118	22,386	4,034	1,907	182.3	35	3.3	51,240
1965	11,908	472	16,098	15,598	23,241	4,202	2,058	196.8	38	3.7	53,343
1966	12,495	498	17,393	16,854	24,394	4,111	2,073	199.0	57	5.5	56,412
1967	12,256	491	18,250	17,685	25,335	4,585	2,344	224.7	80	7.7	58,265
1968	12,659	509	19,580	18,973	27,052	4,902	2,342	225.2	130	12.5	61,763
1969	12,733	516	21,020	20,388	28,421	5,160	2,659	254.5	146	13.9	64,979
1970	12,698	524	22,029	21,367	29,537	5,365	2,650	252.6	229	21.8	67,143
1971	12,043	502	22,819	22,132	30,570	5,553	2,862	273.1	404	37.9	68,698
1972	12,423	523	23,035	22,429	32,966	5,990	2,946	283.9	576	54.0	71,946
1973	13,294	562	22,712	22,245	34,851	6,317	2,998	288.6	888	83.3	74,743
1974	12,889	545	21,732	21,290	33,468	6,066	3,295	317.3	1,215	113.9	72,600
1975	12,813	542	19,948	19,540	32,742	5,934	3,215	309.6	1,839	172.5	70,557
1976	13,752	581	20,372	19,950	35,087	6,359	3,068	295.4	2,037	191.1	74,316

[1] Includes anthracite, bituminous coal, and lignite.
[2] Includes domestically produced crude oil, natural gas liquids, and condensate, plus imported crude oil and products.
[3] Less than 0.1.
[4] Includes coke net imports.

Source: U. S. Department of the Interior, Energy Perspectives 2, p. 63; Dept. of Energy, Monthly Energy Review for 1974, 1975 and 1976.

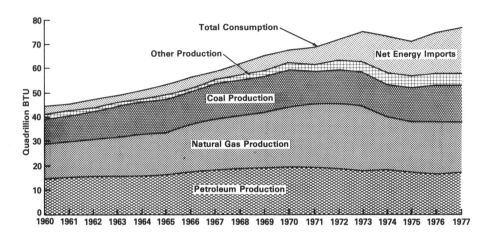

Figure 1-2. U.S. Energy Consumption, Production and Net Imports. Source: U.S. Department of Interior, Energy Perspectives 2, p. 22.

The pattern of energy types used by the U.S. economy has shifted over the last 150 years, as seen in Fig. 1-3. In 1850, wood supplied about 90% of America's energy; however, it was used inefficiently. With the decline in the usage of fuel wood, coal came into prominence, and with it, energy usage efficiency increased. In the 1920s, oil and natural gas usage came into prominence, spurred by its ease of utilization, its cost, and its less severe impact on the environment. With the decline in the availability of oil and natural gas, increased usage of coal must be pursued, and alternative fuel sources, such as nuclear, solar, and synthetic fuels, will become essential.

ENERGY AND GNP

There has been great speculation regarding the relationship of energy consumption and gross national product (GNP). Indeed, the rates of total energy consumed in BTUs to GNP in constant dollars--the energy/GNP rate--has been used extensively in summarizing the relation

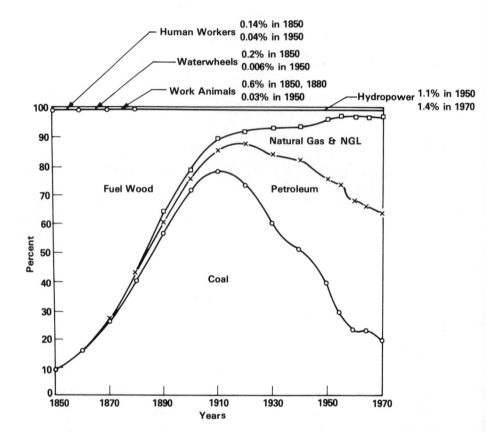

Figure 1-3.   Energy Input Shares, U.S. Economy.   Source:
E. Cook, "Energy Flow Through the United States Economy" (College
Station, Texas: Texas A & M, December 1975), p. 7.

between real economic growth and growth in the consumption of energy inputs. In its "Energy to the Year 1985" report, the Chase Manhattan Bank
expresses this viewpoint forcefully:

> It has been recommended in some quarters that the United
> States should curb its use of energy as a means of alle-
> viating the shortage of supply. However, an analysis of
> the uses of energy reveals little scope for major reduc-
> tions without harm to the nation's economy and its standard
> of living. The great bulk of the energy is utilized for
> essential purposes--as much as two-thirds is for business

related reasons. And most of the remaining third
serves essential private needs. Conceivably, the
use of energy for such recreational purposes as vacation
travel and the viewing of television might be reduced--
but not without widespread economic and political repercussions. There are some minor uses of energy that could
be regarded as strictly non-essential--but their elimination would not permit any significant savings (2).

Figure 1-4 also relates this close correlation between a nation's
GNP per capita and its per capita energy consumption. But the relationship between energy and the GNP is complex. For the last half a century,
the amount of energy required per unit of GNP has generally decreased,
despite the declining real cost of energy during that period (see Table
1-3 and Fig. 1-5). This has been largely due to technological innovation and to shifts in the kind of outputs comprising the GNP. In 1947,
115,600 BTUs were needed per dollar of GNP. By 1960, this figure was
reduced to 92,700, and by 1974 only 79,800 BTUs were associated with
each dollar of GNP.

In a recent study of energy use and economic growth, the Conference
Board found that "the link between them is more elastic than is commonly assumed, provided time is allowed for the necessary adjustments in
production and consumption that will permit less energy to be consumed
per unit of product (3)." Economist John Myers suggests that in light
of the rapid rise in new energy costs these last few years, the rate
could fall by 2% per year for the foreseeable future without adversely
affecting the economy (4).

The conclusion thus drawn is that while studies of the recent past
indicate a tight coupling of energy consumption and GNP, longer range
indications are that their degree of coupling is weakening. We can move
forward in our programs to increase the efficiency of energy use in
industrial processes without fear of economic stagnation.

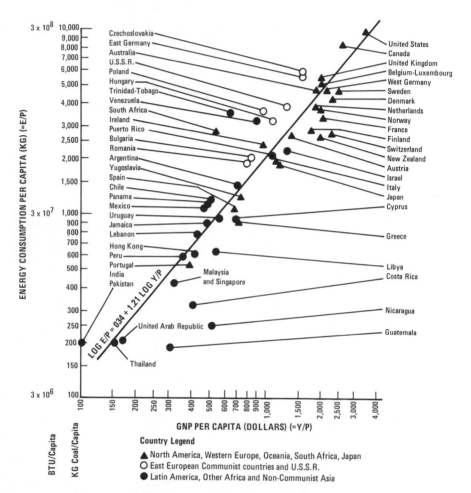

Figure 1-4. Energy Consumption, Population, and GNP.
Source: Leon Glickman and David White, Energy Conservation
Through Effective Energy Utilization, NBS Publication 403
(Washington, D.C.: Government Printing Office, June 1976),
p. 23.

Table 1-3

Industrial Energy Consumption per Dollar of Value
Added to Gross National Product, 1947-74

Year	Value added in manufacturing sector[1] (billion 1958 dollars)	Industrial fuel use of energy (trillion Btu)	Fuel use of energy per dollar of value added (thousand Btu)	Industrial nonfuel use of energy (trillion Btu)	Nonfuel use of energy per dollar of value added (thousand Btu)	Total industrial energy inputs (trillion Btu)	Energy input per dollar of value added (thousand Btu)
1947	114.9	12,224	106.4	1,057	9.2	13,281	115.6
1948	121.5	11,775	96.9	1,047	8.6	12,822	105.5
1949	115.0	10,900	94.8	954	8.3	11,854	103.1
1950	131.3	11,902	90.6	1,007	7.7	12,909	98.3
1951	146.0	13,235	90.7	1,119	7.7	14,354	98.3
1952	150.7	12,754	84.6	1,026	6.8	13,780	91.4
1953	161.2	13,549	84.1	1,019	6.3	14,568	90.4
1954	149.6	12,286	82.1	1,049	7.0	13,317	89.0
1955	165.8	13,912	83.9	1,163	7.0	15,075	90.9
1956	166.9	14,456	86.6	1,245	7.5	15,701	94.1
1957	167.8	14,327	85.4	1,309	7.8	15,636	93.2
1958	153.3	13,341	87.0	1,268	8.2	14,609	95.3
1959	170.7	13,841	81.1	1,414	8.3	15,255	89.4
1960	172.0	14,443	84.0	1,505	8.8	15,948	92.7
1961	171.2	14,430	84.3	1,507	8.8	15,937	93.1
1962	186.2	15,088	81.0	1,563	8.4	16,651	89.4
1963	201.0	15,700	78.1	1,672	8.3	17,372	86.4
1964	215.7	16,533	76.6	1,689	7.8	18,242	84.6
1965	235.1	17,066	72.6	1,744	7.4	18,810	80.0
1966	254.0	17,916	70.5	1,900	7.5	19,816	78.0
1967	254.1	17,891	70.4	2,207	8.7	20,098	79.1
1968	268.4	19,022	70.9	2,385	8.9	21,407	79.8
1969	276.2	19,445	70.4	2,817	10.2	22,262	80.6
1970	260.6	19,788	75.9	2,648	10.2	22,436	86.1
1971	264.1	19,578	74.1	2,810	10.6	22,388	84.8
1972	288.8	20,022	69.3	3,085	10.7	23,107	80.0
1973	309.5	20,016	64.7	3,235	10.5	23,251	75.1
1974	293.1	20,174	68.8	3,215	11.0	23,389	79.8

[1] Bureau of Economic Analysis, Department of Commerce

Source: U.S. Department of the Interior, Bureau of Mines, Division of Interfuels Studies.

PATTERNS OF ENERGY USE IN INDUSTRY

The industrial market is complex and involves the interaction of
the requirements of each type of industry with the characteristics of
each type of fuel. For example, industries such as iron and steel
need special fuels such as metallurgical coal, and industries such as
the glass industry require controlled heat in their heat treating furnaces. These requirements lead to the necessity of using particular

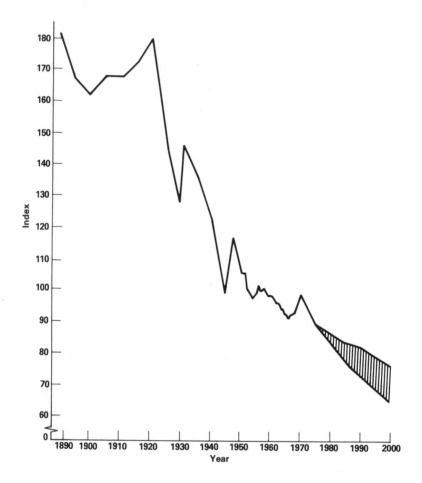

Figure 1-5. Energy/GNP Ratio 1890 to 2000 (1958 = 100).
Source: Richard Williamson, Marcel Grunspan, Charles Mandelbaum,
Robert Bohn, and Martha Snyder, "Analysis of Energy Futures for the
U.S.," ERHQ 18 (Washington, D.C.: ERDA, 1977), p. 16.

fuels, such as low sulfur coal in the case of iron and steel, and gas
in the case of the glass industry. In order to determine what can be
done to improve the effectiveness of energy utilization, we must have
information on the patterns of energy usage by the various industries
within the manufacturing sector.

Energy Consumption by Fuel Type

The manufacturing sector is the leading energy consumer among all economic sectors in the United States. In 1975, about 21 quadrillion BTUs (quads), or 28% of the nation's energy use, was consumed in this sector on fuels, feedstocks, and electric power.

Figure 1-6 shows the change that has occurred in the total fuel, power, and feedstock mix for industry over the last several years. Although this chart only goes back to 1947, we know that fifty years ago approximately 80% of industrial fuel for energy use was derived from coal. By 1947 the use of coal was down to 54% of the total, and subsequently, in the last twenty-five years, there has been an even more rapid decrease in the use of coal.

There are two specific observations from Fig. 1-6 that require comment. The very large increase in natural gas usage from 22% in 1947 to 43% today has come about because of the fact that natural gas is an almost ideal fuel. Moreover, thanks primarily to government regulation, natural gas is the least expensive fuel, both for the general public and

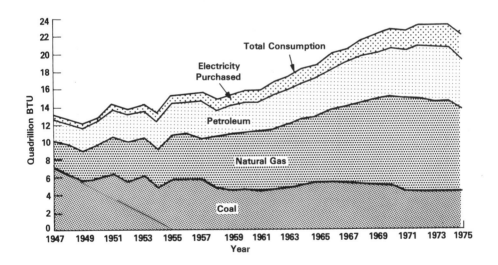

Figure 1-6. Energy Consumption by the Industrial Sector.
Source: U.S. Department of the Interior, Energy Perspectives 2,
p. 66.

for industry. Secondly, the electricity consumption by industry has
more than quadrupled in this twenty-five year period, indicating the
very rapid growth rate that electricity is experiencing.

Between 1947 and 1975, total energy consumption increased at an annual rate of 1.8%; however, the majority of this increased usage was in
oil and natural gas, with coal usage decreasing at an annual rate of
1.9%. Since 1958, energy consumption has increased at a 2.1% annual
rate. Total fossil fuel consumption in 1975 accounted for 86% of energy consumption by the industrial sector for fuel uses.

The large energy users in manufacturing are the chemicals industry,
the primary metals industry, the petroleum refining and coal products
industry, the stone, clay, and glass industry, the paper industry, and
the food industry. Table 1-4 provides a historical summary of energy
usage in these industries, as compiled by the Conference Board.

Table 1-5 presents information by industry and fuel type of 1974
BTU energy consumption in the manufacturing sector. As is evident, the
dependence of these industries on different forms of energy varies considerably. Approximately three quarters of the energy used by the steel
industry is coal. The petroleum industry, on the other hand, uses little coal but is heavily dependent on internally generated process off
gases and natural gas. The primary metals industries, other than steel
and aluminum, depend on natural gas for more than one half of their energy, as do the food processing and stone, clay, and gas industries. The
paper industry is the least dependent on natural gas of all the major
industries but is heavily dependent on fuel oil. The large majority of
this industry's energy consumption, however, is accounted for by forest
fuels (residual pulping liquors, hogged wood, and bark). In the chemicals industry, natural gas, LPG, and heavy liquid feedstocks are the
dominant fuels. Aluminum and many of the remaining small industries rely
most heavily on natural gas and electricity for energy. We can see,
then, that the fuel requirements for the industrial sector is a very
complex equation and cannot be dealt with in simple terms.

Table 1-4

Gross Energy Consumed by High-Energy-Using
Manufacturing Groups, Selected Years 1947-1980
(trillion BTU's)

	1947	1954	1958	1962	1967	1971	1975	1980
Purchased by All Manufacturing Plus Energy Produced and Consumed in the Same Establishment ("Captive Consumption") by SICs 2911 and 3312	10,535	11,934	13,057	14,941	18,264	19,864	24,167	27,763
Purchased by All Manufacturing	8,738	9,766	10,696	12,485	15,463	17,060	20,919	24,383
By Six High Energy-Using 2-Digit Groups								
Food and Kindred Products (SIC 20)	857	883	951	992	1,098	1,284	1,284	1,453
Paper and Allied Products (SIC 26)	635	801	932	1,068	1,367	1,560	1,725	1,771
Chemicals and Allied Products (SIC 28)	1,023	1,753	2,282	2,592	3,257	3,473	4,997	6,020
Petroleum and Coal Products (SIC 29)	550	695	1,018	1,252	1,543	1,820	2,083	2,302
Stone, Clay, and Glass Products (SIC 32)	929	1,032	1,058	1,178	1,341	1,444	1,617	1,864
Primary Metal Industries (SIC 33)	2,547	2,499	2,182	2,833	3,340	3,364	4,031	4,655
Sum of Six Groups	6,541	7,663	8,423	9,915	11,946	12,947	15,737	18,065
By All Other Manufacturing	2,197	2,103	2,273	2,570	3,517	4,113	5,182	6,318
Captive Consumption								
By Petroleum Refining (SIC 2911)	724	1,115	1,081	1,159	1,144	1,223	1,268	1,287
By Blast Furnaces and Steel Mills (SIC 3312)	1,073	1,053	1,280	1,297	1,657	1,581	1,980	2,093

Source: From ENERGY CONSUMPTION IN MANUFACTURING, Copyright
1974, The Ford Foundation. Reprinted with permission of Ballinger
Publishing Company.

Energy Consumption by Process Application

In order to further understand the energy consumption patterns in
the industrial sector, it is important to understand the applications
of energy in the industrial processes. Industrial energy applications
can be broadly classified into the following major types:

- Space conditioning--direct fired, where the fuel is
  directly burned for space conditioning (i.e., not
  burned under a boiler)

Table 1-5

1974 BTU Consumption in the Manufacturing Sector
by Industry and Fuel Type (10^{12} BTU)

Industry/SIC Fuel	Coal	Oil	Natural Gas	Electricity	Other	Total
Food 20	75.3	132.3	475.6	126.7	124.8	934.7
Tobacco 21	5.5	5.2	4.5	3.5	.3	19.1
Textiles 22	22.0	62.7	102.1	91.8	38.1	316.7
Apparel 23	1.0	5.7	15.4	25.0	17.9	64.9
Lumber, Wood 24	2.8	42.4	72.9	50.5	84.7	253.3
Furniture 25	2.8	7.6	25.3	14.2	9.1	59.1
Paper 26	208.8	576.4	414.3	132.7	891.6	2223.8
Pulp, Board 26	204.4	556.7	374.9	103.6	865.2	2104.7
Other Paper 26	4.5	19.7	39.4	29.1	26.4	119.0
Printing 27	.4	7.2	31.4	31.5	18.2	88.7
Chemicals 28	322.2	2085.4	2092.6	436.9	274.0	5211.0
Petroleum 29	5.3	745.3	2154.5	83.7	83.2	3071.9
Rubber 30	29.6	44.0	86.6	64.7	26.0	250.8
Leather 31	1.3	6.3	5.2	5.5	3.5	21.8
Stone & Clay 32	233.0	125.5	696.3	99.6	146.9	1301.3
Cement 32	-1	-1	-1	-1	-1	-1
Other Stone 32	-1	-1	-1	-1	-1	-1
Primary Metals 33	2639.2	418.7	1284.3	536.3	-76.1	4802.3
Steel 33	2533.4	265.5	606.6	50.5	-92.2	3421.4
Aluminum 33	67.2	107.5	229.6	220.1	0	624.4
Other Metals 33	38.6	45.9	403.5	143.8	124.8	756.5
Fabric. Metals 34	11.1	38.9	208.2	88.4	60.7	407.2
Machinery 35	20.1	33.5	164.3	90.7	52.5	361.2
Electric. Mach. 36	13.2	21.4	96.9	84.2	31.0	246.8
Transpor. Equip. 37	47.6	40.5	144.1	97.1	41.1	370.3
Meas. Equip. 38	-1	12.7	15.8	15.6	6.4	50.4
Misc. Indus. 39	.8	8.5	18.8	13.4	9.0	50.5
Other Indus.	-1	-1	-1	-1	-1	-1
Total All Indus.	3642.1	4420.0	8109.0	2091.9	1842.9	20105.9

Source: Federal Energy Administration, Energy Consumption in the Manufacturing Sector, 1977.

- Boiler fuel--further divided into space conditioning and process energy, depending on how the steam from the boiler is utilized

- Direct process heat--for kilns, reheat furnaces, etc., excluding energy used in boiler steam

- Feedstock--where the fuel is used as an ingredient in the process

- Lighting

- Mechanical drive--as in motors, used for crushers, grinders, production lines, etc.

Table 1-6 indicates the functional uses of energy in the manufac-turing sector and the fuel types used. Natural gas, other gases, and

residual oil were the most important fuels used in direct heat applications in manufacturing. Natural gas, residual oil, coal, and "other
energy" (e.g., process residuals) constituted a high proportion of the
over four quads of energy used for process steam in 1974. Over 90% of
the LPG consumed by the manufacturing sector was for feedstock. Other
important feedstocks included coke, natural gas, and heavy liquids.

Industrial Production and Energy Use

The major industrial firms have all achieved a general decline in
energy used per unit of output over the last two decades. The Conference
Board, for example, reports that energy use per unit of product declined
at a 1.6% average annual rate from 1954 to 1967 (5). The Board indicates
that while total manufacturing output rose 87%, total energy use rose
only 53%. Similarly, R. W. Barnes, who has made an analysis of the period 1947 to 1973, reports that industrial production increased in this
twenty-six year period by more than 150% (6). However, the industrial
energy growth was less than 100%, even though during much of this period
of time, the real price of industrial energy, compared to other factors
of production, was decreasing. Analyzing this further, he found that
for the totality of U.S. industry the energy used per unit of production
decreased at a rate of 1.3% per year from 1947, and that for the top six
industries energy use was reduced about 1.5% per unit of production.
Thus, we can see that over this period of time, a very substantial amount
of energy conservation had been achieved in spite of the decreasing real
price of energy. Industry has not been entirely wasteful in its use of
this "cheap" energy.

The principal method by which energy use per unit of output was
reduced was a substitution of capital embodying technology for energy.
There was also a shift within manufacturing away from the energy intensive industries as a group and toward the less energy intensive industries. The former are mainly basic materials producers, so this shift
is part of the long term historical development toward higher degrees
of fabrication and, as such, has contributed to the decline in the energy
output ratio for all manufacturing.

Table 1-6

1974 Btu Consumption in the Manufacturing Sector by Application and Fuel Type (10^{12} Btu)

Fuel	Process Steam	Direct Heat	Indirect Heat (Non-Steam)	Mechanical Drive	Internal Electrical Generation	Electrolytic	Feedstock	Other NSK/NEC	Total
Coal	544.8	237.6	0	0	228.0	0	0	263.9	3642.1
Oil	706.6	425.6	710.7	46.3	114.0	0	1882.5	517.3	4220.0
Natural Gas	1974.1	1727.9	2008.7	137.4	429.7	0	475.4	1582.9	8109.0
Electricity	0	63.7	39.0	809.8	-292.5	373.8	0	1013.0	2091.9
Other	1005.7	65.8	25.3	11.7	22.4	13.0	1518.1	935.3	1842.9
Total	4231.1	2520.6	2783.8	1005.1	501.5	386.9	3876.0	4312.5	20105.9

Source: Federal Energy Administration, Energy Consumption in the Manufacturing Sector, 1977.

One point which must be seriously considered in any evaluation of energy used within industry is the fact that there is a cyclical effect in energy use per unit of output. If capacity utilization is low, energy use per unit will be high. Also, when capacity utilization is very high, energy use per unit of output is usually high. There is an optimum level at which energy use per unit of output is minimized. Thus, if cyclical patterns are ignored, erroneous conclusions regarding energy usage may be easily drawn.

Another important factor in determining energy use per unit of output within an industry is the product mix. Energy use per unit can rise or fall, according to whether energy intensive products become more or less important in the product mix. Rising energy prices may, but will not always, lead to a change in product mix toward less energy intensive products. There is a general belief that recent sharp increases in energy prices, together with present expected interruptions in supplies of energy, will result in an acceleration in energy savings in manufacturing.

WHY ENERGY CONSERVATION?

More than half the current U.S. energy budget is waste. The recent rapid escalation of fuel costs, the instability of supplies and predicted future limitations have focused attention on the need for a complete "rethink" on fuel and energy usage, with particular emphasis on energy conservation. Energy conservation as used here means a reduction of energy waste, the increased efficiency of energy utilization. Conservation should not be confused with curtailment.

Energy conservation will allow the earth's limited resource base of high quality fuels to be stretched further. It will also allow a portion of the fossil fuel base to be reserved for nonenergy purposes: drugs, lubricants, and other materials. The fact is that we may be faced with a continuing shortage of fuel and power unless new sources are developed in the future. Part of the shortage must be compensated for by conservation measures, or major dislocations and even catastrophy may be in store for some companies and institutions.

In addition, a report by the Central Intelligence Agency has indicated that in the absence of greatly increased energy conservation, projected world demand for oil will approach productive capacity by the
early 1980s and substantially exceed capacity by 1985. Thus oil importing countries' ability to obtain oil will be seriously jeopardized. In
addition, the report indicates that prices will rise sharply to ration
available supplies (7).

Our way of life is rife with opportunities for conserving energy.
The U.S. energy budget can be gradually cut significantly without alter-
ing the nation's standard of living. Some experts have indicated that
much of the U.S. energy needs in the next quarter century can be met
simply by improving the efficiency of existing uses. This has yet to
be proved, yet, it can be stated with certainty that "a barrel of oil
saved is more valuable than a new barrel of oil produced."

POTENTIAL FOR INDUSTRIAL ENERGY CONSERVATION

If we look at an industrial plant as a system (see Fig. 1-7), we can
observe that on one side we have inputs of energy raw materials and labor
and on the other side, the output of goods, waste energy, and waste
materials. In order to maximize profits, a manager attempts to keep
the costs of the inputs as low as possible.

In the past in many cases, since the cost of energy was low in relation to the other inputs, it was ignored. However, with today's spi-
raling energy prices, more attention must be given to energy input.
Energy savings can occur either by improving the energy conversion processes, by recycling the waste energy, or by reusing the waste materials.
Many opportunities exist for the application of existing technology to
yield large savings, but in order to identify these areas for savings,
answers must be found to two primary questions:

> (1) What are the areas of activity in which there
> may be significant potential for the better use of energy?

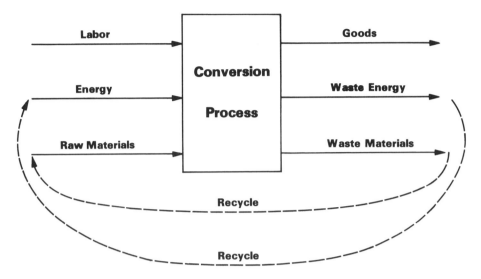

Figure 1-7. Industrial Conversion System.

(2) Within these areas, what are the specific measures of alternative options that could lead to better
and more efficient use of energy?

For the manufacture of any given set of products, a certain minimum
consumption of energy is required, depending on basic factors such as
the raw materials selected, the manufacturing process adopted, and the
level of production to be maintained. Beyond this threshold value, an
economic balance should be struck between the incremental cost of more
energy efficient equipment or techniques and the value of the energy
that can be saved by these means. In an era of cheap, easily available
energy supplies and abundant sources of cooling water, energy economics
and energy conservation have not necessarily been synonymous or even
compatible. Although some of the more intensive industrial users of
energy, including chemicals, paper, and petroleum refining, have long
found it competitively advantageous to design for energy conservation,
instances abound where until recently, the savings realizable from the
use of energy recovery equipment would not offset the cost of its installation. In this era of uncertain energy supply and steeply rising
fuel costs, however, energy conservation must become not only a corporate virtue but an object of national concern.

Many different factors affect energy use for particular end use applications. Among the most important of these are:

- Capital and fuel costs

- Operating and maintenance costs

- Process technology

- Equipment reliability

- Fuel supply availability

- Job requirements

- Space requirements

- Regulatory factors (environmental, safety, etc.)

To determine whether it is possible to reduce our demand for fuels significantly by improving the efficiency of fuel utilization, one must study the processes which are ultimately responsible for fuel consumption and the efficiency with which fuels are used in these processes.

Estimates of the potential for energy conservation in the industrial sector can be described as tenuous, at best. Although we know how much energy is consumed by different types of industry, it is difficult to estimate how much energy could be saved by energy conservation design of industrial process equipment. Even more important, we have no means of predicting the response of industry to energy conservation programs, whether voluntary or mandatory.

Because of the low cost of energy (only about 3% to 5% of producer's price) it is presumed that industry simply does not strive to use energy efficiently in its production processes. Dr. Monty Finnister, Chairman of the British Steel Corporation, stated that less than 50% of all the energy consumed throughout the world is used effectively--the rest is wasted in conversion losses, heat radiation, cooling water, and in other ways. About 55% of the energy used in the steel industry is applied efficiently, which means that 45% is dissipated. The electricity supply industry uses about 30% of the energy inherent in fossil fuels, so that nearly 70% is wasted. In transportation, the position is even worse-- only 25% of the energy intake is effectively used, while 75% is wasted.

Other industries which do not use energy in a primary but in a secondary
form for operating machinery have better records, with an estimated 75%
efficiency (8).

Charles Berg indicates that it would not be unreasonable to assume that energy savings of approximately 30% might be realized
through application of present day energy conservation technologies to
industrial practices (9). The effectiveness with which energy is used
in industry varies greatly, depending upon the nature of the industry
and the size of the plant. He also states that the invention of more
efficient devices, more efficient processes (e.g., cement making, petroleum refining, chemical processing), and especially the institution
of a methodology for waste heat management in plants may be expected
to yield further energy savings in industry beyond the estimated 30%.

A number of projections on our energy conservation potential have
been made. The Petrochemical Energy Group (PEG), a consortium of firms
operating in that area, has made an estimate that the petrochemical industry as a whole can reduce its current level of fuel consumption per
unit of output by perhaps 7-12%, with some further gains of perhaps
5-10% possible over the long term as more efficient processes are developed and new equipment is installed. However, actual reductions
achieved by three of the larger companies--DuPont, Dow, and Union Car-
bide--exceed this estimate.

EXXON has projected that industrial demand reduction will be 19% by
1980 and 30% by 1990 (10). David Freeman has projected major potential
energy savings in the industrial sector in his technical fix scenario
(see Table 1-7). Projected savings in 1985 amount to 10.2 quads, and
in the year 2000, savings could be about 30 quads.

Gyftopoulos et al., in a study for the Ford Foundation Energy project, has determined that with existing technology, it is possible to
reduce the specific fuel consumption in the iron and steel, petroleum
refining, paper, aluminum, copper, and cement industries by one-third
(11). They indicate that if it were realized, such a reduction would
approximately offset the fuel needs for the growth of industry that is
projected for the remainder of the 1970s. Table 1-8 summarizes the specific fuel consumption and the specific fuel savings that might be

Table 1-7

Potential Energy Savings in the Industrial Sector
Technical Fix vs. Historical Growth
(Quadrillion BTU's)

	1985	2000	
Industrial energy use in HG scenario	46	87	
Potential Savings			Conservation Measures
Five energy intensive industries	4.3	13.1	More efficient production processes in paper, steel, aluminum, plastics and cement manufacture.
Miscellaneous process steam	0.5	3.5	Onsite industrial co-generation of steam and electricity.
Miscellaneous direct heat	2.9	5.4	Use of heat recuperators and regenerators with direct use of fuels instead of electric resistive heat.
Other	2.5	7.4	
Total savings	10.2	29.4	
Industrial energy use in TF scenario	36	58	

Note: Only the manufacuring sector's share of energy processing losses is included above.

Source: From A TIME TO CHOOSE: AMERICA'S ENERGY FUTURE. Copyright
1974. The Ford Foundation. Reprinted with permission of Ballinger Publishing Company.

achieved in each of the six industries, assuming widespread application
of the best technology existing today both in the U.S. and abroad.
In addition, Table 1-8 summarizes the overall fuel consumption and fuel
savings that might be achieved in the six industries if the product outputs were those of 1968. It should be recognized that the projected
savings represent an upper limit and that actual savings which can be
realized in practice will be somewhat lower.

Table 1-9 lists values of specific fuel consumption (BTU/ton) for
several industries, based upon 1968 practices, upon suggested practices
using currently demonstrated technology, and upon the theoretical limits

Table 1-8

1968 and Improved Fuel Consumption for Selected U.S. Industries

Industry	Specific Fuel Consumption (10⁶ Btu/ton)		Percentage Improvement in Specific Fuel Consumption Over 1968 Practices	Industry Output (1968) (10⁶ tons/yr)	Total Fuel Consumption (10¹⁵ Btu/yr)	
	With 1968 Practices	With Potential Process Improvements (Technology Existing in 1973)			With 1968 Practices	With Potential Improvements (1973 Technology)
Iron and Steel	26.5	17.2	36%	131	3.47	2.26
Petroleum Refining	4.4	3.3	25%	590	2.6	1.95
Paper and Paperboard	39.0*	23.8*	39%	50	1.95	1.2
Aluminum (Primary and Scrap)	155.0	106.0	32%	4.07	0.63	0.43
Copper	25.8	18.1	33%	3.1	0.08	0.05
Cement	7.9	4.5	43%	72.0	0.57	0.32
TOTALS					9.3	6.2
FUEL SAVING						33%

Source: From POTENTIAL FUEL EFFECTIVENESS IN INDUSTRY. Copyright
1974. The Ford Foundation. Reprinted with permission of Ballinger
Publishing Company.

computed by means of the concept of available useful work. The large
margins that exist between current practices and minimum theoretical
requirements indicate the potential which is available for major long
term reductions in fuel consumption through basic process modifications.

The steady state heat transfer efficiency of individual items of
plant equipment using indirect heat operations ranges from 20-30%, and
some plant systems used in these operations have been reported to operate with efficiencies as low as 5%. Technology applied to the improvement of fuel utilization in these operations might yield significant reductions in the consumption of primary fuels, but the technological com-
munity has not yet responded to this opportunity. Figure 1-8 depicts
the efficiency of industrial processes as a function of the temperature
of the heat required.

Table 1-9

Comparison of Specific Fuel Consumption of Known Processes
with Theoretical Minimum for Selected U.S. Industry

	1968 Specific Fuel Consumption (Btu/ton)	Potential Specific Fuel Consumption Using Technology Existing in 1973 (Btu/ton)	Theoretical Minimum Specific Fuel Consumption Based Upon Thermodynamic Availability Analysis (Btu/ton)
Iron and Steel	26.5×10^6	17.2×10^6	6.0×10^6
Petroleum Refining	4.4×10^6	3.3×10^6	0.4×10^6
Paper	$39.0 \times 10^6*$	$23.8 \times 10^6*$	Greater than -0.2×10^6[†] Smaller than $+0.1 \times 10^6$
Primary Aluminum Production**	190×10^6	152×10^6	25.2×10^6
Cement	7.9×10^6	4.7×10^6	0.8×10^6

*Includes 14.5×10^6 Btu/ton of paper produced from waste products consumed as fuel by
  paper industry.
**Does not include effect of scrap recycling.
[†]Negative value means that no fuel is required.

Source: From POTENTIAL FUEL EFFECTIVENESS IN INDUSTRY. Copyright
          1974. The Ford Foundation. Reprinted with permission of Ballinger
          Publishing Company.

Many opportunities exist for the application of existing technology to the management of waste heat so as to yield large savings. Energy cascading, or utilizing the thermal output of a high temperature process as input to a lower temperature process, is a highly efficient approach. Regeneration or recuperation, where the exhaust energy is recycled back as input energy for the same process, is an example. The waste heat can also be efficiently used for space heating and cooling. It is also possible to convert waste thermal energy into mechanical or electrical energy. Recuperators can cut fuel consumption of radiant tube heat treating furnaces by as much as 30%. Bottoming cycle Rankine engines can extract up to 50 KWH of electricity per million BTUs of enthalpy in waste heat at $700°F$ or higher.

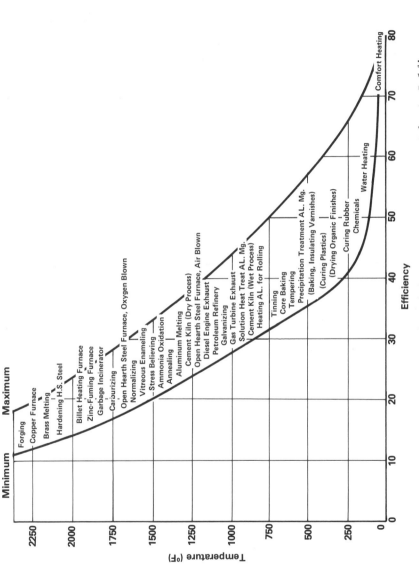

Figure 1-8. Efficiencies for Industrial Processes. Source: John A. Belding and William M. Burnett, From Oil and Gas to Alternate Fuels: The Transition in Conversion Equipment (Washington, D.C.: Energy Research and Development Administration, 1976.

Ideally, every process unit in industry should function near optimum efficiency and fully utilize all waste energy generated within the
process itself. Every major process unit must, therefore, be critically
examined to determine the magnitude and location of energy losses occurring therein and to identify potential process modifications or alternatives which will enhance overall energy utilization efficiency.
Priorities can then be established to provide a proper relationship between the application of energy conservation effort and the potential
benefit of such effort.

INTERNATIONAL INDUSTRIAL ENERGY EFFICIENCIES

Many countries use more energy efficient processes than does the
U.S. for industries ranging from the refining of ores to the manufacturing of hard products. Table 1-10 provides some OECD data related to
energy consumption for selected industries. There are great differences
in energy consumption from country to country in some energy intensive
industries.

The principal causes of the West German and Swedish advantages in
the industrial sector are the somewhat more modern equipment and processes associated with more recent construction, process designs based
on high priced energy, greater use of waste materials for fuel, and
more recycling of used products. West Germany and Sweden have been and
still are more dependent on foreign oil for their total energy needs
than the United States. This a political disadvantage, but it entails
a saving in the energy that is consumed in the production of energy.
There is also a saving in the manner of generating electricity.

In the electric power generation sector, both Sweden and West
Germany enjoy advantages related to the greater use of steam from back-
pressure turbines for process heat and space heating. Some utilities
in Sweden provide both district heat and electricity. In West Germany,
many industries use large amounts of process steam from their own power

Table 1-10

Some OECD Data Related to Energy Consumption

| Country | Sectorial Energy Consumption, 1975 | | Specific Efficiencies | | | | |
| | Industry | Transport | Crude Steel | Pulp and Paper | Cement | Petroleum Products | Aluminum |
	% TPE*		10^4 kcal/ton of Product				
Austria	30.7	19.2	450	344	90	–	1346
Belgium	41.2	11.8	–	–	–	–	–
Canada	26.0	19.1	555	673	95	180	–
Denmark	16.8	17.9	236	353	164	53	–
Italy	34.4	15.4	334	340	96	45	–
Japan	47.3	–	513	512	120	46	1385
Luxembourg	71.0	–	701	–	–	–	–
Netherlands	32.5	10.0	470	–	131	–	1290
Norway	33.9	12.2	189†	659	115	–	1591
Sweden	31.6	12.3	398	489	140	49	1648
Switzerland	18.3	17.8	–	557	–	–	–
United Kingdom	28.5	14.3	478	627	138	73	2107
United States	27.7	25.0	543	579	161	90	947
West Germany	36.1	13.4	326	438	91	80	1481–1503

* TPE = total primary energy

†Electricity only

Source: IGT Highlights, Supplement, April 11, 1977.

plants to generate electricity, often in quantities greater than needed; the excess power is sold to utilities. Little use has been made of co-generation in the U.S.

These statistics generally indicate that there are significantly more energy efficient processes used in foreign countries and that their application in U.S. industry could result in significant energy efficiency improvements.

BARRIERS TO ENERGY CONSERVATION

Although the potential for energy conservation is great, there are many barriers which mitigate against the utilization of the most energy efficient processes. These barriers include the facts that:

- Energy costs continue to be low per unit of product.

- More attractive investments are available.

- Regulations force investments in other areas, e.g., OSHA, environmental regulations.

- The decision process is often fragmented.

- There is a basic conservatism on the part of industry.

- Industry is more willing to satisfice than optimize in energy usage.

- There is a basic lack of knowledge about energy conservation opportunities on the part of industry.

In this text, an attempt will be made to define areas for energy conservation which are economic and which lead to greatly improved energy efficiency in industry.

CONCLUSION

Three thousand years ago, Cassandra told the people of Troy that "the city will fall." No one listened because that was all she said. Prophecy is an idle art unless it stirs us to action.

Many of the ways in which energy is presently used to satisfy the needs of society are not particularly effective. Large quantities of energy are allowed to "leak" out of the industrial sector, and available techniques for utilization of reject heat and waste heat in energy consuming processes are seldom applied. Faced with a shortage of non-polluting fuels and with the recognition that fuels of all types are a non-renewable resource, it is appropriate to give serious attention to improving the effectiveness with which energy is used, as well as to improving the national capacity to supply energy.

If energy utilization effectiveness is not improved, we would require the development of additional energy supply capacity with all of its economic implications. Thus, improving the effectiveness of energy utilization is seen to be necessary both as a measure for conservation (to prevent the waste of a non-renewable natural resource) and as a measure for economic optimization of investments in energy (to eliminate the need for additional energy supply capacity). The implementation of technological improvements in energy consuming processes would require, or would be greatly facilitated by, appropriate price, tax, loan, and regulatory policies, especially as these pertain to new industrial plant equipment.

The remainder of this book will discuss the available methods and technologies which can assist industry to more effectively use energy.

NOTES AND REFERENCES

1. Carrol, Wilson, Energy: Global Prospects 1985-2000, Report of the Workshop on Alternative Strategies (New York: McGraw Hill Book Company, 1977), pp. 3-5.

2. John G. Winger, et al., Outlook for Energy in the United States to 1985 (New York: The Chase Manhattan Bank, June 1972), p. 52.

3. John Myers, "Energy Conservation and Economic Growth: Are They Incompatible?" The Conference Board Record, 12(February 1975), p. 32.

4. Ibid., p. 32.

5. John Myers et al., Energy Consumption in Manufacturing (Cambridge, Massachusetts: Ballinger Publishing Company, 1974), p. 2.

6. R.W. Barnes, "Energy and Industrial Processes: A Look at the Future," Paper presented at the Symposium on Advances in Energy Storage and Conversion, American Chemical Society National. Meeting, San Francisco, California, September 1, 1976, pp. 7-8.

7. Central Intelligence Agency, The International Energy Situation: Outlook to 1985. Report ER 77-10240U (Washington, D.C.: Central Intelligence Agency, April 1977). p. 1.

8. Cedric Beatson, "Recapture the Heat that is Escaping from Your Factory," The Engineer, 222(October 10, 1974), p. 58.

9. Charles Berg, Energy Conservation Through Effective Utilization, NGSIR 73-102 (Washington, D.C.: National Bureau of Standards, February 1973), p. 11.

10. EXXON, Energy Outlook 1977-1990.

11. Elias Gyftopoulos, Lazaros Lazaridis, and Thomas Widmer, Potential Fuel Effectiveness in Industry (Cambridge, Massachusetts: Ballinger Publishing Company, 1974), p. 1.

Chapter 2

IMPROVING ENERGY MANAGEMENT

As energy is used more effectively, product costs can be reduced
and profits improved. This can be accomplished even in the face of
sharply increasing energy costs. The ways in which energy can be saved
may be classified into three general categories:

(1) Housekeeping measures. Industrial firms can
achieve significant reductions in energy usage by better
maintenance and housekeeping (e.g., shutting off standby furnaces, practicing better space conditioning,
eliminating steam and heat leaks) and greater emphasis
on the optimization of energy usage. Other savings can
be achieved through improved operating practices (e.g.,
operating at lower but acceptable temperatures without
affecting productivity).

(2) Equipment and process modification. Equipment
modifications can either be applied to existing equipment (retrofitting) or be incorporated in the design
of new equipment, or both. The improvements could be
the result of better quality control, of the use of
more durable or more efficient components, or of the
implementation of a novel or formerly neglected, more
efficient design concept. Changes for realizing greater
efficiency can also be made in a process or in the replacement of a process, producing the desired amount
and quality of goods while using less energy.

(3) Integrated operations. Better utilization of
equipment can be achieved by carefully examining the
production processes, schedules, and operating practices. Typically, industrial plants are multi-unit,
multi-product installations, which in the past have
often been designed and built as a series of independent operations with minimum consideration of overall plant energy efficiency. Improvements in plant
efficiency can be achieved through a number of ways:

- Proper sequencing of process operations. For
  example, using high pressure steam or gas for
  power generation or motivating purposes before
  extracting the heat content of the steam or
  burning the gas will make maximum use of the
  available energy content.

- Rearranging schedules to utilize process
  equipment for continuous periods of operation, thus avoiding numerous short runs
  and minimizing heat-up losses.

- Scheduling process operations during off-
  peak periods to level electrical energy
  demand and conserving the use of energy
  during peak demand periods.

The sequential use of energy and materials, starting at their
highest performance level, should be practiced. This involves minimizing the degradation of use after each level of performance is completed. Energy conservation can be achieved by aggregating sequential
and interacting activities in both space and time so that energy systems may supply requirements in a continuous manner from the highest
temperature level to the lowest temperature heat sink. Separation of
process activities in space or time tends to force large temperature
changes and consequent energy waste, as well as increase energy consumption for transportation and handling.

All of these categories of changes require capital additions and
represent the substitution of capital for purchased energy. There are,
however, many housekeeping changes that only require more careful management. Establishment of an effective energy management program involves the following basic elements:

- Understanding the basic principles of energy and
  its use in the plant.

- Conducting comprehensive surveys to measure all
  energy input and output during a given period.
  Equipment associated with large energy consumption
  should be identified.

- Creating and communicating a plan of action.

- Setting targets for unit energy consumption.

- Managing and controlling energy use against the
  targets.

No company program, however, is ever stronger than the commitment
which top management puts behind it or the resources allocated to make
it go. Only management can provide an energy conservation program with
enough authority to overcome such common obstacles as historic operating practices and employee indifference.

Most of the energy consuming facilities and processes now in use
in the U.S. were designed and built in an era of cheap, abundant ener-
gy--in particular, fossil fuels. A consequence was that given high
capital costs and increasing labor costs, there was little incentive
for energy efficient designs. A great deal of technology does exist
for more energy conservative designs, but these have seen more use
abroad, where energy has continually been higher priced than in the
U.S. With rapid increases in fuel cost, changes in fuel availability,
and the ecological situation, old "rules of thumb" and "accepted practices" will have to be critically re-examined. Many of the known
methods must be modernized and perfected. For example, in many processes in which the feedstock is also fuel, plant designs that optimize feedstock consumption will indirectly reduce overall energy requirements.

The quantity of fuel saved through the measures cited in current
reports is impressive, especially in view of the simplicity of the measures themselves. Simple, straightforward steps, such as adjusting
combustion equipment and controlling plant ventilation, have yielded
fuel savings of 10% or more. It appears that those who earlier expressed their belief that substantial quantities of fuel might be

conserved through more effective use of energy in industrial processes
have been vindicated.

Another thing industry might do to improve fuel efficiency is to
re-examine technical measures that have been available in the past.
This is, in fact, taking place today at a vigorous pace. This chapter
will discuss specific practices and capital costs which can assist in
the improvement of energy efficiency in an industrial operation.

OPTIMIZING STEAM USAGE

One area of large potential saving is better utilization of steam,
since approximately half of the energy consumed in the industrial sector goes for steam generation. Reduction in energy use can be accomplished either by increasing the efficiency of the industrial boiler,
by improving the utilization patterns for steam, or by reducing steam
losses. This section will deal with the latter two aspects, and Chapter 5 will deal with improving the efficiency of industrial boilers.

Stop Steam Leaks

One of the largest steam losses in an industrial plant is the
blowing through of steam traps. Since initial inspections commonly
reveal that as high as 7% of the traps in any one system leak, all
steam traps should be checked at least monthly for leakage. Proper
application of steam traps is also necessary for heat conversion. It
is the only way that condensate and air are automatically removed as
fast as they accumulate, without waste of steam.

Leakage can be determined by sound, by trial and error isolation
for eliminating a pressure build-up in both exhaust steam and condensate lines, and by a test valve on the downstream side of the trap,
which can be opened to the atmosphere with the downstream shutoff
valve in the closed position. It has been determined that by careful maintenance and frequent inspection, steam losses can be reduced
to 1%.

Large energy savings can be accrued through an aggressive program for the inspection and repair of steam traps. One method for determining the savings which can be gained by repairing steam traps has been developed by R. D. Glenn (1).

Example: In a plant where the value of steam is $1.50/MMBTU, an inspection program revealed that a trap on a 100 psig steam line was stuck open. The orifice in the trap was 1/8 inch. Figure 2-1 shows that steam loss was indicated to be at the rate of 540 MMBTU/year. By repairing the trap,

Annual savings = 540 MMBTU/year x $1.50/MMBTU

= $810 per year

In addition, William Sisson has developed a nomograph (Fig. 2-2) which provides a simple way to estimate the energy loss resulting from steam leaks through various sized openings (2). The nomograph is based on the Grashof equation, using an approximated average value for the factor K:

$$W = K(0.01654A)P^{0.9696}$$

where:

W = steam flow in lb/sec

K = flow coefficient - 0.95

A = areas of opening, in^2

P = steam pressure, psig

Sample Problem: If the steam pressure is 400 psig, what is the steam loss per month (720 hours) through a 1/8 inch opening? If 13,000 BTU/lb coal is used to produce the steam, what is the approximate loss in tons of coal per month?

Solution: (1) Align 400 on P scale with 1/8 on A scale. Extend to W scale and read steam loss as 175,000 lb/month.

(2) Connect 175,000 on W scale with 13,000 on H scale and read coal loss at 8 tons/month where line crosses C scale.

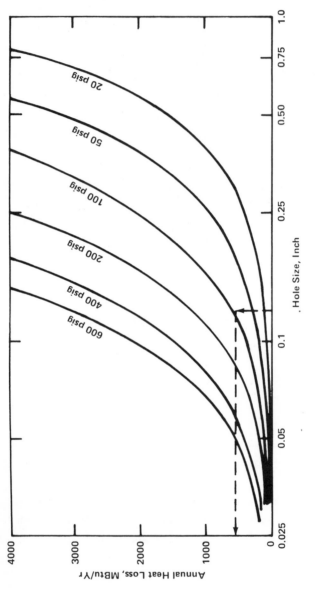

Figure 2-1. Heat Loss from Steam Leaks. Source: U.S. Department of Commerce, Energy Conservation Program Guide for Industry and Commerce, NBS Handbook 115 (Washington, D.C.: Government Printing Office, 1974), p. 3-24.

Losses from Steam Leaks (Per Month of 720 Hours)

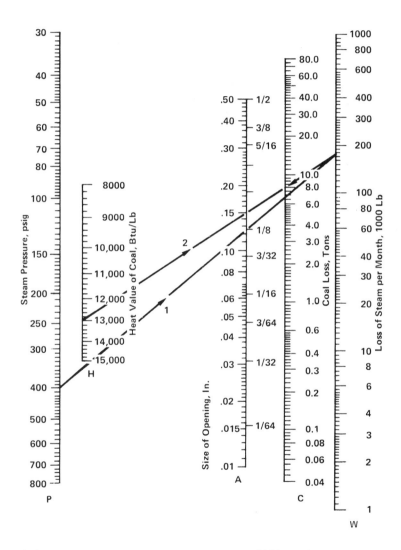

*Figure 2-2. Steam Loss Nomogram. Source: William Sisson,
"Nomogram Determines Loss of Steam and Fuel Due to Leaks," Power Engineering 79 (September 1975), 55.*

Steam Condensate Recovery

Clean condensate is desirable for boiler feedwater, since it contains heat and is practically free of dissolved minerals. Therefore,
all condensate leaks should be repaired. The pumps of condensate receiver and pumping sets should be kept in good running order so that
condensate is always pumped back to the boiler house and not permitted
to overflow from the receiver to the sewer. In cases where clean condensate is normally discharged to the sewer, an evaluation should be
made to determine whether a receiver and pumping set and piping can
be justified.

The amount of fuel used for steam generation can be reduced
10-30% by returning steam condensate to the boiler plant for use as
feedwater (3). For example, in a plant where the value of steam is

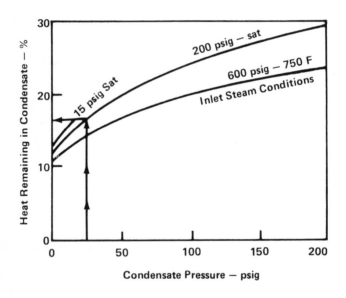

*Figure 2-3. Heat in Steam Condensate (calculated from steam
tables). Source: U.S. Department of Commerce, Energy Conservation
Program Guide for Industry and Commerce, NBS Handbook 115, 1974,
p. 3-22.*

$1.50/MMBTU, saturated steam was delivered to one building at a pressure of 200 psig, at an average rate of 30,000 lb/hr, and for an average of 7500 hr/yr. The steam was reduced through control valves and condensed in heating coils at an average pressure of 25 psig. The condensate was returned to the boiler plant and used as feedwater. The amount and value of the heat recovered is calculated below.

From Fig. 2-3, it is determined that 17% of the heat remains in the 25 psig condensate from the 200 psig saturated steam.

From the steam table, the heat value
    for 200 psig saturated steam   = 1198 BTU/lb
    for 70°F of make up water if
        condensate is not returned to boiler = __38__ BTU/lb
                   Net Heat Value = 1160 BTU/lb

The heat recovered in condensate
    = 17%/100% x 1160 BTU/lb x 30,000 lb/hr x 7500 hr/yr
    = 44,370 MMBTU/yr

The value of the heat recovered
    = 44,370 MMBTU/yr x $1.50/MMBTU
    = $66,555 per year

These values only represent heat saving potential, because no heat loss has been considered for returning the condensate to the boiler plant. Heat loss is dependent on the length of return lines and on the amount of insulation on them.

In addition to energy savings, returning of condensate to the boiler plant will accomplish the following:

• Reduce requirement for treated makeup boiler feedwater.

• Reduce energy and chemical requirements in the water treating operation.

• Reduce water pollution.

• Reduce (but not eliminate) the cost of losses due to steam trap leakage.

Another energy saving technique is to flash high pressure steam condensate to a lower pressure system, in order to produce significant quantities of steam (4). Consider the example of a building so remote from the boiler plant that a condensate return system cannot be economically justified so that all condensate is discharged to the sewer. Assume that an average of 30,000 lb/hr of condensate at 150 psig is produced for an average of 7500 hr/yr, with the value of steam being $1.50 per MMBTU. There is, at the same time, a requirement for a 15 psig lower pressure steam header, and the load on this header could utilize the entire amount of recoverable heat.

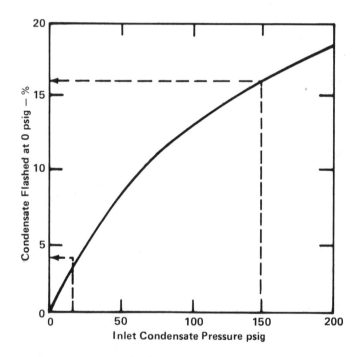

Figure 2-4. Flashing Steam Condensate (calculated from steam tables). Source: Department of Commerce, Energy Conservation Guide for Industry and Commerce, NBS Handbook 115, 1974, p. 3-23.

The quantity of 15 psig steam that would be produced by flashing condensate is estimated from Fig. 2-4 as follows:

Flashing from 150 psig to 0 psig = 16.4%

Flashing from 15 psig to 0 psig = 4.0%

                            Difference   12.4%

The potential amount of 15 psig steam available from flashing the 150 psig condensate

= 30,000 lb/yr x 12.4%/100%

= 3,720 lb/yr

The heat in 15 psig saturated steam, less the heat in water at $70^{\circ}F$

= 1126 BTU/lb

Heat available in steam flashed

= 3,720 lb/hr x 1126 BTU/lb x 7500 hr/yr

= 31,415 MMBTU/yr

Maximum value of recoverable heat

= 31,415 MMBTU/yr x $1.50/MMBTU

= $47,123 per year

In addition to saving heat, the flashing of condensate will:

- Reduce the requirement for treated makeup boiler feed water.

- Reduce the requirement for energy and chemicals used in the water.

- Reduce the cost of losses due to steam trap leakage, as a result of eliminating a separate long 15 psig steam line.

- Reduce the amount of boiler plant capacity required.

Other Conservation Suggestions

Insulation

- For surface temperatures over $140^{\circ}F$, the installation of piping and vessel insulation gives good payback. The many uninsulated hot lines in older plants, which have a side effect of lowering the ambient temperature for personnel comfort, should be insulated.

- Continue maintenance of insulation for steam lines and equipment. Give particular attention to maintenance of insulation covering. Poor covering contributes to steam load increases during rainstorms.

Steam Traps

• Consider a steam trap survey to determine the most efficient type of trap for future replacement.

• Remove or block off unnecessary traps, such as those encountered in high pressure single flow direction steam lines.

Miscellaneous

• Review long steam lines to distant single services; consider relocating or converting equipment requiring steam. An example is remote tanks with steam coils and reciprocating and turbine driven pumps.

• Stop idling standby turbines. This can be done in most cases. Idling can cost up to 15% of full load water rate and could waste 15,000 lb/day of steam for a 100 hp turbine.

• Tracing is a subject that deserves discussion. During cold weather, tracing is commonly used to protect piping and equipment from freezing. However, improper operation and maintenance of tracing systems will waste energy. Winter tracing should not be used until necessary, and steam should be turned off when the temperature does not require its use. Traced lines should be insulated, and traps should be repaired in the autumn, both to ensure their functioning when needed and to save steam.

• If a unit has waste heat steam generating facilities, consider preheating boiler feed water with process streams which are currently air or water cooled.

• Review all steam piping systems for possible replacement of underground lines and lines passing through areas subject to flooding or water accumulation, such as those encountered in valve boxes.

WASTE HEAT MANAGEMENT

Many areas around an industrial plant are prime considerations for heat recovery. Process streams, for example, can be equipped with heat exchangers for heat recovery. There may be heat discharged from one stream which may be useful elsewhere. The primary thing to remember is to look at all discharges, whatever the type. Compute the energy release above a reasonable recoverable level, and the economics will become evident immediately.

Various uses for waste heat are available. These include electricity generation, process steam generation, hot water, process heating, preheating combustion air, and space heating. Devices which can
be used for recovering waste heat include recuperators, regenerators,
waste heat boilers, and bottoming cycle engines. Recuperators and
regenerators are both used to preheat combustion air with exhaust
gases. A recuperator is merely a direct heat exchanger for exhaust
gases and combustion air. A regenerator uses a thermal storage material, such as the checkerworks in glass furnaces, which receives heat
from exhaust gas and transfers it to incoming air.

Waste heat steam boilers have also proven to be very effective
waste heat recovery devices. A paper mill operating a rotary kiln for
incinerating a pollutant and exhausting gases at $1200^{\circ}F$ could use a
heat recovery steam boiler which allows the generation of 3,420 pounds
of 2000 psi steam per hour for a two year total of 60,000,000 pounds
of steam. The investment required for the boiler is $20,000, which
would result in a payback period of less than one year.

In many industrial processes waste heat at relatively low tempera-
ture levels are generated (e.g., $300^{\circ}F$ to $600^{\circ}F$). Very often this heat
is not recovered. An organic rankine bottoming cycle engine can be
used for converting this waste heat into electricity. Although this
device is still in the developmental stage, small (100 KW) units, which
could prove to be extremely valuable because of their availability to
use low temperature heat sources, are presently available on the market.

The effectiveness of fuel usage could be improved in many industrial processes by recovery of useful energy that is now lost, either
through exhaust gas or materials in process. However, the design of
heat recovery systems requires considerable skill and experience and
should not be attempted without professional assistance. Consideration must be given to the type of energy that is available, to the
quality and quantity of that energy, and to the requirements for the
recovery systems.

This topic will be covered in greater detail in Chapter 4.

ELECTRICITY CONSERVATION

Electric energy has historically been so cheap that none but the most energy intensive industries have ever made any effort to control its use. And even for them, the goal has frequently been just to pare down demand charges. But the days of cheap power are gone. Energy prices are soaring, and utilities are beginning to overhaul their rate structures. More and more, the idea of controlling energy consumption is catching on in all sizes of plants across the nation.

Electricity normally constitutes a plant's largest utility bill and is involved with nearly every piece of plant energy consuming equipment. Therefore, special attention should be given to the use of electricity for energy conserving possibilities, as well as for potential cost reductions. Table 2-1 depicts energy consumption in the industrial sector by equipment type. As is readily evident, the major energy consumers are electric drives.

Means for improving energy efficiency and reducing energy costs can be achieved through reduction in maximum demand load, due to improved equipment demand control, power factor correction, improved power factor, improved motor selection and utilization, and improved lighting practices. These practices can result in reductions in electricity usage of 20-50%.

Production Equipment Demand Control

A utility bill is comprised of three key elements:

1. The user's actual consumption of energy, expressed in KWH. This charge tends to offset the cost of the fuel that is converted to electricity.

2. The user's demand on the system, with respect to his average connected load, expressed in KW or KVA. This charge covers fixed costs of interest and the depreciation of investment in the equipment necessary to meet the maximum power requirements of that particular user.

3. Fuel adjustment charges, which permit the utility to pass along increases in fuel costs to the user.

Table 2-1

Total Industrial Electrical Consumption (1972)
(billions kW-hr)

Industrial Motor Drive (except HVAC)		458
Pumps	143	
Compressors	83	
Blowers and fans	73	
Machine tools	40	
Other integral HP applications	52	
DC drives	47	
Fractional HP applications	20	
Other Industrial Electrical Usage		142
Electrolytic		
Direct heat		
HVAC		
Transportation		
Lighting		
Total Industrial		600

Source: Federal Energy Administration, Energy Efficiency
and Electric Motors, Conservation Paper 58,
August 1976, p. 26.

At the end of the billing period, usually 30 days, the total number of KWH accumulated forms the basis for the energy charge. To determine the peak demand that occurred during the billing period, the utility establishes shorter periods of time, usually 15, 30, or 60 minutes,
called the demand interval. Demand is defined as being the average
energy used during the demand interval, or KWH/hr, expressed as KW.
Peak demand is the highest KW load obtained during any demand interval
within the billing period. Normally, billing demand remains in effect
for one month, but it may be levied for an entire season, or even for
a full year, depending on the contract the user has with his utility.

The efficiency of load utilization is called the load factor,
which is the ratio of the average KW load over the entire billing period to the highest demand over any demand interval in the period. If
the plant's load efficiency is running at 60%, a 40% energy peak exists
that, ideally, can be deferred to low demand periods. The ideal load
factor is 100%. To approach this, maximum demand has to be lowered.

Careful attention by management to the more ordered use of electrical power throughout a plant can improve the load factor and reduce costs. The most practical measure required for improving the load factor consists of reducing the maximum demand to the practicable minimum. Some of the ways for accomplishing this are to ensure that:

- A starting schedule be prepared and adhered to, so that the whole of the plant starting load does not occur within the same half hour period of the demand meter.

- Maximum advantage is taken of night usage, i.e., wherever operationally possible, to accomplish such tasks as operating intermittent equipment, recharging industrial truck batteries, etc.

- Heavy machinery, which takes a high starting load and is used intermittently, is never started twice within one half hour recording period of the maximum demand meter.

- No two units of heavy machinery that are used only during short periods of the working day are ever used at the same time.

- Wherever possible, clutches are fitted so that motors are started on no load. It is worth remembering that star/delta starters will only take approximately twice the normal running current, whereas direct on-line starters will take five to six times the normal current.

- Large production machinery are scheduled as uniformly as possible among all shifts to minimize the demand charges for purchased power.

Automatic means of demand control can also be instituted. The devices can be relatively simple or can be large scale computer systems. They all perform similar functions, i.e., they measure the energy being consumed in a facility, and they limit demand by turning off non-essential loads. Three techniques can be used to control demand: ideal rate control, predictive control, and instantaneous rate control (5).

Ideal Rate Control. The energy consumed during a demand interval is measured and accumulated. The accumulation is compared with a theoretical rate of energy consumption, which is based on the demand interval and the maximum allowable KW during that interval. When the

difference between the ideal rate and actual comparison reaches a preset minimum, a load is shed. When the difference increases, due to a
drop in the rate of consumption, a load is restored or added. See
Fig. 2-5.

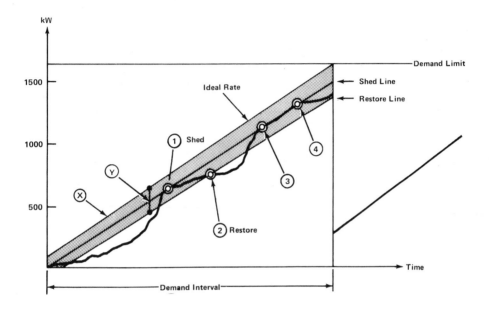

Figure 2-5. Ideal Rate Control. Source: Thomas C. Elliott,
"Demand Control of Industry Power Cuts Utility Bills, Points to
Energy Savings," Power, 120 (June 1976), 21. Reprinted with per-
mission from POWER, June 1976.

In a hypothetical facility which has a demand limit of 1600 KW
and a demand interval of fifteen minutes, as actual consumption begins
to approach the ideal rate (point 1), the difference between the two
accumulators reaches X, and a load is shed. During the next measurement, if the rate of consumption is such that the difference is still
less than X, another load is shed. When the difference between actual
consumption and the ideal rate reaches a larger difference, Y, as is
shown at point 2, a load is restored. If the rate increases, loads are
shed again, as can be seen at points 3 and 4. This process continues
until the end of the demand interval, at which point a new measurement
is started.

Ideal rate control is recommended for applications in which only
one or two loads are being controlled, or in which all the loads being
controlled have the same KW rating. This technique requires that (1)
the controllers be synchronized to the demand interval prescribed by
the power company, and that (2) the devices normally put all loads on
line at or near the start of each demand interval. Unfortunately, this
produces large transient surges at the beginning of each interval,
which is undesirable to user and utility alike.

Predictive Control. In predictive control, controllers are synchronized to the utility's demand interval, as shown in Fig. 2-6. At
the beginning of the interval, the devices add all loads onto the line
and start to accumulate actual energy consumption. At each measurement
point, the units measure the instantaneous rate of energy consumption
and predict whether or not the pre-established limit will be exceeded.
The control system achieves this by adding the accumulated energy to
the product of the rate and the time remaining in the demand interval.

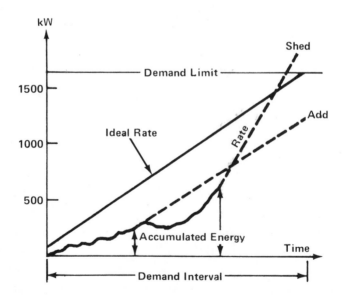

*Figure 2-6. Predictive Control. Source: Elliott, "Demand Con-
trol of Industry Power," Power, p. 21. Reprinted with permission from
POWER, June 1976.*

If this calculated value exceeds the preset limit, the system begins
to shed loads; if the calculated value is less than the preset limit,
it resets loads.

Because of the added calculations, the predictive technique is
generally a more complex and expensive approach to demand control. It
requires that units be synchronized to the utility's demand interval,
and since several types of controllers add all loads at the beginning
of each demand interval, serious power surges can be created.

Instantaneous Rate Control. Instantaneous or continuous integral
control is based on measuring the rate of power consumption at short
time intervals and comparing this rate with a preset limit. See Fig.
2-7. If at any time the instantaneous rate exceeds the preset rate,
load will be shed. As long as the instantaneous rate is below the preset rate, load will be added.

Because this approach is based on an instantaneous rate of consumption, it is independent of the actual demand interval. For this reason, it does not require synchronization to the utility's demand interval; thus, surges are avoided. Another advantage is that the technique

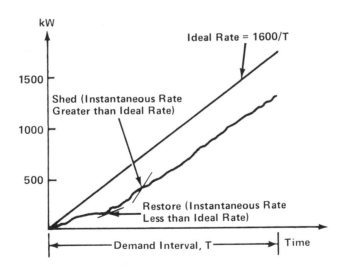

Figure 2-7. Instantaneous Rate Control. Source: Elliott, "Demand Control of Industry Power," Power, p. 21. Reprinted with permission
from POWER, June 1976.

attempts to hold the rate of power consumption at a predetermined value which produces a more evenly distributed rate of power consumption. The approach is expensive, however, because computer assistance is needed, but it is effective, especially where loading is complex.

In the installation of load controllers in the facility, it is essential to recognize that the most important aspect is the setting up of the program for load shedding. Only the user with a thorough knowledge of how his facility operates can do this. Judgments must be made, establishing which loads can be shed, in what priority they are to be shed, and for how long during each demand interval. If done well, installation of load controllers can result in significant reduction in demand charges and leveling of electrical load, the latter of which are advantageous to the user as well as to the utility.

Power Factor

Power factor is the percentage of current in an AC circuit which can be used as energy, and is the ratio of true power in kilowatts to apparent power in kilovolt amperes. The kilowatts, or "real power," performs the "real work" done by the electricity. The out of phase component, kilovolt ampere reactive, or "phantom power," provides the magnetizing force necessary for operation of the work performing device.

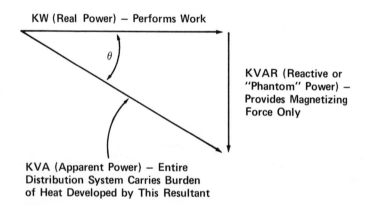

Figure 2-8. Power Factor.

Figure 2-8 provides the relationship between KW, KVAR, and KVA.
The angle between the KW and KVA vectors is known as the "phase angle"
and is used as a measure of the relative amount of KVAR of the system.
The quantity known as "power factor" is simply the cosine of this angle.

As the amount of KVAR is decreased, the phase angle is diminished,
and the magnitude of KVA approaches that of the KW. When the phase
angle decreases to zero, power factor becomes 1.00 or 100%. At 100%
power factor, KVA is equal to KW, and all of the heating developed in
the system is a function of current that is actually performing productive work. Low or "poor" power factor is caused by use of inductive
or magnetic devices, such as induction motors, transformers, chokes
for fluorescent lights, and welding plants. The reactive current uses
part of the capacity of the distribution network, although it does no
useful work.

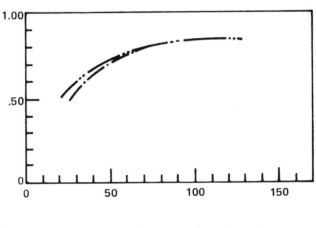

—·—— = 600 Volt, 30, 20 HP Induction Motor, 1800 RPM
—··—— = 600 Volt, 30, 100 HP Induction Motor, 1800 RPM

*Figure 2-9. Power Factor VS Percent Full Motor Load. Source:
P.N. Silverthorne, "Power Factor Correction for Energy Conservation,"
ASHRAE Journal, Vol. DGW-13 (May 1975), 29. Reprinted with the permission of the American Society of Heating, Refrigerating and Air Conditioning Engineers, Inc.*

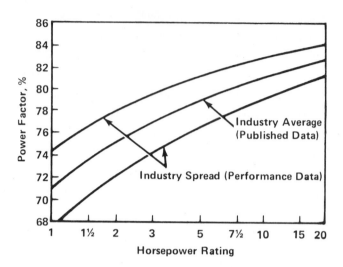

Figure 2-10. Power Factor VS Horsepower Relationships of Commercially Available Motors. Source: Gould, Inc., Century Electric Division.

Generally speaking, the power factor for inductive motors falls as the mechanical load on the motor decreases; therefore, the aim should be to keep all motors almost fully loaded. Figure 2-9 demonstrates this for a 20HP and 100HP motor by relating power factor to percent full load on the motor. Of more interest is the relationship of power factor to operating mechanical equipment. The figure shows that at very low load on a motor the power factor drops to as low as 0.5 at 25% load.

Additionally, there is a substantial range of power factors in comparable commercially available motors, and larger units have better power factors than their smaller counterparts. This is shown in Fig. 2-10.

Penalties for Poor Power Factor

Poor power factor penalizes the industrial user of electricity in a variety of ways. First of all, utilities typically penalize industrial customers for low power factors. In fact, it is the rule, rather than the exception, for industrial power contracts to specify minimum

permissable power factor, Users having power factors below 80-90%
typically have surcharges added to their bills. If the customer does
not bring this power factor up to the minimum in a specified period
of time, the power company will, under some power contracts, perform
the correction, bill the customer for the installation, and charge a
stiff monthly "rental" for the capacitors on the electrical power bill.
These penalties may amount to thousands of dollars a month for large
industrial plants.

In addition, poor power factor robs the distribution system of
capacity that could be used to handle work performing load and results
in currents higher than necessary to perform a given job, thereby contributing to excessive voltage drop and high system losses.

How to Improve Power Factor

The losses due to low power factor are largely correctable in the
following ways:

- Use of high power factor equipment (e.g., high power factor
  lighting ballasts) and electric resistance devices (e.g.,
  boilers, furnaces, etc.). Operation of equipment at close
  to full loads (e.g., induction motors) will improve the power factor.

- Use of synchronous motors instead of or along with induction
  motors. The characteristics of synchronous motors are such
  that they can be operated at unity or leading power factor.
  The leading power factor is very simply adjusted, and properly balanced systems could theoretically use one horsepower
  of synchronous motor to correct approximately four inductive
  horsepowers. Good in theory, but practice shows that synchronous motors are generally only available in 500 HP or
  larger sizes. To aid in power factor correction, moreover,
  they must be applied to continuously running loads.

- Use of a synchronous condenser. A synchronous condenser is
  a large piece of rotating equipment which is similar to a
  synchronous motor. The synchronous condenser drives no
  load; it merely functions to improve the power factor. Synchronous condensers are not a solution in industrial plants
  and are used mainly by electric utilities.

- Use of capacitors, the only practical method of power factor
  correction. For a given cycle of power supply, during the
  first quarter of the cycle, a capacitor absorbs power, while

an inductor emits power. This sequence reverses during the
next quarter of the cycle, reverses again in the next quarter
cycle, and so on. Hence, a capacitor matched in capacity to
the magnetic field of an inductive motor can supply the reactive power required by a motor through the phenomenon of continuous interchange of energy between an inductor and a capacitor. This technique is the simplest and the most versatile
for industrial power factor improvement. Capacitors can be
bought in blocks and combined to provide the required amount
of capacitive reactance.

Benefits of Power Factor Enhancement

The major benefit of power factor correction is the power company's
providing attractive billing incentives, since the utility is faced
with the same problem of reduced capacity as is the customer. The
utility's facilities--distribution, transmission, and generation--must
be oversized to accomodate the reactive component. As a general rule,
where power factor is poor, capacitors can be installed for about one-
fifth of the cost of distribution facilities. What this means is that
for every dollar spent for power factor improvement capacitors, five
dollars worth of distribution system KVA might be released. Capacitors
should be installed at the terminals of the offending loads, rather
than for the entire system, reducing current from this point all the
way back through the plant distribution system to and including the
utility's generator.

Additional benefits include the reduction in system component siz-
ing, including transformers, major feeders, substations, and bus ducts.
Power factor correction can also increase system capacity, and it can
reduce power system losses, resulting in improved voltage throughout
the system. Furthermore, there is likely to be less load on transformers serving motors, potentially increasing transformer life and decreasing the likelihood of transformer failure. And power factor correction can offer longer motor life for those motors operating close
to their nameplate current.

Electric Motors

Electric motors are a major user of the electricity consumed in the U.S. The Edison Electric Institute reports that in 1972 more than 548 billion KWH per year were used to power electric motors in industrial and commercial process equipment. This accounted for nearly 33% of the U.S. electrical energy demand, or 9% of the total energy demand.

Energy Consumption Patterns. Table 2-2 shows the electrical energy consumed by electric motors in the individual two digit SIC codes. The top five industries, which account for slightly over 60% of the industrial sector's total, are essentially continuous process industries.

Table 2-3 shows the estimated population of motors, the estimated average drive motor size in kilowatts, and the average usage in hours per year. This information was compiled by Arthur D. Little by taking past data from the Census of Manufacturers and multiplying it by the number of years expected as the useful life for that type of equipment.

Motor Efficiency. Motors can range in efficiency from a low of 10% in small devices to a high exceeding 90% in large, special purpose industrial motors. Most industrial motors are rated in the 50-70% efficiency range.

Figure 2-11 provides published motor efficiency data as compiled by Arthur D. Little. As is evident, the data for a given size and type of motor are highly variable between manufacturers and often even between different models of the same manufacturer. Also evident is the trend in the manufacture of commercially standardized electric motors, over the last few decades, in compromising higher efficiency ratings in favor of other factors, such as lower initial cost and conservation of materials.

Conservation Potential. Industrial end users have traditionally been uninterested in electric motor efficiency because in the past, such economies have been relatively unimportant. Compared to other potential savings, such as improving product line interruptions and better inventory management, motor efficiency has just not been a very visible economic issue. Today, improvement in motor efficiency can come about in several ways:

Table 2-2

Electric Energy Consumed by Electric Motors in the Industrial Sector
(Corrected for Non-Electric Drive Component)

SIC Code	Industry Group	1971 Rank	Electric Energy Consumed by Electric Motors (kW-hr x 10^9)	Percent of Total	Cumulative Percent of Total
28	Chemicals and Allied Products	1	80.9	18	18
33	Primary Metal Industries	2	76.1	17	35
26	Paper and Allied Products	3	59.1	13	48
20	Food and Kindred Products	4	34.9	8	56
29	Petroleum and Coal Products	5	28.8	6	62
32	Stone, Clay, and Glass Products	6	24.5	5	67
37	Transportation Equipment	7	24.2	5	72
22	Textile Mill Products	8	23.7	5	77
36	Electrical Equipment and Supplies	9	20.4	4	81
35	Machinery, Except Electrical	10	19.2	4	85
34	Fabricated Metal Products	11	17.8	4	89
30	Rubber and Plastics Products	12	16.0	3	93
24	Lumber and Wood Products	13	9.3	2	95
27	Printing and Publishing	14	7.5	2	97
23 25, 38, 31 21, 39 and 19	Remaining Industries	15–21	15.9	4	100
	Total		458.3	100%	

Source: Federal Energy Administration, Energy Efficiency and Electric Motors, Conservation Paper
58 (Washington, D.C.: Government Printing Office, August 1976), p. 22.

1. design equipment so that motors always operate at full load;

2. using variable speed drives;

3. substituting more efficient motors.

Table 2-4 shows the potential for improved efficiency in electric
motors. It depicts very vividly the fact that the potential for improved efficiency varies considerably, depending on the size of the
motor, with the smaller sizes having the greatest potential for improvement. It shows that with proper attention placed on selection
of motors, greater energy efficiencies and economies can be achieved.

Table 2-3

Industrial Equipment Usage Profile (1971)
(Integral HP Motor Drive Only)

Equipment Type	Estimated Population (000's)	Estimated Average Drive Size (kW)	Average Usage (hr/yr)	Estimated Electricity Consumption (kW-hr x 10^6)
Centrifugal Pumps	6,310	11.42	3,000	216,000
Rotary Pumps	3,934	7.55	3,000	89,000
Reciprocating Pumps	383	7.5	3,000	8,600
Turbine and Other Pumps	545.6	8.0	3,000	13,400
Subtotal – Pumps	11,172.6	-	-	327,000
Air Compressors	705	13.94	3,000	29,400
Refrigeration Compressors	50.3	175	4,000	35,200
Vacuum Pumps	1,256.0	3.75	4,000	18,900
Subtotal – Compressors	2,011.3	-	-	83,500
Centrifugal Blowers and Fans	1,808.0	20.0	2,500	90,400
Axial and Propeller Fans	2,352.0	11.16	2,500	65,600
Subtotal – Blowers and Fans	4,160.0	-	-	156,000
Metal-Cutting Machine Tools	2,362.0	9.0	1,000	21,300
Metal-Forming Machine Tools	703.0	26.25	1,000	18,500
Subtotal – Machine Tools	3,065.0	-	-	39,800
Subtotal	20,408.9			606,300
Adjustment for Usage in Other Sectors				(267,300)
				339,000
Other AC Integral HP				63,300
Fractional AC				20,000
DC Motors and Drives				36,000
Total				458,300

Source: Federal Energy Administration, Energy Efficiency and Electric Motors, Conservation
Paper 58, p. 23.

Facility managers should become more aware of and should install more efficient motors when replacements or new equipment must be installed. It should be realized, however, that the efficiency rating is but one of many performance parameters to be considered in the selection of an electric motor, and in many cases a low efficiency rating can be, and is, entirely justified. This is particularly true where a motor has relatively infrequent usage, making an inexpensive but inefficient motor economical. But it would cost little for industrial

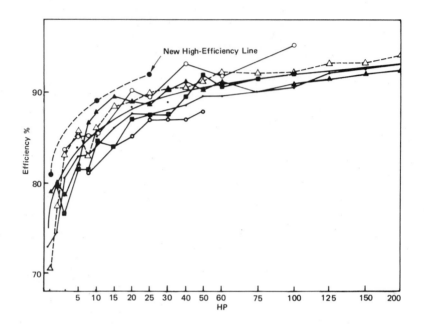

Figure 2-11. Published Motor Efficiencies of Principal Manufacturers (open, drip proof, 1800 RPM, NEMA Design B). Source: Federal Energy Administration, Energy Efficiency and Electric Motors, Conservation Paper 58, p. ES-9.

users to select an optimally efficient motor, if the data were readily available and believable. Life cycle cost analysis from the industrial user's point of view (including first cost, operating and maintenance costs, salvage, etc.) should dictate the selection of more efficient, if more expensive, models.

Lighting

We use lighting systems to help us perform our jobs. The relationship between light, vision, visibility, and the task to be performed and the worker's productivity is very complex. We know that if too little light is used, increased employee errors and reduced output may result. On the other hand, if too much lighting is used, operating costs are needlessly high, and profits again suffer. There is

Table 2-4

**Current and Future Motor Efficiencies in Integral
HP AC-Polyphase Motors**

	Current			Future
	Worst	Best	Average	Improved Efficiency Models
1 HP	68	78	73	85.5
5 HP	78	81.5	80	89
10 HP	81	88	85	90
50 HP	88.5	92.0	90	92.5
100 HP	90.5	92.5	91.5	93
200 HP	94	95	94.5	95

Source: Federal Energy Administration, Energy Efficiency and
Electric Motors, Conservation Paper 58, p. 43.

a point at which the amount and kind of lighting used will help to
maximize profits by helping to maximize productivity.

Energy used for illumination is a major candidate for conservation measures. These can be achieved either through improved housekeeping or through better lighting design. In most cases, simple
changes can result in improvement of efficiency and in the quality of
illumination. Of course, one must consider the benefits versus costs
of these modifications. For example, a $100 modification that pays
for itself in one year is much more effective than a $10 modification
that pays in four years.

The following are some housekeeping or maintenance techniques
which can be used to improve the efficiency of one's present lighting
system:

- Turn out lights when they are not required.

- Reduce lighting by removing lights or installing lower
  wattage bulbs. Guidelines to be followed are 50 footcandles at inspection stations, 30 footcandles in work
  areas, and 10 footcandles in corridors, stairways, etc.

- Install light switches (especially on incandescent lights
  and in coolers and freezers) to permit employees to turn
  out lights not being used.

- Make greater use of natural light by turning off lights near windows when there is sufficient light from the outside.

- Increase the illumination, where necessary, by painting the room or area with lighter and more reflective paints.

- Reduce exterior and parking lot lighting levels. In some instances, this may require increasing security provisions. Consider time switches or photoelectric cells to limit the burning time to the shortest term possible.

- Turn off or remove supplementary lighting, such as decorative office lamps and drafting table lamps, unless it is required by the specific task.

- Reschedule janitorial work to regular shift hours, in order to reduce lamp burning hours.

- Keep lighting fixture clean.

- Reorient work stations. Group tasks which require the same lighting levels, and adjust the light output of the system accordingly. Place work stations requiring the highest level of illumination near windows to take advantage of natural daylight.

- Modify illumination level by changing the type of lamps used, by changing the type of luminaires used, or by adding new controls and associated equipment.

- Use more efficient lamps. In general, the lamp that provides the most lumens per watt will be the most cost effective in the long run. The more lumens a lamp produces from each watt of input, the more efficient it is.

Various types of lamps are available. For most indoor commercial and industrial applications, lamps can be divided into three categories: incandescent, fluorescent, and high intensity discharge (mercury vapor, metal halide, high pressure sodium, and low pressure sodium). Their basic characteristics are shown in Table 2-5. At one end of the spectrum is the incandescent lamp which has the characteristics of low unit cost, familiar color traits, large variations in shape, wattage, input voltage, small size, simplicity of wiring and control. At the other extreme is the low pressure sodium lamp. Although it is not normally used for indoor applications, the sodium lamp has excellent

Table 2-5

Interior General Lighting (Commercial and Industrial) Lamp Selection Guide

Lamp	Incandescent		Fluorescent				High intensity discharge			
Type	Standard	Tungsten halogen	Cool white	Warm white	Deluxe cool white	Deluxe warm white	Deluxe white mercury	Warm deluxe white mercury	Metal halide	High pressure sodium
Application	Local lighting accent, display, Low initial cost general lighting	Accent, display, general lighting in lobbies, theaters, etc.	Most commercial and industrial general lighting systems	Many commercial and industrial general lighting systems	General and display lighting where good color daylight-like atmosphere important	General and display lighting where good color Incandescent-like atmosphere important	Store and other commerical and industrial-general lighting	Store and other commercial-general lighting	Industrial, commercial general lighting, pools, arenas	Industrial general lighting. Some commercial applications
Lamp optical controllability potential	Excellent	Excellent	Fair	Fair	Fair	Fair	Good	Good	Excellent	Excellent
Glare control with typical luminaire	Good	Excellent	Good	Good	Good	Good	Excellent	Excellent	Good	Good
Color effects: accents	Warm colors	Warm colors	Yellow, blue, green	Yellow, green	None-excellent color balance	Yellow, orange, red-like incandescent	Red, yellow, blue, green	Red, orange, yellow, green	Yellow, geen, blue	Yellow, orange
Color effects: grays	Blues	Blues	Reds	Red, blues	None	Blues	Deep reds	Deep reds, blues	Reds	Deep reds, greens, blues
Range of lamp wattages for typical applications	30-1000	200-1500	20-215	20-215	20-215	20-110	50-1000	175-1000	400-1000	150-1000
Range of initial lamp lumens for typical applications	210-23740	3460-34730	1300-16000	1300-15000	850-11000	820-6550	1575-63000	6500-58000	34000-100000	16000-130000
Initial lamp lumens/watt	7-24	17-24	65-85	65-85	40-60	40-60	30-65	35-60	85-100	100-130
Average rated lamp life (hrs)	750-2000	1500-4000	9000 (3 hours/start)-30000 (continuous burning for F40 types)				16000-24000+	24000+	8000-15000 (10 or more hours/start)	12000-20000 (10 or more hours/start)
Light output depreciation characteristics	Good	Excellent	Good	Good	Fair	Fair	Fair	Fair	Fair	Excellent

Source: Terry McGowan. "Lightning Design Materials and Methods: All About Sources," Reprinted from the September 1973 issue of Progressive Architecture, Copyright 1973, Reinhold Publishing Company.

life and efficiency. Other lamps, such as fluorescents, fall in-between.

While efficiency alone should not be the prime consideration in
the selection of a lamp type, it is a very important factor. In most
cases, a more efficient light source can be substituted for a less
efficient source with little, if any, loss in task visibility or color
rendition. The total annual energy and operating cost (life cycle
cost) saving achieved will help to lower any electric utility bill.

INSULATION

In insulation systems design, the first question raised is usually
"Why insulate?" The answer is obvious. Energy conservation through
the use of optimal economic insulation has large benefits for industry
and for the U.S. economy. From the industrial manager's point of view,
insulation will reduce the loss of energy from a surface operating at
a temperature other than ambient, will reduce plant operating expenditures for fuel and power, will improve process efficiency, and will
increase system output capacity. It may even reduce the required capital cost.

The benefits of increased thermal insulation usage are not con-
fined to a reduction in energy consumption. Insulation can reduce
ventilation and air conditioning requirements for indoor facilities and
can reduce equipment corrosion by reducing the condensation of corrosive agents in gaseous streams. Reduced fuel consumption will result
in reduction in combustive air pollutant emissions (primarily the sulfur oxides and suspended particulates). In this way, greater insulation use can help to improve air quality.

Conservation Potential

The amount of energy rejected as heat in various temperature ranges
per year for six manufacturing industries has been computed by Reding
and Shepard and is shown in Table 2-6. This rejected heat represents

Table 2-6

Estimated Heat Rejection in the Six Biggest Fuel Consuming
Industries (1972-1973)

Industry (or Process) and Operation	Radiation, Convection, Conduction, and Unaccounted Losses[2]	Rejected Heat, (10^{12} Btu/year)[1]				
		Below 100°C	100–250°C	250–800°C	800–1800°C	Total, All Ranges
Chemical						
Chlorine/Caustic Soda	40	254	51			
Ethylene/Propylene	20	151	79	40		
Ammonia	32	115	139	40		
Ethylbenzene/Styrene	1	24	28			
Carbon Black	2		32			
Sodium Carbonate	2	36	16			
Oxygen/Nitrogen	1	171	147			
Cumene		4	4			
Phenol/Acetone		8	8			
Other	150	628	290	79		
TOTAL	248	1391	794	159		2,345
Primary Metals						
Steel	298	79	397	596	795	
Aluminum	119	378	79	20	79	
Other	119	397	119	40	79	
TOTAL	536	854	595	656	953	3,058
Petroleum	318	994	497	695		2,186
Paper	199	1391	755	119		2,265
Glass–Cement–Other						
Cement	119	79	40	219		
Glass	99	60	12	87	36	
Other	99	79	40	159	20	
TOTAL	317	218	92	465	56	1,049
Food	60	795	159	119		1,073
GRAND TOTALS	1679	5643	3110	2213	1009	11,975
Ratio to Grand Total Σ_T		0.47	0.26	0.18	0.09	

[1] To convert to 10^{15} J/year, multiply by 1.0543. The estimated accuracy of the values is ±20%.
[2] Radiation etc. losses are distributed within other losses shown in the table.

Source: J. T. Reding and B. P. Shepard, Energy Consumption Fuel Utilization and Conservation in Industry, EPA 650/2-75-032-d (Washington, D.C.: Government Printing Office, August 1975).

energy potentially recoverable through better use of thermal insulation. Analysis of the data provides an estimate of the total rejected
energy in the form of heat from each of the major industries in each
temperature range. These estimates show that a very large fraction
(47%) of the rejected heat is available at a temperature less than
100°C; that about three fourths of the rejected heat is available at
less than 250°C; and that 27% of the rejected heat is available in
the temperature range 250-1800°C. The biggest potential for immediate
energy savings is in those manufacturing process operations in which
relatively low quality heat is rejected (i.e., at temperatures below
250°C).

An estimate of potential energy savings in the six big fuel consuming industries on the basis of short term conservation measures
(less than five years) is shown in Table 2-7. Research and development on new processes, increasing product yields, new and improved
thermal insulations, and improved systems may result in additional
conservation of energy, but the near term estimates were made by assuming use of only presently available materials and techniques.

A Federal Energy Administration study estimates that over 1400
trillion BTU's, the equivalent of over 122 million barrels of oil,
could have been saved in 1974, had industry alone installed economic
thicknesses of insulation. This is in addition to the energy presently
being saved with existing insulation. The potential additional energy
conservation available to industry through the economic use of insulation through 1990 is estimated to be the equivalent of 3.5 billion
barrels of oil, or 250 million barrels per year (6).

Longer term conservation through the use of thermal insulations
can be increased through improved or more effective thermal insulations
than those presently available to reduce heat flows for a given insulation thickness; through improving hot wall refractories which are
used to allow lower effective furnace wall heat transfer, permitting
higher inner wall temperatures without adverse shortening of the life;
through cast and plastic mix refractories having higher service temperatures, lower thermal conductivities, and improved mechanical properties, in order to allow improved reactor vessel and furnace thermal

Table 2-7

Estimated Energy Conservation Potential in Immediate Period in
Six Largest Energy-Consuming Industries, 1972

Industry	Estimated Energy Savings by Insulation and Maintenance*		Estimated Total Industry Process Energy Usage**	
	(Btu/Year)	(J/Year)	(Btu/Year)	(J/Year)
Chemical	252×10^{12}	266×10^{15}	2662×10^{12}	2810×10^{15}
Primary Metals	500	528	5900	6230
Petroleum	343	362	3044	3210
Paper	208	219	2563	2700
Glass, Cement, Structural Clay Products, Others	190	201	2000	2110
Food	47	50	1283	1350

*Estimated precision, ±30%.
**Process energy only; does not include feedstock energy usage. Estimated precision, ±10%.

Source: R. G. Donnelly, V. J. Tennery, D. L. McElroy, R. G. Godfrey, and J. O. Kolb, Industrial Thermal Insulation: An Assessment, Report TID-27120 (Washington, D.C.: Energy Research and Development Administration, August 1976), p. xiv.

conductivities, and improved mechanical properties, in order to allow improved reactor vessel and furnace thermal insulation without significant shortening of the service life of the refractory linings (7).

Insulation Materials

A large variety of thermal insulation materials are manufactured for application in the various industrial temperature ranges and environments. Donnelley et al. have classified industrial thermal insulations on the basis of temperature (8). These temperature ranges are:

1. cryogenic, $-270^{\circ}C$ $(-454^{\circ}F)$ $\leq T \leq -100^{\circ}C$ $(-148^{\circ}F)$

2. low temperature, $-100^{\circ}C$ $(-148^{\circ}F)$ $\leq T \leq 100^{\circ}C$ $(212^{\circ}F)$

3. intermediate temperature, $100^{\circ}C$ $(212^{\circ}F)$ $\leq T \leq 500^{\circ}C$ $(932^{\circ}F)$

4. high temperature, $T \geq 500^{\circ}C$ $(932^{\circ}F)$.

An insulation material's applicability is limited by an upper temperature at which, due to high thermal conductivity, it becomes structurally unstable or non-competitive. But insulations also have a lower temperature limit, due to undesirable properties. Table 2-8 provides an indication of the types of insulations available at the various temperature ranges.

In many cases, materials design and selection are based on properties other than thermal conductivity. Some of these properties are listed in Table 2-9. Standard methods for determining these properties are not generally available. Hence, one must be careful when comparing a specific property of various insulations and using it as the only basis for materials choice in a given system design. Many of these property values for various types of commercially available insulations are unknown.

Economic Thickness

Various procedures have been developed for calculating the economic thickness of insulation. The cost factors involved in the traditional analysis of economic thickness and which determine it consist of two types. The first type of costs decreases with additional insulation thickness, which reflects cost savings from additional insulation use. These decreasing costs are:

- Energy costs

- Capital investment in heat producing equipment

- Maintenance costs

- Interest costs on investment

The second type increases with additional insulation thickness. Increasing cost items are:

- Insulation costs

- Interest for insulation investment

- Maintenance cost for insulation

Table 2-8

Industrial Insulations by Temperature Range of Application

Cryogenic and Low Temperature [−270°C (−454°F) ⩽ T ⩽ 100°C (212°F)]

  Evacuated

    Multifoil
    Opacified powders

  Mass type

    Glass foams
    Organic foams
    Fiber glass
    Loose fill
    Balsa wood

Intermediate Temperature [100°C (212°F) ⩽ T ⩽ 500°C (932°F)]

  All inorganic

    Perlite
    Calcium silicate
    Foam glass
    Mineral wool
    Reflective
    Loose fill
    Insulating firebrick

High Temperature [T > 500°C (932°F)]

  All inorganic, carbon, or metallic

    Loose fill
    Reflective
    Insulating firebrick
    Ceramic foams
    Ceramic fiber
    Pyrolytic carbon
    Carbon fibers

Source: Donnelly, et al., Industrial Thermal Insulation: An Assessment, p. 12.

Optimal economic insulation thickness can be derived through the minimum total cost method and the marginal cost method. Each of these procedures should yield the same solution.

The minimum total cost method involves the actual calculations of lost energy and insulation costs for each insulation thickness. The thickness producing the lowest total cost is the optimal economic solution.

Table 2-9

Selected Properties of Thermal Insulations Often Used in Design

Property	Insulation Type							
	Rigid and Semi-Rigid	Flexible	Blanket, Felt, or Batt	Loose Fill	Reflective	Cryogenic	Insulating Fire Brick	Sprayed and Foamed in Place
Thermal conductivity	X	X	X	X			X	X
Thermal diffusivity	X	X	X	X			X	X
Maximum temperature limits	X	X	X	X	X	X	X	X
Density	X	X	X	X	X	X	X	X
Abrasion resistance	X	X			X		X	X
Alkalinity	X	X	X					X
Capillarity	X	X	X	X				X
Flame spread index	X	X	X	X				X
Smoke density index	X	X	X	X				X
Combustibility	X	X	X					X
Compressive strength	X	X	X			X	X	X
Hydroscopicity	X	X	X	X			X	X
Linear shrinkage at maximum temperature	X	X	X		X	X	X	X
Specific heat	X	X	X				X	X
Tensile strength	X	X		X				
Water absorption during submersion	X	X		X				X
Water vapor transmission	X							X
Heat transmission	X				X	X	X	

Source: Donnelly, et al., Industrial Thermal Insulation: An Assessment, p. 13.

The marginal cost method involves a solution for the lowest cost thickness without having to calculate total annual costs. Using this method, the optimum thickness is determined to be the point at which the last dollar invested in insulation results in exactly one dollar in energy cost savings on a discounted cash flow basis.

The marginal cost is the change in installed cost between two successive insulation thicknesses. The marginal method is demonstrated in Fig. 2-12 by comparing the change in insulation cost (ΔC) with the savings in cost of lost energy (ΔS) resulting from the addition of an insulation thickness (ΔL). At thickness L_1, the addition of ΔL causes a large saving in lost energy (ΔS_1) at a small increase in insulation cost (ΔC_1). At thickness L_3, the addition of ΔL is rather expensive (ΔC_3) and saves only a small amount in energy costs (ΔS_3).

The optimal economic thickness is arrived at when the last dollar invested for insulation results in one additional dollar in energy cost savings. This condition is met at thickness L_2, where ΔS_2 and ΔC_2 resulting from the addition of ΔL are equal, ($m_c = m_s$).

Another effort in the calculation of economic thickness was a joint effort by Union Carbide Corporation and West Virginia University, who developed an insulation assessment manual which was published by the National Insulation Manufacturers Association (NIMA) in 1961 (9). This methodology is simplified by observing that insulation cost at a specific time and location increases linearly with thickness.

After the NIMA publication in 1961, an updated economic thickness manual, ECON-I, was published in 1973 by the Thermal Insulation Manufacturers Association (TIMA) (10). It followed the same methodology that was used in the NIMA manual, with the following differences. The NIMA manual required the user to determine the average annual heat and maintenance costs. ECON-I, on the other hand, internalized a 7.5%/year interest rate and an annual insulation maintenance cost of 1% of the total plant investment. A heat cost escalation rate of 4%/year was also included in ECON-I, with the capability to adjust for escalation rates up to 10%/year. Insulation cost factors were updated to 1973 conditions in ECON-I.

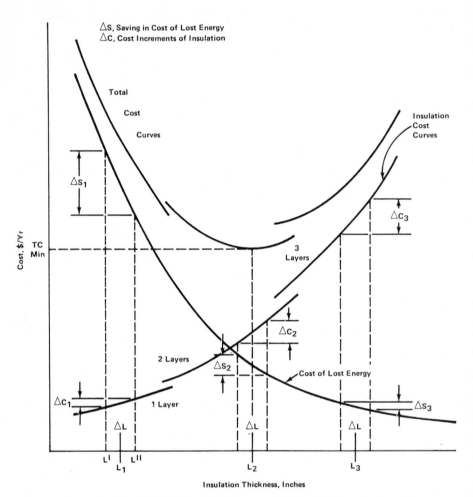

Figure 2-12. *Insulation Thickness VS Insulation Cost.* Source: Federal Energy Administration, *Economic Thickness for Industrial Insulation,* Conservation Paper 46 (Washington, D.C.: Government Printing Office, August 1976), p. 6.

In 1975, TIMA published an updated economic thickness manual called R-ECON for adding or retrofitting insulation thickness (11). It used the same methodology but updated some of the economic data to more current values and at the same time restricted the range of several important cost factors.

York Research Company has recently developed a computer analysis
of economic insulation thickness in a design manual prepared for the
Federal Energy Administration (12). The methodology incorporates several features that will make it useful to industrial users. First,
the discounted cash flow method is used to obtain the economic thickness and net present value basis. Second, insulation cost factors input by the user allow for single, double, and triple layers and also
for three complexities of pipe fittings. Third, cost of money and
other cost factors can be varied independently over a wide range. A
sample problem using the York technique follows (13):

High Temperature Steam Pipe. A new electric utility installation
will have a 1000 foot long, 16 inch nominal pipe size high pressure
steam line leading from the steam generator to the turbogenerator
building. The steam temperature will be $1050^{\circ}F$. The line will be
outside, where the average ambient temperature is $62^{\circ}F$.

The utility will have a net continuous output capacity of 565
megawatts. Total electrical generation per year is expected to be
2,772 million KWH (plant factor of 56%), and the line is expected to
be in service 8500 hours per year. Total expected capital investment
for the steam plant (boiler, piping, condenser, etc.) is $55 million.
The first year fuel cost will be $18 per ton of coal, which has a
heating value of 12,500 BTU per pound. The plant will have a depreciation period of thirty years; however, the insulation is expected
to have a useful service life of only fifteen years in the outdoor
environment. The boiler efficiency is 92%, and it requires 10,100
BTU of fuel to produce each KWH of electricity.

The return on investment requirement for the company is 15%. The
bonds issued by the company to finance the plant pay a 9% dividend,
with flotation and administrative costs adding another 15% to this
cost over the thirty year life. Fuel cost is expected to increase at
an average annual rate of 7%.

The insulation to be used on the pipe is calcium silicate, with
an alumium jacket. The average of insulation contractor unit price
estimates for this insulation on a 16 inch pipe at the time of construction is as follows:

Thickness	Dollars Per Linear Foot
2 in. single layer	11.86
4 in. single layer	17.29
5 in. double layer	27.36
7 in. double layer	36.50

The worksheets that follow display the input data and resulting economic thickness, which is 6 inches, for the above problem. As is shown, the calculation is first made by using the single layer incremental cost (m_{c1} = $2.72 per inch per linear foot). The thickness using m_{c1} is in the double layer range, so the procedure is repeated, using the double layer slope (m_{c2} = $4.57 per inch per linear foot). This incremental cost produces the economic thickness of 6 inches, which is the double layer range and is, therefore, the correct solution.

Cost of Heat Worksheet

STEP 1. Find the Multiplier for the average annual heat cost, A, using Fig. 2-13.

    a. Enter insulation project service life, years.

        n_1 = 15

    b. Enter annual fuel price increase, percent.

        i_1 = 0.07

    c. Find multiplier, A.

        A = 1.67

STEP 2. Find the first year cost of heat C_h, using Fig. 2-14.

    a. Enter heating value of fuel, BTU per lb.

        H = 12,500

    b. Enter efficiency of conversion fuel to heat, percent.

        E = 0.92

    c. Enter first year price of fuel, dollars per ton.

        P_c = 18

    d. Find first year cost of heat, dollars per million BTU.

        C_h = 0.78

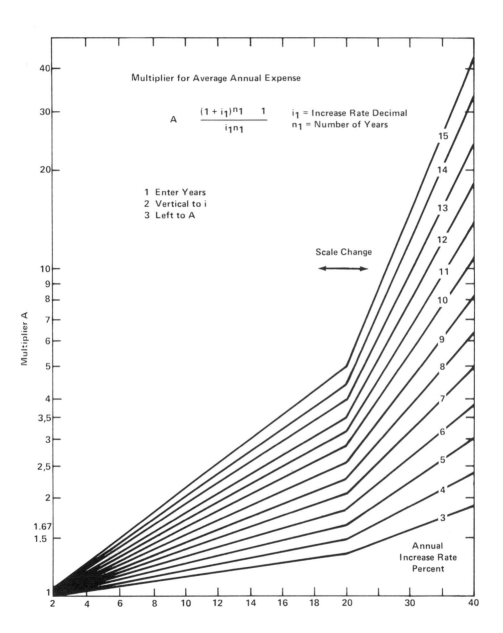

Figure 2-13. Multiplier to Apply to Present Costs for Determining The Average Annual Costs When Uniform Cost Increases Occur in Future Years.

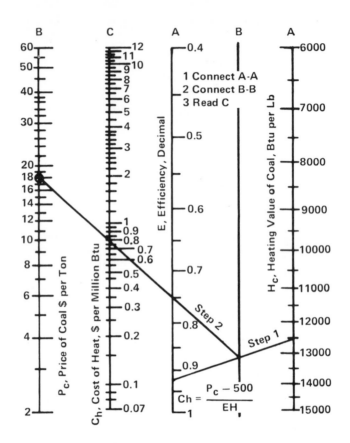

Figure 2-14. The Cost of Heat, C_h, for Coal as the Heat Source.

STEP 3. Find the average annual heat cost, using Fig. 2-15.

    a. Find the average annual cost of heat for purchased steam and electric heat plants (no operating or maintenance costs), dollars per million BTU.

       $A_{Ch} =$

    b. Find average annual heat cost for coal, oil, and gas plants (10% operation and maintenance costs), dollars per million BTU.

       $A_{Ch} = 1.40$

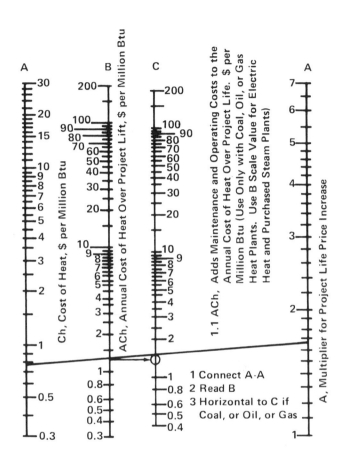

Figure 2-15. The Average Annual Value of Heat Cost Including Operating and Maintenance Costs at the Heat Producing Facility.

STEP 4. Find the compound interest factor, $(1 + i_2)^{n_2}$, using Fig. 2-16.

    a. Enter life of facility, years.

        $n_2 = 30$

    b. Enter annual cost of money to finance plant, decimal.

        $i_2 = 0.10$

    c. Find compound interest factor.

        $(1 + i_2)^{n_2}$, from Step 4.

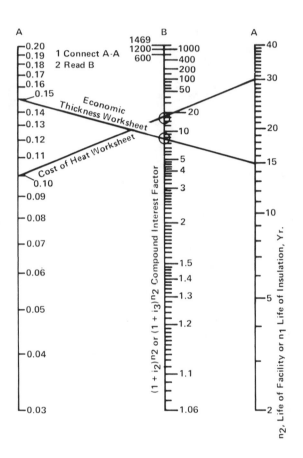

Figure 2-16. Compound Interest Factors.

STEP 5. Find the annual amortization multiplier for capital investment, B, using Figure 2-17.

    a. Find B using i_2 and $(1 + i_2)^{n_2}$, from Step 4.

       B + 0.105

STEP 6. Find the annual capital cost of heat, C_k, using Fig. 2-18.

    a. Enter expected average annual heat production, Q, millions of 10^6 BTU (10,000 BTU per KWH x 2,777 x 10^6 KWH per year x 0.92) x 10^{-6}.

       Q = 25.76

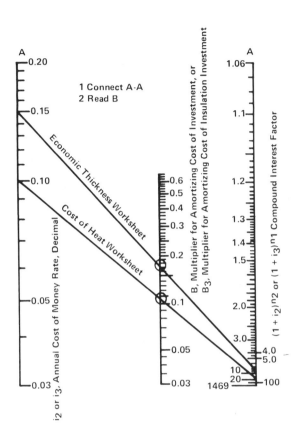

Figure 2-17. The B Multiplier.

b. Enter capital investment in heat plant, millions of dollars.

P_f = 55

c. Find annual capital cost of heat, dollars per million BTU.

C_k = 0.23

STEP 7. Find the projected cost of heat, M.

$M = A_{Ch} + C_k$

a. From Step 3, we know that A_{Ch} including maintenance and operation is, dollars per million BTU.

A_{Ch} = 1.40

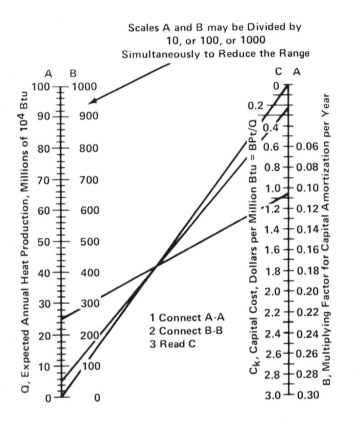

Figure 2-18. The Distribution of the Heat Production Capital Costs, C_k, Over the Energy Output on an Annualized Basis.

b. From Step 6, we know that C_k is, dollars per million BTU.

$C_k = 0.23$

c. Find the projected cost of heat, M, dollars per million BTU.

$M = 1.63$

Now proceed to the economic thickness determination worksheet.

Economic Thickness Determination Worksheet

STEP 1. Calculate mean insulation temperature, $^\circ$F.

$$a.\ t_m = \frac{t_p + t_a}{2} = \frac{1050 + 62}{2}$$

$$= 556$$

STEP 2. Enter insulation thermal conductivity using Fig. 2-19, which is for high temperature insulation materials.

    a. Using the mean temperature found in Step 1, find thermal conductivity, k, BTUH per sq ft per deg F per inch thickness.

       k = 0.6

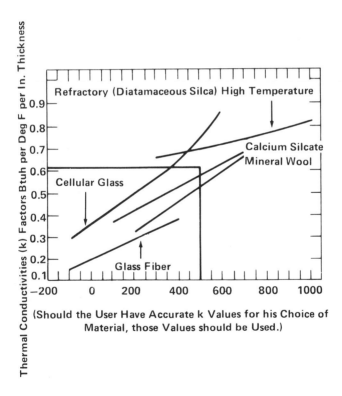

Figure 2-19. Thermal Conductivity Factors for High Temperature Insulation Materials.

STEP 3. Calculate temperature difference, $^\circ$F.

    a. $\Delta T = (t_p - t_a)(1050 - 62)$

        $= 998$

STEP 4. Enter annual hours of operation, hr.

    a. From the sample problem description, we know the plant operates, hr.

        $Y = 8500$

STEP 5. Find D_s for a flat insulation installation of D_p for a pipe insulation installation.

    a. The example problem deals with pipe insulation, so find the value for D_p on Fig. 2-20.

        $D_p = 4.3$

STEP 6. Find B_3, the multiplier for amortization cost of insulation investment, using Figs. 2-16 and 2-17.

    a. i_3, the annual cost of money rate, decimal

        $i_3 = 0.15$

    b. n_1, the life of the insulation, years

        $n_1 = 15$

    c. The compound interest factor

$$(1 + i_3)^{n_1} = 8.1$$

    d. The multiplier for amortizing cost of investment, B_3

        $B_3 = 0.171$

STEP 7. Use the incremental cost, m_{c1}

    a. m_{c1}, single layer price is determined by

        $(P_2 - P_1/L_2 - L_1) = (17.29 - 11.86/4 - 2)$

        $m_{c1} = 2.72$

        m_{c2}, double layer price, dollar per in. thickness per linear ft is determined by $(36.50 - 27.36/7 - 5)$

        $m_{c2} = 4.57$

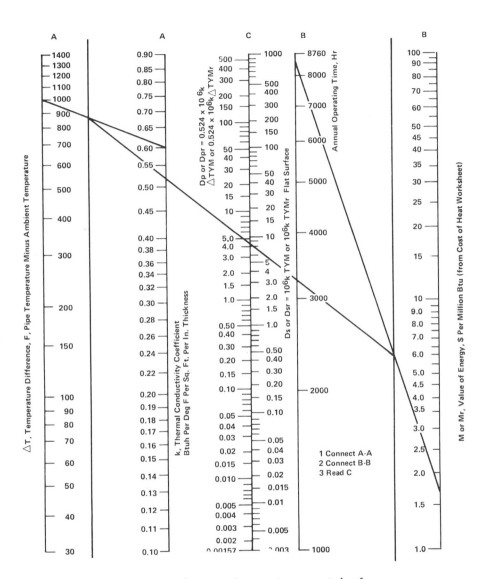

Figure 2-20. Annual Cost of Heat Lost or Gained.

STEP 8. Find Z_p, the factor for round pipe surface, using Fig. 2-21.

    a. For a single layer of insulation, m_{c1}

$$Z_{p1} = 8.5$$

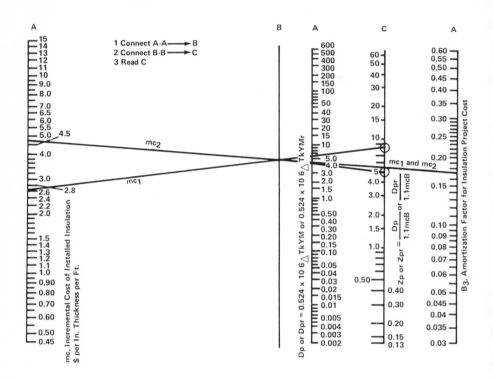

Figure 2-21. Z_p Factor for Pipe Surfaces.

b. For a double layer of insulation, m_{c2}

$Z_{p2} = 5.0$

STEP 9. Calculate kR_s.

a. k x 0.7 (R_s value of 0.7 is typical)

$kR_s = 0.42$

STEP 10. Use Fig. 2-22 to determine the economic thickness.

a. Use the value found for Z_{p1} in Step 8, in.

$W_1 = 8$

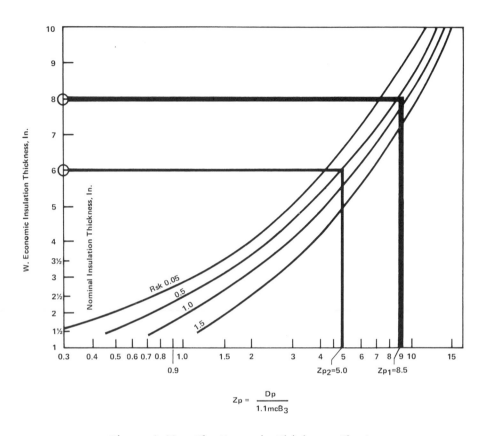

Figure 2-22.   The Economic Thickness Chart.

STEP 11. If the economic thickness found in Step 10 is within the
single layer range (corresponding to the single layer slope, m_{c1}, used
in Step 7), the thickness is correct.   If the thickness is beyond the
single layer range, repeat the procedure from Step 7 on, using the
double layer slope, m_{c2}.

   a. You would use the value for Z_{p2}, 5 in.

        $w_2 = 6$

STEP 12. If the economic thickness, using the double layer slope,
is in the triple layer range, repeat the procedure from Step 7 on,
using the triple layer slope, m_{c3}.

Conclusion

In past years, first cost analysis was the primary criterion for
capital expenditures; hence, using more than minimal amounts of insulation was not justified. This rationale is no longer acceptable.
Life cycle costing should become the predominant industrial design
criterion. Insulation of all heated surfaces not presently insulated
is economically justified. Analysis of economic insulation thickness
should be performed to determine the potential for energy and cost
savings.

SPACE CONDITIONING

Space conditioning requires a substantial amount of energy in the
industrial and commercial sectors. Figure 2-23 depicts commercial
building energy consumption data for 1976. As can be seen, space
conditioning accounts for 48% of the energy used. Energy usage depends
also upon the end use of the building, which is shown in Table 2-10.

Significant potential exists for energy conservation in commercial buildings. In a study performed by Arthur D. Little, the potential for energy conservation was estimated for existing buildings and
new construction. These estimates are shown in Table 2-11. Each index
in the table represents the probable energy consumption across all new
or existing buildings within the same category. The factors shown
represent the extent of conservation which could be achieved using
practical methods and existing materials. In addition, the estimates
allow for more technological improvements in selected HVAC and electrical components between now and 1990. The areas most susceptible to
energy conservation appear to be lighting and space heating and cooling
in the commercial sector.

Existing means for improving space conditioning energy efficiency
include the following:

Table 2-10

1970 Energy Demand Building Type by End-Use
(trillions of Btu's)

	Space Heating	Air Cond.	H₂O Heating	Cooking	App & Lighting	Other	Total
Residential							
Mobile Homes	137	6	24	18	23	0	209
Single Family Detached	6,219	132	1,297	283	635	0	8,567
Low Density	1,199	10	253	84	125	0	1,670
Multi-Family Low Rise	346	12	123	44	64	0	589
Multi-Family High Rise	244	3	72	25	32	0	376
Total Residential	8,146	163	1,769	454	879	0	11,411

	Space Heating	Air Cond.	H₂O Heat & App	Refrigeration	Lighting	Unallocated	Total
Commercial							
Office Buildings	480	22	53	0	62		617
Retail Establishments	335	31	82	127	83		658
Schools	603	28	89	0	80		799
Hospitals	217	12	95	0	55		379
Other	572	57	146	0	147		922
Total Commercial	2,206	150	464	127	427	510	3,885
Total	10,352	313	2,233	582	1,306	510	15,296

Note: Electricity measured at point-of-entry and does not include power plant losses.

Source: Arthur D. Little, Inc., Energy Use Patterns.

Table 2-11

Energy Conservation Factors for
Residential and Commercial Buildings
(1970 = 1.00)

	Existing Buildings	New Construction
Mobile Home		
Hot Water Heating	.95	.75
Space Heating – Gas/Oil (Elec)	.80 (.83)	.60 (.65)
Cooling – Gas/Oil (Elec)	.86 (.88)	.69 (.74)
Heat Pump (Heating & Cooling)	–	.45
Single Family, Attached & Detached		
Hot Water Heating	.95	.75
Space Heating	.71 (.74)	.50 (.55)
Cooling	.87 (.89)	.59 (.61)
Heat Pump (Heating & Cooling)	–	.45
Multi-Family, Low-Rise & High-Rise		
Hot Water Heating	.95	.75
Auxiliary Equipment	.95	.90
Space Heating	.76 (.80)	.55 (.60)
Cooling	.85 (.87)	.56 (.62)
Heat Pump (Heating & Cooling)	–	.45
Office Buildings		
Lighting	.80	.50
Auxiliary Equipment	.95	.90
Space Heating	.78	.60
Cooling	.82	.53
Hot Water Heating	.95	.90
Retail Establishments		
Lighting	.70	.50
Auxiliary Equipment	.95	.90
Space Heating	.76	.50
Cooling	.76	.54
Hot Water Heating	.95	.90
Schools, Educational		
Lighting	.80	.50
Auxiliary Equipment	.95	.90
Space Heating	.79	.50
Cooling	.81	.59
Hot Water Heating	.95	.90
Hospitals		
Lighting	.80	.60
Auxiliary Equipment	.95	.90
Space Heating	.84	.60
Cooling	.91	.67
Hot Water Heating	1.00	.90

Source: Arthur D. Little, Inc., estimates.

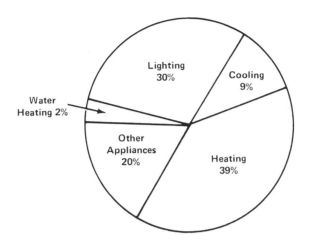

Figure 2-23. Commercial Energy Consumption Data (1976). Source: Eric Hirst, Oak Ridge National Laboratory.

• Where possible, the maximum temperature in occupied buildings should be maintained at a temperature between 65° and 68°F during the normal work day, and 55°F on weekends and holidays during the heating season.

• The maximum temperature in unoccupied areas, such as storage areas, should be lowered to 50°F during the heating season.

• The minimum temperature in air conditioned areas should be in a range between 80° and 82°F during the cooling season.

• Plant heating and air conditioning systems should be studied to determine if they are correctly designed; many are not. Equipment selection is an important aspect.

• Equipment suppliers or qualified engineers should determine the extent to which the heating, ventilating, and air conditioning equipment is optimally suited to performance. In many cases, HVAC equipment is oversized, and occasionally one boiler in a multiboiler system can be eliminated by more efficient use of the others.

• Humidifiers should be utilized to maintain higher humidity levels which provide comfort at lower temperatures.

- Under most conditions, proper insulation of ceilings and walls will pay for itself in heating and cooling costs. For existing buildings, expert technical advice should be sought.

- All windows should be closed, locked, and sealed at the beginning of the heating season. Keep a constant lookout for opened windows and doors, as large amounts of heated or cooled air can escape from unregulated openings.

- Stratification of air in the facility should be eliminated, since this will cool the ceiling and warm the floor. This can be accomplished by simply mixing the air, placing vertical ductwork from the floor to the ceiling and installing a fan which can pull the air off the floor and discharge it at the upper end of the duct.

- Spot heaters in low density work areas can be used to reduce the general heating load in the building.

- Where safety permits, reduction in ventilating air can result in substantial power savings. In some cases, reducing ventilating air quantities by 50% can reduce the power load by 12.5%.

- The use of infrared heaters in high bay areas and warehouses should be considered. Infrared heat provides comfort and prevents condensation on stored materials. It requires less fuel than other kinds of heating, because it heats solid objects in the radiation path without having to heat the air in the room. Fuel savings of more than 20%, when compared to conventional heating, are common.

- The burners in the boilers should be checked regularly, in order to insure proper combustion. At least once a month, during winter, it would be wise to have a qualified person with necessary instruments adjust the boilers for maximum efficiency.

- The use of exhaust air to precondition incoming air should be considered.

- During winter months, the recirculation of air should be maximized wherever possible to reduce fresh make-up air heating requirements.

- Usage of gasoline powered equipment should be limited to one area of the facility, in order to allow maximum reduction of ventilation in remaining parts of the facility.

- Operations requiring higher rates of ventilation should be separated from other operations and should be accorded their own air supply and exhaust system.

- High efficiency filters should be installed in high contaminant process areas (such as welding shops) to allow greater recirculation of heated plant air.

- Truck doors, loading docks, and vestibules should be regulated to prevent the loss of conditioned air. Control methods might include:

  (1) interlocks to turn off air handling units in the area when the doors are open;

  (2) separate openings from other areas by partitions or curtains;

  (3) installation of air curtains for smaller openings.

- An economizer cycle can be incorporated in HVAC installations to utilize 100% outdoor air for cooling purposes wherever possible, in lieu of operating the refrigeration equipment.

- The installation and use of smaller packaged steam generating units or hot water boilers should be considered for the end of the heating season. Operating large boilers at low load and low efficiency wastes energy and is costly.

- The use of air conditioning reheat systems for humidity control should be reduced, except where absolutely necessary.

NOTES AND REFERENCES

1. R. D. Glenn, "Energy Conservation Opportunity in Steam Use," Energy Conservation Through Effective Energy Utilization, NBS Special Publication 403 (Washington, D.C.: Government Printing Office, 1974).

2. William Sisson, "Nomogram Determines Loss of Steam and Fuel Due to Leaks," Power Engineering, 79 (September 1975), p. 55.

3. U.S. Department of Commerce, Energy Conservation Program Guide for Industry and Commerce, NBS Handbook 115 (Washington, D.C.: Government Printing Office, 1974), p. 3-22.

4. Ibid., p. 3-23.

5. Thomas C. Elliott, "Demand Control of Industry Power Cuts Utility Bills, Points to Energy Savings," Power, 120 (June 1976), p. 21. Reprinted with permission from POWER, June 1976.

6. Federal Energy Administration, Economic Thickness for Industrial Insulation, Conservation Paper 46 (Washington, D.C.: Government Printing Office, August 1976), p. 1.

7. R. G. Donnelly, V. J. Tennery, D. L. McElroy, R. G. Godfrey, and J. O. Kolb, Industrial Thermal Insulation: An Assessment, Report No. TID-27120 (Washington, D.C.: Energy Research and Development Administration, August 1976), p. 28.

8. Ibid., p. 11.

9. How to Determine Economic Thickness of Insulation (New York: National Insulation Manufacturers Association, 1961).

10. ECON-I, How to Determine Economic Thickness of Thermal Insulation (Mt. Kisco, New York: Thermal Insulation Manufacturers Association, 1973).

11. R-ECON, A Method for Determining Economic Thickness of Add on Thermal Insulation (Mt. Kisco, New York: Thermal Insulation Manufacturers Association, 1975).

12. Federal Energy Administration, Economic Thickness for Industrial Insulation.

13. Ibid., p. 100.

Chapter 3

THERMODYNAMIC AVAILABILITY ANALYSIS

Thermodynamic principles dictate that all energy, regardless of its source, has a cost and that the measure of effectiveness with which fuel is used in industrial processes requires consideration of properties other than energy alone. The boundaries of energy discussions are being expanded to cover energy "quality" as well as quantity. Thus, discussions are being broadened, using J. W. Gibbs' concept of "free energy."

Every plant engineer knows that one BTU of energy in the form of $2000^{o}F$ heat has more value that one BTU of energy in the form of $150^{o}F$ coolant water. He also knows that the energy value of one BTU of electrical energy at the plant has a higher quality than one BTU of coal. The point is that quality, as well as quantity, must be considered in evaluating energy sources, forms, and use. The laws of thermodynamics indicate that the relevant quantity is a property called available useful work, which is related to a property called entropy. The essence of the issue is the proper discrimination between energy and thermodynamic availability--between the quality of an energy form and the quality of the energy offered by that form (1).

WHAT IS AVAILABLE WORK?

In order to understand the concepts regarding available work and "second law" analysis, which will be discussed in this chapter and in the remainder of this book, one must have an understanding of what is meant by the first and second laws of thermodynamics.

First Law Efficiency

The first law of thermodynamics is simply the law of conservation of energy: that energy can neither be created nor destroyed. In symbolic terms:

$$Q - W' = E_{in} - E_{out}$$

where:

Q is the net heat flowing into the system during the process.

W' is the net work done by the system during the process.

E_{in} is the internal energy of the system at the start of the process.

E_{out} is the internal energy of the system at the end of the process.

This means that the net energy content of a system in a given period is equal to the energy content of the material leaving the system, plus the heat added to the system.

Conventional thermodynamic analysis of energy use commonly consists of conducting heat and work balances around the process and is based on the first law of thermodynamics. To determine energy use efficiency in terms of the first law,

$$\eta = \frac{E_o}{E_i}$$

where:

η = energy efficiency

E_o = energy transfer achieved by a system

E_i = energy input to the system

When the theoretical maximum value of this ratio is greater than 1, it is usually called a coefficient of performance. When $\eta_{max} \leq 1$, it is usually called an efficiency.

As a general figure of merit, the first law efficiency has some drawbacks: First, its maximum value depends on the system and on temperatures and may be greater than, less than, or equal to 1. Second, it does not adequately emphasize the central role of the second law in governing the possible efficiency of energy use. Finally, it cannot readily be generalized to complex systems in which the desired output is some combination of work and heat.

The first law of thermodynamics does, however, give us some insight into process efficiency, in that we can compare the total energy leaving a process as useful product to the total energy entering the process. Energy which escapes the useful streams leaving the process altogether is thereby quantified.

Second Law Efficiency

The use of the first law of thermodynamics is not adequate for considering minimum task energy. We know that energy is not lost, and that, in any process involving heat, the constraints of the second law of thermodynamics usually insures that not all of the energy can be made available in useful form. The second law allows us to define a quantity, available work, that has the dimension of energy yet is actually consumed in a process. There is an upper limit to the amount of work that can be produced from thermal energy at a temperature T_1 with an ambient temperature of T_o. This upper limit is the available work, or the maximum work, that can be provided by the system. Work is the highest quality (lowest entropy) form of energy and, consequently, the most valuable.

In order to understand and calculate second law efficiencies, an important concept is available work. Its definition is given as

B = maximum work that can be provided by a system as it proceeds to a specified final state in thermodynamic equilibrium with the atmosphere.

Thus, a more meaningful measure of process efficiency is derived from the second law of thermodynamics. The second law efficiency parameter compares the amount of work available from a given quantity of energy to the amount of work needed to operate the process. The concept of available work is illustrated by several examples given in Table 3-1. An assessment based on availability simultaneously accounts for both the quantity and quality of energy passing through the system. Properly done, this assessment will tell us how much work is lost in a given process, how much of the lost work is wasted, and where it is wasted.

Table 3-1

Available Work Provided by Sources and Need by End Users

Note: T_1 (hot) $> T_2$ (warm) $> T_0$ (ambient) $> T_3$ (cool)

	Work W_{in}	Fuel with Heat of Combustion H	Heat Q_1 from Hot Reservoir at T_1
Sources	(e.g., water power, wind power, raised weight) or electricity if the wall socket is treated as the source	(e.g., coal, oil, gas)	(e.g., geothermal source, solar collector source) also fission reactors and fossil-fuel plants if alternatives to thermal operation are excluded
	$B = W_{in}$	$B \equiv |\Delta H|$* (Usually to within 10%)	$B = Q_1 \left(1 - \dfrac{T_0}{T_1}\right)$
	Work W_{out}	Heat Q_2 Added to Warm Reservoir at T_2	Heat Q_3 Extracted from Cool Reservoir at T_3
End Uses	(e.g., turning shafts, pumping fluids, propelling vehicles)	(e.g., space heating, cooking, baking, drying)	(e.g., refrigerating, air conditioning)
	$B_{min} = W_{out}$	$B_{min} = Q_2 \left(1 - \dfrac{T_0}{T_2}\right)$	$B_{min} = Q_3 \left(\dfrac{T_0}{T_3} - 1\right)$

*An exact consideration is required for each fuel.

Source: Water Carnahan, et al., Efficient Use of Energy: A Physics Perspective (Princeton, New Jersey: The American Physical Society, January 1975), p. 25.

In terms of the available work concept, the definition of second law efficiency can be stated as

$$\varepsilon = \frac{B_{min}}{B_{actual}}$$

This states that efficiency is equal to the ratio of the least available work that could have done the job to the actual available work used to do the job.

Table 3-2 gives a set of first law and second law efficiencies for single source-single output devices.

Loss of Thermodynamic Availability

An absolute measure of the performance of a given process can be given by the concept of "lost work." Lost work is the difference between the available work flowing into a process and the available work flowing from the process. The total annual lost work for a given process is an absolute measure of the annual energy consumption burden contributed by that process; it can represent a good basis for the establishment of energy conservation priorities.

The primary benefit of the concept of "lost work" is that it is an accurate and easily applied measure of the total burden imposed by any given process unit on available energy resources. This measure is universally applicable to all process units, regardless of type or of the industrial sector in which it is found. The measure of effectiveness of an industrial process, then, will be the increase in available useful work between raw materials entering and industrial products leaving the process, divided by the available useful work of the fuel consumed. A detailed analysis of the elements of work loss and the process work requirements will reveal potential areas for improvement which may otherwise go undetected.

Table 3-2

First Law and Second Law Efficiencies for
Single Source–Single Output Devices

End Use \ Source	Work W_{in}	Fuel: Heat of Combusiton $|\Delta H|$ Available Work B	Heat Q_1 From Hot Reservoir at T_1
Work W_{out}	1. $\eta = W_{out}/W_{in}$ $\epsilon = \eta$ (e.g., electric motor)	2. $\eta = W_{out}/|\Delta H|$ $\epsilon = \dfrac{W_{out}}{B} \; (\equiv \eta)$ (e.g., power plant)	3. $\eta = W_{out}/Q_1$ $\epsilon = \dfrac{\eta}{1 - (T_0/T_1)}$ (e.g., geothermal plant)
Heat Q_2 added to warm reservoir at T_2	4. $\eta(COP) = Q_2/W_{in}$ $\epsilon = \eta\left(1 - \dfrac{T_0}{T_2}\right)$ (e.g., electrically driven heat pump)	5. $\eta(COP) = Q_2/|\Delta H|$ $\epsilon = \dfrac{Q_2}{B}\left(1 - \dfrac{T_0}{T_2}\right)$ (e.g., engine-driven heat pump)	6. $\eta(COP) = Q_2/Q_1$ $\epsilon = \eta\,\dfrac{1 - (T_0/T_2)}{1 - (T_0/T_1)}$ (e.g., solar hot water heater)
Heat Q_3 extracted from cool reservoir at T_3	7. $\eta(COP) = Q_3/W_{in}$ $\epsilon = \eta\left(\dfrac{T_0}{T_3} - 1\right)$ (e.g., electric refrigerator)	8. $\eta(COP) = Q_3/|\Delta H|$ $\epsilon = \dfrac{Q_3}{B}\left(\dfrac{T_0}{T_3} - 1\right)$ (e.g., gas-powered air conditioner)	9. $\eta(COP) = Q_3/Q_1$ $\epsilon = \eta\,\dfrac{(T_0/T_3) - 1}{1 - (T_0/T_1)}$ (e.g., absorption refrigerator)

Source: Carnahan, et al., Efficient Use of Energy: A Physics Perspective, p. 27.

POTENTIAL FOR CONSERVATION IN INDUSTRY

Our efforts to increase the efficiency with which we use energy
resources could be greatly enhanced by the adoption of thermodynamic
availability as a measure of the effectiveness of our energy utilization. If we analyze the theoretical potential for energy conservation, we should note that there has been considerable discussion in
technical circles as to whether we should be concerned with the first
law of thermodynamics or the second law of thermodynamics (2). Some
work has been done toward evaluating the efficiency of various industrial processes according to both of these viewpoints (3).

In the Battelle study, an attempt was made to calculate the minimum amount of energy and availability required in each of seven industries for the converting of raw materials into the products of that
industry (4). In the steel, container glass, and aluminum industries,
minimum energy and availability requirements were calculated. In the
copper, rubber, plastics, and paper industries, such minimum values
could not be calculated because the energy and availability of the
raw materials are greater than they are for the products. Theoretically, these four industries do not require energy to convert their raw
materials into their products. In fact, however, the processes currently in use for converting the raw materials into products do require
significant energy.

A brief summary of the results of the analyses of the seven industries is given in Table 3-3. As seen, the efficiency and effectiveness of energy use in industrial processes vary widely: some processes
waste relatively little of their input energy, while other processes
waste nearly all their input energy.

The high efficiencies of rubber, plastics, and, to some extent,
paper reflect the fact that these industries are essentially processing fuels. The energy values of the feedstock and the product in
these processes lead to a falsely high value of efficiency.

Availability destruction for the process and/or the ratio of
availability destruction to the availability of purchased fuel and
electricity is another measure of process performance. Table 3-4 lists
these quantities for each industry studied. The fact that the availability destruction is greater than the total purchased availability
for some industries indicates that some of the feedstock availability
is being destroyed. Thus, although the rubber, paper, and plastics
manufacturing processes have high efficiencies, these processes also
have large availability destructions, relative to their purchased
availability. This illustrates the point that a high first law efficiency does not always indicate good process performance or even low
availability destruction.

Table 3-3

Efficiency and Effectiveness Summary
for Seven Industries

	Percent			
	Direct Equivalent		Fuel Equivalent	
Industry	Efficiency	Effectiveness	Efficiency	Effectiveness
Steel	42.0	40.1	41.0	39.0
Copper	3.5	9.2	2.7	7.10
Aluminum	29.6	32.6	14.2	15.1
Container glass	23.9	22.1	21.3	19.5
Rubber High temperature	92.4	96.2	92.4	96.2
Selected plastics Polyethylene				
Low-density	83.5	93.1	79.0	88.0
High-density	75.1	87.9	73.4	85.5
Polystyrene	82.7	97.9	82.2	97.2
Polyvinyl chloride	65.5	90.3	65.0	89.5
Paper	44.3	44.5	44.3	44.5

Source: E. H. Hall, et al., Evaluation of the Theoretical Potential for Energy Conservation in
Seven Basic Industries (Columbus, Ohio: Battelle Columbus Laboratories, July
1975), p. 65.

A study by Ross and William provides second law efficiencies of
selected energy intensive industries, as shown below (5):

Iron and Steel	35%
Petroleum Refining	12%
Paper	0.4%
Primary Aluminum Production	17%
Cement Manufacturing	17%

They found that a goal of 25-50% is reasonable for practical systems,
the low end being appropriate for general purpose systems with wide
applicability and the high end for sophisticated special purpose
systems. Electricity generation exhibits second law efficiencies of
30-35%.

Table 3-4

Availability Destruction, Purchased Availability,
and Their Ratio

Industry	Availability Destruction[a]	Purchased Availability[b]	A.D./P.A. Percent
	(x 10^6 Btu/ton of Product)		
Steel	12.0	17.8	67
Copper	62.2	36.6	170
Aluminum	61.8	89.6	69
Glass containers	8.9	10.7	83
High Temp rubber	2.1	1.5	140
Low density polyethylene	3.3	1.9	174
High density polyethylene	6.3	2.8	225
Polyvinyl chloride	2.7	1.6	169
Polystyrene	1.9	1.8	106
Paper	25.8	1.6	1616

(a) Based on direct equivalence of electrical input.
(b) Includes direct equivalent of electrical input; does not include feedstock availability.

Source: Hall, Evaluation of the Theoretical Potential for Energy Conservation, p. 66.

Gyftopoulos et al., in a Ford Foundation study, calculated the efficiencies of five selected industries and determined the potential for efficiency gains based on currently available technology and for theoretical limits computed by means of the concept of available useful work (see Table 3-5) (6). The large variations that exist between current practices and minimum theoretical limits indicate the potential which is available for major long term reductions in fuel consumption through basic process modifications.

On the basis of the above studies, Table 3-6 presents estimates of the energy efficiency potentials of seven industries. It should be realized that the technological minimum level may not be obtainable in practice. There are inescapable heat losses in a practical sense, since it is uneconomical to invest capital for perfect insulation or for heat recovery at low temperature differences. It does seem reasonable that overall improvements in energy use per unit of production

Table 3-5

Comparison of Specific Fuel Consumption of Known Processes
with Theoretical Minimum for Selected U.S. Industries

	1968 Specific Fuel Consumption (Btu/ton)	Potential Specific Fuel Consumption Using Technology Existing in 1973 (Btu/ton)	Theoretical Minimum Specific Fuel Consumption Based Upon Thermodynamic Availability Analysis (Btu/ton)
Iron and Steel	26.5×10^6	17.2×10^6	6.0×10^6
Petroleum Refining	4.4×10^6	3.3×10^6	0.4×10^6
Paper	$39.0 \times 10^{6*}$	$23.8 \times 10^{6*}$	Greater than $-0.2 \times 10^{6\dagger}$ Smaller than $+0.1 \times 10^6$
Primary Aluminum Production**	190×10^6	152×10^6	25.2×10^6
Cement	7.9×10^6	4.7×10^6	0.8×10^6

Source: From POTENTIAL FUEL EFFECTIVENESS IN INDUSTRY. Copyright
1974. The Ford Foundation. Reprinted with permission of Ballinger
Publishing Company.

Table 3-6

Energy Conservation Potential
for Selected Industries
(1972 = 100)

Industry	Savings Attainable by 2000	Technological Minimum	Second Law Minimum
Paper	75	69	0
Cement	75	60	10
Chemicals	80	50	0
Aluminum	60	45	29
Steel	75	50	30
Glass	75	35	20
Petroleum Refining	80	70	10

of about 25-35% throughout industry is possible, with some industries
attaining greater savings than others.

Chapters 7-13 will evaluate the potentials and technologies available for achieving these savings in selected industries.

NOTES AND REFERENCES

1. Charles A. Berg, "A Technical Basis for Energy Conservation,"
    Technology Review, 76 (February 1974). See also George
    Hatsopoulos and J. H. Kernan, Principles of General Thermody-
    namics (New York: John Wiley, 1961).

2. See Berg, "A Technical Basis for Energy Conservation"; H. C.
    Hottel and T. B. Howard, New Energy Technology--Some Facts
    and Assessments (Cambridge, Massachusetts: MIT Press, 1971);
    and Elias Gyftopoulos, Lazaros Lazaridis, and Thomas Widmer,
    Potential Fuel Effectiveness in Industry (Cambridge, Massachu-
    setts: Ballinger Publishing Company, 1974).

3. E. H. Hall et al., Evaluation of the Theoretical Potential for
    Energy Conservation in Seven Basic Industries (Columbus, Ohio:
    Battelle Columbus Laboratories, July 1975). See also Marc
    Ross and Robert Williams, Assessing the Potential for Energy
    Conservation (Albany, New York: The Institute for Public
    Policy Alternatives, July 1, 1975; and Gyftopoulos et al.,
    Potential Fuel Effectiveness in Industry.

4. E. H. Hall et al., Evaluation of the Theoretical Potential for
    Energy Conservation.

5. Ross and Williams, Assessing the Potential for Energy Conser-
    vation.

6. Gyftopoulos et al., Potential Fuel Effectiveness in Industy,
    p. 8.

Chapter 4

WASTE HEAT RECOVERY

While there are significant differences between individual industries, approximately 50% of the energy consumed by all U.S. industry
for heat and power is rejected as waste heat to air and water. The
generation of electric power consumes about 25% of the total primary
energy consumed in the U.S., and approximately 60-70% of such energy
is rejected as waste heat. In many instances, this wasted energy is
not only costly but can produce undesirable environmental damage. The
problem addressed in this chapter is how to convert waste heat into
useful work in such a way as to increase profits. Various types of
heat recovery devices will be discussed, and case studies of applications will be provided.

Waste heat is defined as heat which is rejected from a process
at a temperature high enough above the ambient temperature to permit
the extraction of additional heating value from it. Waste energy value
can be categorized according to three temperature ranges. The high
temperature range refers to temperatures above $1200^{\circ}F$. The medium
temperature range is between $450^{\circ}F$ and $1200^{\circ}F$, and the low temperature
range is below $450^{\circ}F$ (1).

High and medium temperature waste heat can be used to produce
process steam, to generate electricity, to preheat combustion air,

or used for low quality process requirements. In the low temperature
range, waste energy, which would be otherwise useless, can be used for
space heating and cooling or can be made useful by application of a
heat pump.

The essential quality of heat is not amount but value. This distinction is apparent when one compares the value of recovering similar
quantities of heat at 250°F or 1000°F, or of a given amount of heat
in a corrosive or in an inert environment. For example, if process
steam is needed, waste heat in clean flue gas at 1200°F is quite useful; the same heat in a dirty flue at 300°F might not be worth bother-
ing with and certainly would be much more difficult to use.

WHY WASTE HEAT RECOVERY?

There are three basic reasons for recovering waste heat:

1. Economic. Energy costs are skyrocketing, and recovery
can reduce overall costs. Many investments are cost effective,
and investment paybacks of one year or less are not unusual.

2. Heat availability. Scarce energy supplies can cause
plant shutdowns. Heat is readily available in most plants in
the form of waste heat.

3. Conservation of the nation's natural resources. We are
using up our scarce energy supplies, and severe shortages will
be prevalent in a few decades. With heat recovery, we can prolong this time of reckoning.

The economic recovery of waste heat depends upon a number of factors. First, there must be a use for the waste heat within the plant
or facility. Second, an adequate quantity of waste heat must be available. An estimate of the quantity of waste heat available can be made
using the first law of thermodynamics. Third, the heat must be of adequate quality for the intended use; for example, heat available at
250°F cannot be used for a process requiring 500° heat. Heat quality
and availability can be determined by using the second law of thermodynamics. Fourth, the heat must be transported from the waste stream

to the process or material where it is to be used. This is a problem
of heat transfer. Fifth, the process must be economic or have a fairly short payback time.

POTENTIAL FOR WASTE HEAT RECOVERY

    Waste heat is normally available in the form of sensible heat
being exchanged from a combustion process or of heat being released
from a material in process. This waste heat is released to the sur-
roundings and loses its value very rapidly. In order for it to be recovered, one must identify the sources of this waste heat and determine how the heat can be utilized.

    Table 4-1 provides an indication of high quality waste heat from
various process equipment which are usually directly fired. This
waste heat can still be directly used, e.g., for drying, where no
heat exchanger need be employed.

Table 4-1

High Temperature Waste Gas Sources

Type of Device	Temperature F
Nickel refining furnace	2500 - 3000
Aluminum refining furnace	1200 - 1400
Zinc refining furnace	1400 - 2000
Copper refining furnace	1400 - 1500
Steel heating furnaces	1700 - 1900
Copper reverberatory furnace	1650 - 2000
Open hearth furnace	1200 - 1300
Cement kiln (Dry process)	1150 - 1350
Glass melting furnace	1800 - 2800
Hydrogen plants	1200 - 1800
Solid waste incinerators	1200 - 1800
Fume incinerators	1200 - 2600

Sources: W. M. Rohrer and K. Kreider, "Sources and Uses
of Waste Heat," Waste Heat Management Guidebook,
NBS Handbook 121 (Washington, D.C.: Government
Printing Office, January 1977), p. 5.

Table 4-2 provides temperature ranges of waste gas from industrial process equipment in the medium quality range. Waste heat in this temperature range is of sufficient quality to be used to extract mechanical work, e.g., steam generation or operation of a gas turbine.

Table 4-3 provides heat sources in the low temperature range. Low temperature heat can be used for preheating process materials, for generating hot water, for space heating or cooling, or as the heat source for a heat pump.

Waste heat recovery can be divided into two categories, depending on the ultimate use of the recovered energy. For example, if combustion air is preheated by using hot waste gases, the waste energy is returned directly to the process, with the primary effect of reducing the specific energy consumption of that process. On the other hand, if the hot waste gases are used to generate steam or hot water which would be used somewhere else in the facility, the waste heat recovery is considered as a secondary operation. This is due to the fact that while there is an overall reduction in specific energy consumption for the facility, there is no energy reduction for the process from which the waste heat originates.

Table 4-2

Medium Temperature Waste Gas Sources

Type of Device	Temperature F
Steam boiler exhausts	450 – 900
Gas turbine exhausts	700 – 1000
Reciprocating engine exhausts	600 – 1100
Reciprocating engine exhausts (turbocharged)	450 – 700
Heat treating furnaces	800 – 1200
Drying and baking ovens	450 – 1100
Catalytic crackers	800 – 1200
Annealing furnace cooling systems	800 – 1200

Source: Rohrer and Kreider, "Sources and Uses of Waste Heat," p. 6.

Table 4-3

Low Temperature Waste Gas Sources

Source	Temperature F
Process steam condensate	130 – 190
Cooling water from:	
Furnace doors	90 – 130
Bearings	90 – 190
Welding machines	90 – 190
Injection molding machines	90 – 190
Annealing furnaces	150 – 450
Forming dies	80 – 190
Air compressors	80 – 120
Pumps	80 – 190
Internal combustion engines	150 – 250
Air conditioning and refrigeration condensers	90 – 110
Liquid still condensers	90 – 190
Drying, baking and curing ovens	200 – 450
Hot processed liquids	90 – 450
Hot processed solids	200 – 450

Source: Rohrer and Kreider, "Sources and Uses of Waste Heat," p. 6.

Methods for utilizing the waste heat can be classified into the following categories:

- Direct utilization, e.g., for drying or preheating process materials when no external heat exchanger is employed.

- Recuperation, in which waste gases and air or other gas for preheating are separated by a metallic or, in cases of very high temperatures, a refractory heat exchange surface. Transfer of energy from one fluid to another occurs continuously.

- Regeneration, in which heat from waste gas is conducted to and stored in a heat exchange medium, in a refractory, or in metallic materials and subsequently heats air for preheating. The gas flow alternately shares the same heat transfer surfaces and is switched either by means of a flow reversing valve or by rotating the heat storage matrix.

- Waste heat boiler, a form of recuperation in which hot waste gases generate process steam or hot water. Both water tube or fire tube designs can be used.

- Co-generation, in which electricity and process steam are generated together (e.g., a steam turbine driving an electrical generator). After removing the necessary energy for doing work, the steam turbine exhausts partially spent steam at a lower pressure than that of the inlet pressure. The energy in the turbine exhaust steam can then be used for process heat in the usual ways. This topic will be covered in greater detail in Chapter 6.

- Energy cascading, in which the energy is used at its highest quality first and then used at lower qualities in other associated processes, until the energy is of such low quality that it is no longer useful. The chemical industry has been proficient in the use of energy cascading, in which the heat condensation from one distillation column operates a second unit. The food industry also has many opportunities for energy cascading.

To use the waste heat from the sources described in Tables 4-1, 4-2, and 4-3, heat transfer must occur from the waste stream to the process stream. The waste recovery systems available will be discussed later in this chapter. Potential applications for the waste heat are described below (2):

- Medium to high temperature exhaust gases can be used to preheat the combustion air for:

      Boilers using air preheaters
      Furnaces using recuperators
      Ovens using recuperators
      Gas turbines using regenerators

- Low to medium temperature exhaust gases can be used to preheat boiler feed water or boiler makeup using economizers, which are simply gas to liquid water heating devices.

- Exhaust gases and cooling water from condensers can be used to preheat liquid and/or solid feedstocks in industrial processes. Finned tubes and shell and tube heat exchangers are used.

- Exhaust gases can be used to generate steam in waste heat boilers for the production of electrical power, mechanical power, process steam, and any combination of the above.

• Waste heat may be transferred to an intermediate fluid by
  heat exchangers or waste heat boilers, or it may be used
  by circulating the hot exit gas through pipes or ducts.
  Waste heat can be used to operate an absorption cooling
  unit for air conditioning or refrigeration.

The essential considerations in making optimal choice of waste
heat recovery devices are:

• Temperature (quality) of the waste heat

• Flow rate of waste heat

• Chemical composition and pollutants in the waste heat

• Temperature requirements of the heated fluid or material

DIRECT UTILIZATION OF WASTE HEAT

If the physical and chemical nature of the waste gases are
such that they can be used directly, either in the original process
or in a related process, there will be a great monetary saving to the
industrial firm, due to the elimination of or great reduction in the
requirement for capital equipment. Direct fired tunnel kilns are a
particularly good example of this approach in both heating and cooling
applications. Cold air entering the exit end of the kiln passes countercurrent to the material which it cools and becomes preheated combustion air and/or air for dryers. Combustion gases from the firing
area flow countercurrent to the material towards the exhaust stack,
thereby preheating the material as it approaches the heating area.
The glass industry and the cement industry can use this process in the
preheating of the charge. Other ways in which it can be used are producing blast furnace coke from coal and preheating ferrous scrap
charges to electric furnaces or to basic oxygen furnaces.

In addition to tunnel kilns, other forms of preheaters include
shaft furnaces, fluidized beds, and batch devices, each designed with
a particular application in mind.

RECUPERATORS AND REGENERATORS

Recuperators and regenerators are the most common methods for preheating combustion air, for preheating boiler feed water, and in some
cases, for preheating the fuel. In essence, recuperators and regenerators reduce fuel consumption by returning waste energy back to the process. From a thermodynamic standpoint, recuperators and regenerators
serve the same purpose, but each uses different heat transfer mechan-
isms to accomplish this. The recuperator is a direct heat exchanger
between exhaust gas and combustion air. The regenerator is a direct
heat exchanger between exhaust gas and combustion air. The regenerator, on the other hand, is a device that contains a thermal storage
material which receives heat from exhaust gases and transfers it to
the incoming combustion air.

Recuperators (3)

The simplest configuration for a heat exchanger is the metallic
radiation recuperator which consists of two concentric lengths of metal
tubing as shown in Fig. 4-1. The inner tube carries the hot exhaust
gases, while the external annulus carries the combustion air from the
atmosphere to the air inlets of the furnace burners. The hot gases
are cooled by the incoming combustion air, which now carries additional
energy into the combustion chamber. This is energy that does not have
to be supplied by the fuel; consequently, less fuel is burned for a
given furnace loading. The saving in fuel also means a decrease in
combustion air, and therefore, stack losses are decreased, not only
by lowering the stack gas temperatures but also by discharging smaller
quantities of exhaust gas.

This particular recuperator gets its name from the fact that a
substantial portion of the heat transfer from the hot gases to the
surface of the inner tube takes place by radiative heat transfer. The
cold air in the annulus, however, is almost transparent to infrared
radiation, so that only convective heat transfer takes place to the
incoming air. As shown in the diagram, the two gas flows are usually

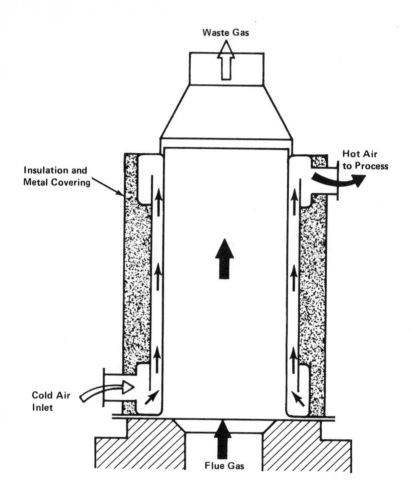

Figure 4-1. Metallic Radiation Recuperator.

parallel, although the configuration would be simpler and the heat
transfer more efficient if the flows were opposed in direction (counterflow). The reason for the use of parallel flow is that recuperators frequently serve the additional function of cooling the duct carrying away the exhaust gases, which consequently, extends its service
life.

The inner tube is often fabricated from high temperature materials, such as stainless steels of high nickel content. The large temperature differential at the inlet causes differential expansion, since the outer shell is usually of a different and less expensive material. The mechanical design must take this effect into account. More elaborate designs of radiation recuperators incorporate two sections: the bottom operating in parallel flow and the upper section using the more efficient counterflow arrangement. Because of the large axial expansions experienced and the stress conditions at the bottom of the recuperator, the unit is often supported at the top by a free standing support frame with an expansion joint between the furnace and recuperator.

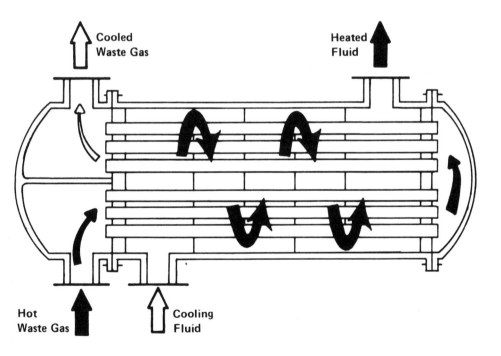

Figure 4-2. Convective Type Recuperator.

A second common configuration for recuperators is called the tube
type of convective recuperator. As seen in Fig. 4-2 the hot gases are
carried through a number of parallel small diameter tubes, while the
incoming air to be heated enters a shell surrounding the tubes and
passes over the hot tubes one or more times in a direction normal to
their axes.

If the tubes are baffled to allow the gas to pass over them twice,
the heat exchanger is termed a two pass recuperator; if two baffles
are used, a three pass recuperator, etc. Although baffling increases
both the cost of the exchanger and the pressure drop in the combustion
air path, it increases the effectiveness of heat exchange. Tube type
recuperators are generally more compact and have a higher effectiveness
than radiation recuperators, because of the larger heat transfer area,
made possible through the use of multiple tubes and multiple passes of
the gases.

The principal limitation on the heat recovery of metal recuperators is the reduced life of the liner at inlet temperatures exceeding
$2000^{\circ}F$. At this temperature, it is necessary to use the less efficient
arrangement of parallel flows of exhaust gas and coolant, in order to
maintain sufficient cooling of the inner shell. In addition, when
furnace combustion air flow dropped back because of reduced load, the
heat transfer rate from hot waste gases that preheat combustion air
becomes excessive, causing rapid surface deterioration. Then, it is
usually necessary to provide an ambient air bypass to cool the exhaust
gases.

In order to overcome the temperature limitations of metal recuperators, ceramic tube recuperators have been developed, whose materials
allow operation on the gas side to $2800^{\circ}F$ and on the preheated air
side to $2200^{\circ}F$ on an experimental basis, and to $1500^{\circ}F$ on a more or
less practical basis. Because early ceramic recuperators were built
of tile and joined with furnace cement, thermal cycling caused cracking of joints and rapid deterioration of the tubes. Later developments
introduced various kinds of short silicon carbide tubes, which can be
joined by flexible seals located in the air headers. Figure 4-3 illustrates the type of design which maintains the seals at comparatively

low temperatures and has reduced the seal leakage rates to a small percentage. Earlier designs had experienced leakage rates from 8-60%. The new designs are reported to last two years with air preheat temperatures as high as 1300°F and with much lower leakage rates.

An alternate arrangement for the convective type recuperator, in which the cold combustion air is heated in a bank of parallel vertical tubes that extend into the flue gas stream, is shown in Fig. 4-4. The advantage claimed for this arrangement is the ease of replacing individual tubes, which can be done during full capacity furnace operation. This minimizes the cost, the inconvenience, and possible furnace damage that might be due to a shutdown forced by recuperator failure.

For maximum effectiveness of heat transfer, combinations of radiation type and convective type recuperators are used, with the convective type always following the high temperature radiation recuperator. Figure 4-5 shows this arrangement.

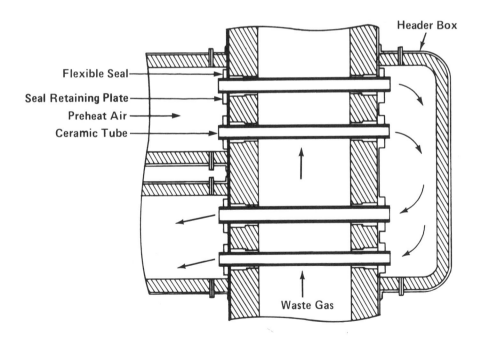

Figure 4-3. Ceramic Recuperator.

Although the use of recuperators conserves fuel in industrial furnaces, and although their original cost is relatively modest, the purpose of the unit is often just the beginning of a somewhat more extensive capital improvement program. The use of a recuperator, which raises the temperature of the incoming combustion air, may require purchase of high temperature burners, larger diameter air lines with flexible fittings to allow for expansion, cold air lines for cooling

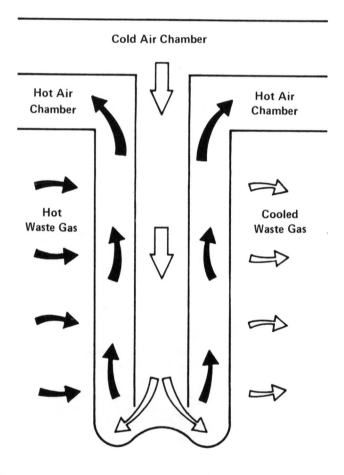

Figure 4-4. Vertical Tube Within Tube Recuperator.

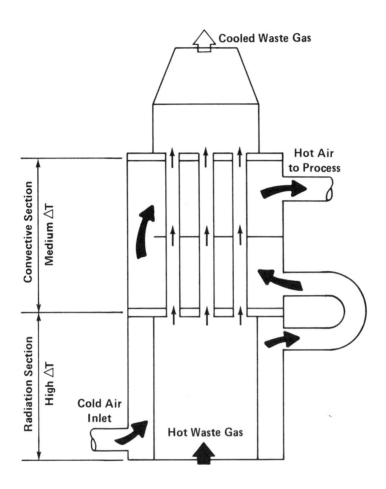

Figure 4-5. Combined Radiation and Convective Type Recuperator.

the burners, modified combustion controls to maintain the required
air/fuel ratio (despite variable recuperator heating), stack dampers,
cold air bleeds, controls to protect the recuperator during blower
failure or power failures, and larger fans to overcome the additional
pressure drop in the recuperator. It is vitally important to protect
the recuperator against damage caused by excessive temperatures, since

the cost of rebuilding a damaged recuperator may be as high as 90% of the initial cost of manufacture and since the drop in efficiency of a damaged recuperator may easily increase fuel costs by 10-15%.

Figure 4-6 shows a schematic diagram of one radiant tube burner fitted with a radiation recuperator. With such a short stack, it is necessary to use two annuli so that the incoming air can achieve reasonable heat exchange efficiencies.

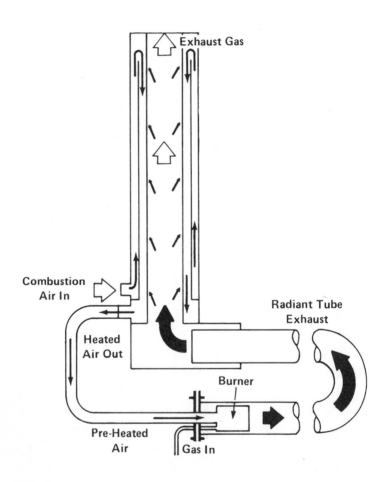

Figure 4-6. Small Radiation Type Recuperator Fitted to a Radiant Type Burner.

Recuperators are used for recovering heat from exhaust gases to heat other gases in the medium to high temperature range. Some typical applications are soaking ovens, annealing ovens, melting furnaces, afterburners and gas incinerators, radiant type burners, reheat furnaces, and other gas to gas waste heat recovery processes in the medium to high temperature range.

Switched Regenerators

The switched regenerator employs two or more chambers filled with refractory checkers which absorb heat from one fluid and then transfers it to another fluid by alternating the flow of exhaust gas and combustion air through the chambers. See Fig. 4-7. Checked regenerators, being refractory, are used in high temperature corrosive environments, such as those in coke ovens, hot blast stoves, glass furnaces, open hearth furnaces, and soaking pits. Switch regenerators, as compared with recuperators, are more disadvantageous because of higher cost and lower effectiveness caused by high leakage rates, increased space requirements, inconsistent operation, and a requirement for more sophisticated methods of control.

Heat Wheels

A heat wheel is a rotary type of regenerator which, using the Ljungstrom principle, is finding increasing applications. Heat wheels can be built in two configurations: as a rotary disk and as a rotary drum. Figure 4-8 illustrates these two configurations.

The heat wheel is fabricated from a material which is porous and which has a high heat capacity. As the disk or drum rotates slowly, heat is transferred to it by the hot gases; and as the disk or drum continues to rotate, it gives up its heat to the cooler intake air. Heat transfer efficiencies of 75-80% have been achieved.

Heat wheels are available in four types (4). The first consists of a metal frame that is packed with a core of knitted mesh, stainless steel or aluminum wire resembling that found in the common metallic

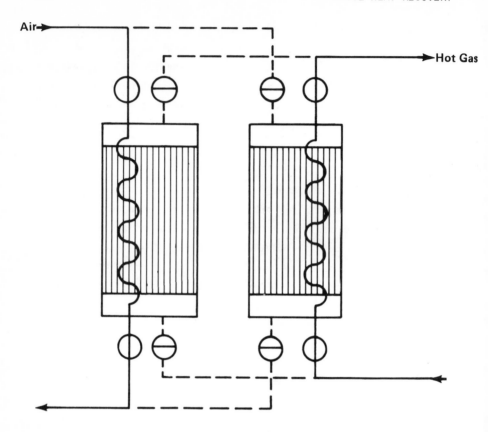

Figure 4-7. Switched Regenerator.

pot scraper. The second, called a laminar wheel, is fabricated from
corrugated metal and is composed of many parallel flow passages. The
third variety is also a laminar wheel but is constructed from a ceramic matrix of honeycomb configuration. This type is used for higher
temperature applications with a present day limit of about 1600°F.
The fourth variety is of laminar construction, but the flow passages
are coated with a hygroscopic material so that latent heat may be recovered. The packing material of the hygroscopic wheel may be any of
a number of materials.

Because of the construction of the heat wheel, small amounts of
gas can be transferred from the exhaust to the intake duct, resulting
in contamination. If this contamination cannot be tolerated, it can
be reduced by the addition of a purge section, where a small amount

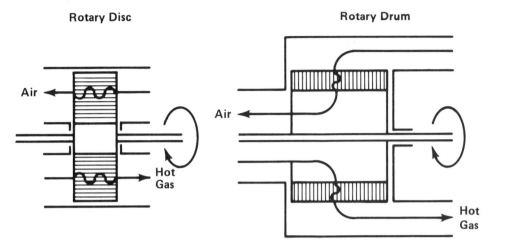

Figure 4-8.   Heat Wheels.

of clean air is blown through the wheel and then exhausted to the atmosphere. Common practice is to use about six air changes of clean
air for purging. This limits gas contamination to as little as 0.4%
and particle contamination to less than 0.2% in laminar wheels, and
it reduces cross contamination to less than 1% in packed wheels.

Heat wheel applications include those for space heating, process
heat recovery in low and moderate temperature environments, curing or
drying ovens, and air preheaters.

HEAT PIPES

Heat pipes can be used effectively to recover heat in a variety
of processes. They show promise as an industrial waste heat recovery
option because of their high efficiency and compact size. They are
also free from cross contamination.

The basic heat pipe is a closed container with a capillary·wick
structure that is saturated with a vaporized fluid. See Fig. 4-9.

The heat pipe uses an evaporation-condensation cycle, with the
capillary wick serving as the pump to move the condensed fluid back
to the heat input area. The heat absorbed from hot exhaust gases

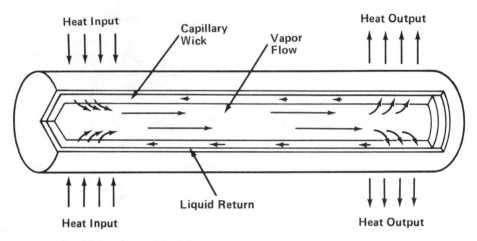

Figure 4-9. The Basic Heat Pipe.

evaporates the entrained fluid, causing the vapor to collect in the
center core. The latent heat of vaporization is carried in the vapor
to the cold end of the heat pipe, which is located in the cold gas
duct. Here the vapor condenses, giving up its latent heat. The condensed liquid is then carried by capillary action back to the hot end,
where it is recycled.

The following are three examples of how the unique features of
the heat pipe can be used to recover waste heat (5).

Process Control Application. Constant temperature heat pipes can
be applied to control temperature and utilize waste heat in process
control applications. Candidates are chemical reactors in which the
temperature of an exothermic or an endothermic reaction must be controlled. For example, in a methanation reaction, heat pipes can be
used to control the methanation temperature, while the waste heat is
used to produce steam. This system is shown in Fig. 4-10. In the
lower chamber, $3H_2$ and CO gas react in the presence of a Ni catalyst,
to produce CH_4 synthetic pipeline gas. The reaction is extremely exothermic, and as a result, a tremendous amount of waste heat is generated. This heat is transferred via heat pipes to the upper chamber,
where steam is generated. This steam can then be used either to drive
a turbine or in the gasifier. In either case, waste heat is recovered.
The major advantage is the precise temperature control of the reaction,

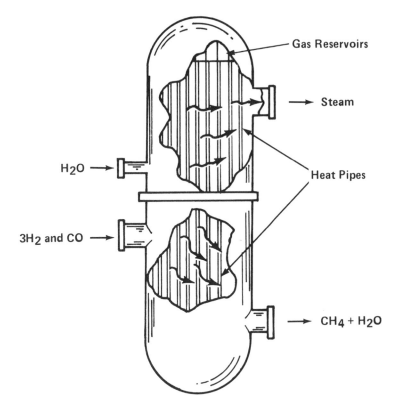

Figure 4-10. Heat Pipes for Temperature Control and Waste Heat Recovery in Methanation Process.

which is obtained automatically with constant temperature heat pipes.

Plume Control in Sulfur Scrubbers. Most sulfur scrubbers emit wet exhaust, which produces a visible plume. This plume is highly undesirable and must be controlled. Conventionally, steam lines are used to preheat flue gases or to inject hot combustion products into the scrubbed gas, in order to increase the scrubber's exhaust temperature. A bank of heat pipes can recover waste heat from hot flue gas entering the sulfur scrubber and then use this heat to preheat flue gases leaving the scrubber, thus alleviating the plume problem.

In the simplest scheme (Fig. 4-11), heat transfer from the heat pipes will transfer from the flue gas duct, directly to the scrubber. Another configuration can be designed which removes waste heat from the flue gas via heat pipes, as shown in Fig. 4-12. The waste heat

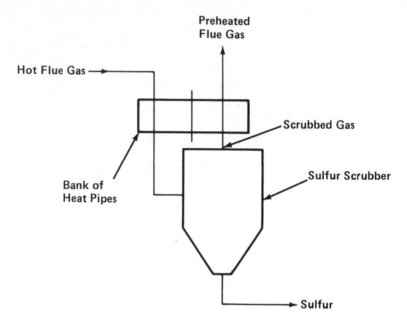

Figure 4-11. Direct Flue Gas Preheating.

is used to preheat the air, and the hot air is then mixed with the
scrubber gas. Preheated flue gas disperses into the atmosphere without creating a visible plume. A similar technique could also be employed to reduce or eliminate plumes in wet cooling towers. In either
case, the plume control can be accomplished with energy that normally
would be wasted.

Improvement of Thermodynamic Efficiency of Gas Turbines Through
Heat Recovery. In the past, expelling the exhaust from a gas turbine
into the atmosphere was a common practice, simply because fuel was
inexpensive and the effects of thermal discharge were of minor importance. However, present day conditions are responsible for a reversal:
fuel is becoming more expensive, and the effects of thermal pollution
are a major consideration. Heat pipes can be used to recover waste
heat, to improve thermodynamic efficiency of the turbine, and to reduce
temperatures of the exhaust gases.

A simple heat pipe system is illustrated in Fig. 4-13. When the
exhaust gas passes over a bank of heat pipes, the exhaust temperatures
are reduced. The recovered waste heat is then used to preheat incoming
air. The potential advantages include reduction of negative thermal

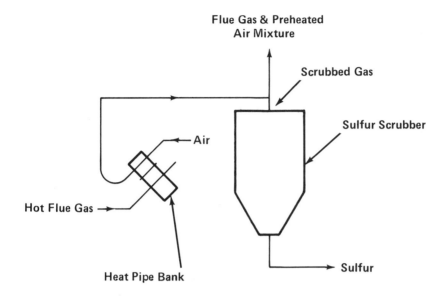

Figure 4-12. Indirect Flue Gas Preheating.

effects and utilization of waste heat to improve the thermodynamic efficiency of the gas turbine.

Other applications of heat pipes include:

• Air dryers

• Heating, ventilation, and air conditioning systems

• Air preheaters

The use of heat pipes can provide a viable alternative to existing heat recovery techniques.

WASTE HEAT BOILERS

Waste heat boilers are designed to use the heat in furnace flue
gases, thereby increasing overall efficiency. Figure 4-14 shows a
typical arrangement of such installations. Most waste heat boilers
are water tube boilers in which the hot exhaust gas passes over a number of parallel tubes containing water. The water is vaporized in the
tubes and collected in a steam drum for use as heating or process

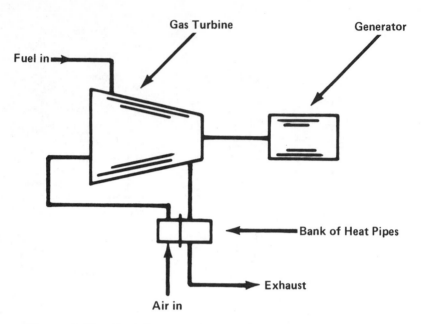

Figure 4-13. Heat Recovery System for an Open Cycle Gas Turbine.

steam. A more compact boiler can be produced if the water tubes are
finned to increase the effective heat transfer area on the gas side.

BOTTOMING CYCLE ENGINES

Another attractive process for the recovery of waste heat at relatively low temperatures is being developed by companies such as Sundstrand. This is the organic Rankine bottoming cycle engine for power
generation or for motivating pumps, compressors, fans, etc. An application of the bottoming cycle engine is shown in Fig. 4-15. Table
4-4 provides an indication of the system performance when it is coupled
to a GE recuperated gas turbine. Efficiency increases of 29% can be
achieved.

Advantages of an organic Rankine cycle engine are its simple
single stage turbine design, its non-corrosive working fluids, its
relatively low capital costs, and its low maintenance costs and requirements. Table 4-5 shows the incremental power costs for a typical

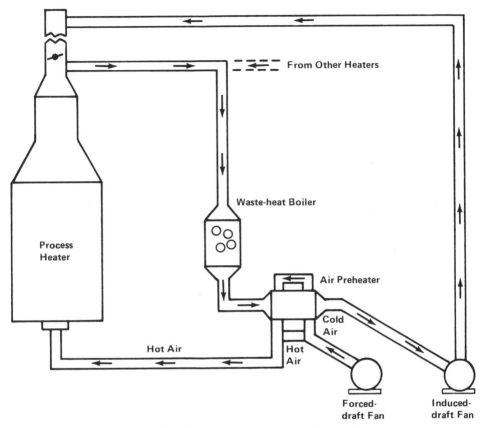

Figure 4-14. Typical Waste Heat Boiler Installation.

10,000 KW bottoming cycle plant recovering 700°F waste gases. These power costs are substantially lower than conventional or nuclear plant costs.

HEAT PUMPS

In all of the previously discussed heat recovery devices, fluids with temperatures less than 250°F are of little value. Such waste heat can be used economically for space heating and other temperature applications, such as for food processing, through a device called a heat pump.

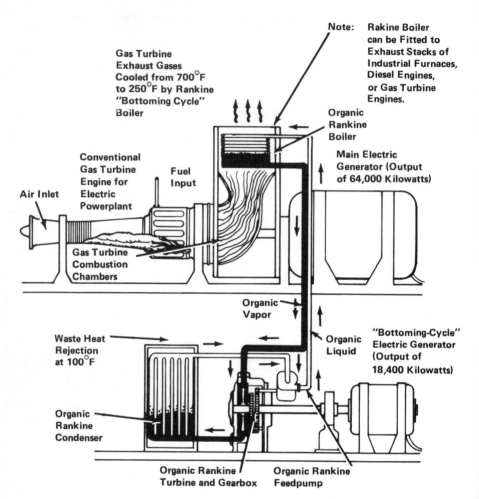

Figure 4-15. Organic Rankine Bottoming Cycle Engine.

A heat pump is a device which consists of two heat exchangers, a compressor, and an expansion device. Figure 4-16 is a schematic flow diagram of an industrial electric heat pump designed to deliver 180°F hot water using heat available from 90°F water. The performance of a heat pump is described in terms of the coefficient of performance, or COP, which is defined as:

Table 4-4

Waste Heat Recovery from Gas Turbine

Gas Turbine	
Model number	GE PG7791R
Power (59°F, 14.7 psia)	64,000 kwe
SFC	0.48 lb/kwhre
Heat rate (HHV)	9,300 Btu/kwhre
η_{OA} (HHV)	37%
Exhaust gas flow rate	1.90×10^6 lb/hr
Exhaust gas temperature	715°F
Organic Rankine Cycle Bottoming Plant	
Exhaust gas temperature from boiler	250°F
Fluorinol - 85 flow rate	7.9×10^5 lb/hr, 1190 GPM
Turbine	
Type	Single Stage, Axial Impulse
RPM	3600
Tip diameter	6.2 ft
Blade height	6 inches
Power	18,400 kwe
Overall Plant Characteristics	
Gas turbine power	64,000 kwe
Organic Rankine cycle power	18,400 kwe
Total binary plant power	82,400 kwe
% Increase in power output	29%
η_{OA} (HHV) – Gas turbine	37%
– Binary plant	47%
Heat rate (HHV)	
– Gas turbine	9300 Btu/kwhre
– Binary plant	7250 Btu/kwhre

Source: From POTENTIAL FUEL EFFECTIVENESS IN INDUSTRY. Copyright 1974. The Ford Foundation. Reprinted with permission of Ballinger Publishing Company.

$$COP = \frac{\text{Heat transferred in Condenser}}{\text{Compressor Work}} = \frac{T_H}{T_H - T_L}$$

where T_L is the waste heat temperature and T_H is the temperature leaving the heat pump.

Figure 4-17 shows a typical industrial heat pump COP for an installation having a 90°F heat source. As is evident, if hot water at 190°F is desired, a COP of 3.0 will be achieved. If 230°F steam is required, the COP drops to about 2.5. As greater delivery temperatures are required, the COP continues to fall. At this time, there is a

Table 4-5

**Incremental Power Cost from
Binary Gas Turbine Rankine Cycle Plant**

Capital Cost Contribution	
Installed plant cost	$150/kwe
Fixed charges on capital	15%
Load factor	0.80
Equivalent hours/years	7000
Capital charge contribution	0.32 cent/kwhr
Fuel Cost Contribution	0.00 cent/kwhr
Operation and Maintenance	0.05 cent/kwhr
Cost of Power Leaving Plant	0.37 cent/kwhr

Source: From POTENTIAL FUEL EFFECTIVENESS IN INDUSTRY. Copyright
1974. The Ford Foundation. Reprinted with permission of Ballinger
Publishing Company.

limit of 230°F for industrial heat pumps, due to refrigerant limits.
Research and development of new refrigerants may raise the limit of
the heat pump use to as high as 400-500°F.

Fuel savings which may be achieved by installation of air preheaters can be calculated from overall heat balances of the heater
after fixing the stack temperature and excess air level. The amount
of fuel which can be saved is determined by the kind of fuel used, and
by the furnace waste gas and preheat temperatures. Figure 4-18 shows
the potential energy savings that are possible when air is preheated
for combustion with heavy fuel oil.

For a waste gas temperature of 1000°C, a combustion air temperature of 700°C would result in a fuel savings of approximately 34%,
when compared to a similar process without waste heat recovery. Similarly, for the same waste gas temperature, an improvement in the degree of air preheat by 400-700°C would result in fuel savings of about
12%. Similar relationships showing comparable savings are available
for other fuels of interest.

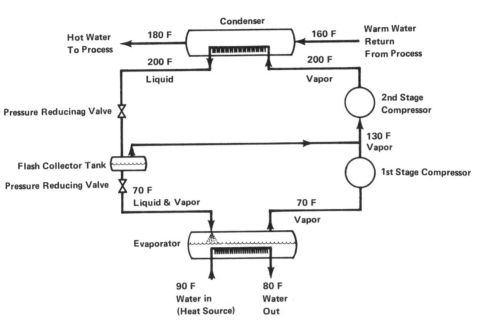

Figure 4-16. Heat Pump Schematic Diagram. Source: Kenneth Kreider and Michael McNeil, eds., Waste Heat Management Guidebook, NBS Handbook 121 (Washington, D.C.: Government Printing Office, January 1977), p. 122.

The following is an example of the savings potential of preheating combustion air to 400°C, using heat recovery from flue gases at 600°C. Assume the following:

• Furnace operations = 7000 hr/yr

• Fuel costs   = $1.00, $2.00, $3.00 per MMBTU

• Boiler loading   = 70%

• Fuel Consumption = 130 MMBTU/hr at full loading

According to Fig. 4-18, fuel savings of 16% is achievable. Thus, the annual fuel savings would be

.16 x 7000 x .70 x 130 = 101,920 MMBTU

which is worth the following:

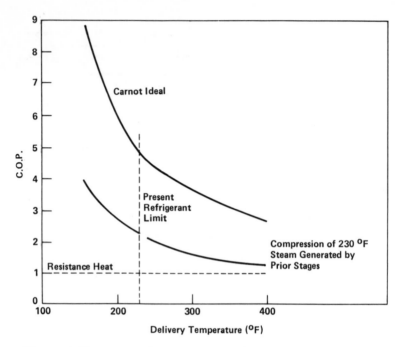

Figure 4-17. *Typical Heat Pump Performance*. *Source: Kreider and McNeil, <u>Waste Heat Management Guidebook</u>, p. 124.*

at $1/MMBTU = $101,920
at $2/MMBTU = $203,840
at $3/MMBTU = $305,760

If the costs for installing the air preheater are assumed to be:

cost of preheater $ 50,000
cost of installation   70,000
                Total $120,000

the payback would be:

at $1 = 1.18 years
at $2 = .59 years
at $3 = .39 year

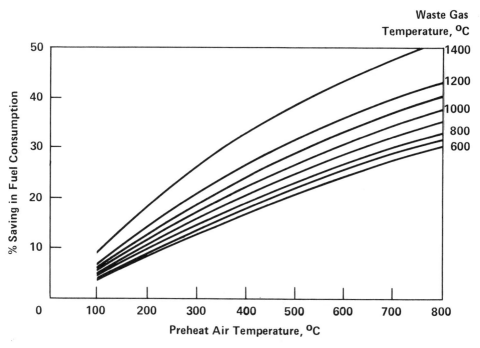

Figure 4-18. Fuel Oil Savings from Heated Combustion Air.
Source: H.R. McChesny, "State of the Art--Regenerative and Recuperative Heat Recovery," Efficient Use of Fuels in the Metallurgical Industries (Chicago, Illinois: Institute of Gas Technology, 1974), p.
86.

Sisson has developed a nomogram that makes it possible to quickly
estimate potential yearly savings (6). In constructing the nomogram,
it was assumed that each 40°F reduction in flue gas temperature would
decrease fuel consumption by 1%. Load factor is defined as the ratio
of the actual amount of steam generated over a given period of time to
the amount that could have been generated by operating the boiler at
full capacity for that same period.

Sample Problem: A 100,000 lb/hr industrial boiler has an input of
125 MMBTU/hr at full load and a flue gas temperature of 650°F. If
the load factor is 80% and fuel cost is $2.80/MMBTU, find the annual
saving possible by reducing the flue gas temperature to 320°F.

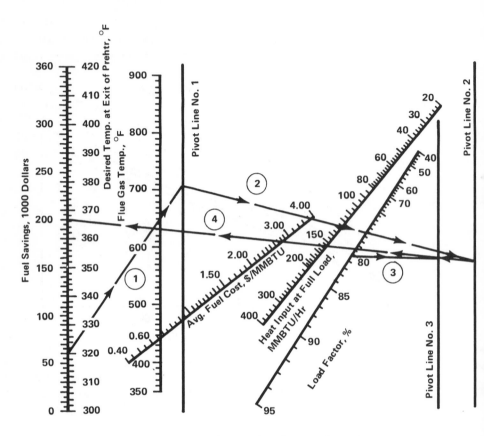

Figure 4-19. Nomogram for Estimating Fuel Savings. Source:
William Sisson, "Nomogram Estimates Fuel Savings by Economizer or
Air Preheater," Power Engineering, 80 (April 1976), 97.

Solution: (1) Connect 320°F on "Desired Temperature" scale with
650°F on "Flue Gas Temperature" scale and extend to pivot line
No. 1.

(2) Connect this point with 125 million BTU/hr on
"Heat Input" scale and extend to pivot line No. 2.

(3) Connect this point with 80% on "Load Factor" scale
and mark crossing point with pivot line No. 3.

(4) Connect marked point with $2.80 on "Average Fuel
Cost" scale and extend to "Fuel Savings" scale. Read answer as
$202,000 per year.

SUMMARY

Waste heat recovery is an important facet of managing energy in
an industrial firm. The energy exhausts to the atmosphere which has
already been paid for and should not be discarded until all possible
economic use of it has been achieved. It is necessary to identify the
prospective uses for the waste energy, to make an economic analysis
of the costs and savings of the options, and finally, to determine
which among these options has the greatest return on investment.

In the evaluation for additional heat recovery equipment, caution must be observed in the physical loading of existing structures.
Heat savings from reduction of stack temperatures must be calculated
individually to determine the feasibility of equipment installation.
Heat recovery equipment manufacturers can provide precise cost data,
together with projected savings as a result of the equipment modifications.

Table 4-6 presents, in matrix form, a summary of a number of significant attributes of the most common types of industrial heat recovery devices. This matrix allows rapid comparisons to be made in selecting competing types of heat exchangers. The characteristics given
in the table for each type of heat exchanger are: allowable temperature range, ability to transfer moisture, ability to withstand large
temperature differentials, availability as packaged units, suitability
for retrofitting, and compactness and the allowable combinations of
heat transfer fluids.

Although these large savings are available, waste heat recovery
equipment have not been installed in every plant that can benefit from
it. Basically, the reason is that when fuels were cheap and readily
available, waste heat recovery was not important, and payback periods
for such equipment were long. The increases in the prices of fuels
in recent years and the prospects for further increases, together with
fuel allotments, have fundamentally changed the economics of waste heat
recovery. The engineer or manager must learn about available options
in waste heat recovery to reduce energy costs and improve profitability.

Table 4-6

Operation and Application Characteristics of Industrial Heat Exchangers

Commercial Heat Transfer Equipment	Low Temperature Sub-Zero - 250°F	Intermediate Temperature 250°F - 1200°F	High Temperature 1200°F-2000°F	Recovers Moisture	Large Temperature Differentials Permitted	Packaged Units Available	Can Be Retrofit	No-Cross Contamination	Compact Size	Gas-to-Gas Heat Exchange	Gas-to-Liquid Heat Exchanger	Liquid-to-Liquid Heat Exchanger	Corrosive Gases Permitted With Special Construction
Radiation Recuperator			●		●	1	●	●		●			●
Convection Recuperator		●	●		●	●	●	●		●			●
Metallic Heat Wheel	●	●		2		●	●	3	●	●			●
Hygroscopic Heat Wheel	●	●		●		●	●	3	●	●			
Ceramic Heat Wheel		●	●		●	●	●	●	●	●			●
Passive Regenerator	●	●			●	●	●	●	●	●			●
Finned-Tube Heat Exchanger	●	●			●	●	●	●	●		●		4
Tube Shell-and-Tube Exchanger	●	●			●	●	●	●	●		●	●	
Waste Heat Boilers	●	●					●	●			●		4
Heat Pipes	●	●			5	●	●	●	●	●			●

1. Off-the-shelf items available in small capacities only.

2. Controversial subject. Some authorities claim moisture recovery. Do not advise depending on it.

3. With a purge section added, cross-contamination can be limited to less than 1% by mass.

4. Can be constructed of corrosion-resistant materials, but consider possible extensive damage to equipment caused by leaks or tube ruptures.

5. Allowable temperatures and temperature differential limited by the phase equilibrium properties of the internal fluid.

Source: Kreider and McNeil, Waste Heat Management Guidebook, p. 153.

NOTES AND REFERENCES

1. Kenneth Kreider and Michael McNeil, eds., Waste Heat Management Guidebook, NBS Handbook 121 (Washington, D.C.: Government Printing Office, January 1977), p. 5.

2. Wesley Rohrer and Kenneth Kreider, "Sources and Uses of Waste Heat," Waste Heat Management Guidebook, pp. 6-7.

3. This material is abstracted from W. M. Rohrer, "Commercial Options in Waste Heat Recovery Equipment," Waste Heat Management Guidebook, pp. 142-45.

4. Ibid., p. 146.

5. A. Basiulis and M. Plost, "Waste Heat Utilization Through the Use of Heat Pipes," Paper delivered at the Annual Winter Meeting of the Heat Transfer Division of the American Society of Mechanical Engineers, Houston, Texas, November 30 - December 4, 1975, pp. 3-4.

6. William Sisson, "Nomogram Estimates Fuel Savings by Economizers or Air Preheater," Power Engineering, 80 (April 1976), p. 97.

Chapter 5

IMPROVING BOILER EFFICIENCY

The use of process steam by industry accounts for some 16% of the total U.S. energy consumption and for approximately 40% of total industry fuel and energy consumption. Process steam has myriad applications in industry and is the largest consumer of industrial fuel and energy.

Due to the relatively low cost of fuel, most boiler installations prior to the Arab oil embargo were not fully equipped to operate at or near optimum performance levels. A Federal Energy Administration study indicates that based on 1975 boiler populations, the efficiency of industrial boilers can be improved by approximately 4% on the average, for an equivalent savings of 80 million barrels of oil per year (1). According to another study, industrial boilers that are not equipped with heat recovery equipment are operating at between 3-8% below maximum attainable efficiency (2). And S. G. Duke reports that in 1973 only about 30% of the 900 new industrial steam boilers installed in the U.S. were equipped with fuel conserving control systems (3).

With the rapidly rising cost of energy, several alternatives are now available for increasing the efficiency of industrial boilers. The purchase of more efficient boilers, the retrofit of existing

boilers and the adoption of improved boiler operating practices can
result in significant energy conservation and increased profitability.

INDUSTRIAL BOILER INVENTORY

Boilers are built in various sizes to serve particular markets.
Figure 5-1 provides an indication of boiler uses in relation to boiler
size. Division by size allows us to group boilers with similar physical characteristics, auxiliary equipment, and degree of combustion control sophistication, all of which affect total fuel consumption and

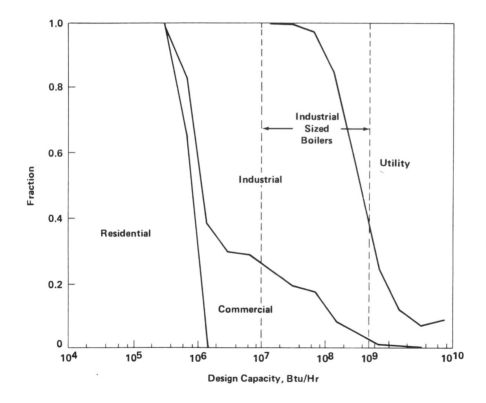

*Figure 5-1. Boiler Use Related to Boiler Size. Source: KVB,
Inc., Industrial Boiler User's Manual, Vol. 1, Report No. FEA-D-77/026
(Washington, D.C.: Federal Energy Administration, January 1977), p.
197.*

operating efficiency. It is common practice to equate 10^3 lb/hr
(k lb/hr) saturated steam flow with 10^6 BTU/hr (MMBTU/hr) heat capacity based on 1000 BTU/lb of steam generated. Industrial boilers have
therefore been defined to include the size range between 10^4 to 5×10^5
lb/hr steam flow capacity or 10^7 to 5×10^8 BTU/hr heat output capacity.

Table 5-1 provides a summary of industrial boilers by size and
type. In 1975, there were 62,000 industrial sized boilers of over
10,000 pounds per hour steam output in operation, with a total capacity in excess of 2300 billion BTU's per hour. The total annual
energy consumption by these boilers was in excess of 8800 trillion
BTU's. Note that water tube boilers constitute 87% of the total industrial boiler capacity, of which 70% is in the size range of 10,000-
250,000 lb/hr.

In order to assess the energy conservation potential of industrial
boilers, it is necessary to obtain a complete inventory of these boilers. An analysis of boiler sales data by the American Boiler Manufacturers Association shows the following breakdown of boiler users by
industry in 1976:

Chemical and Allied Products	19.0%
Paper and Allied Products	17.5%
Petroleum Refining and Related Industries	13.3%
Food and Kindred Products	12.3%
Electric Utility--Non-Generating Use	9.4%
Miscellaneous Manufacturing	8.0%
Primary Metal Industries	6.0%
Textile Mill Products	5.0%
Transportation Equipment	4.5%
Lumber and Wood Products	3.0%
Rubber Products	2.0%

This tabulation indicates the industries in which energy conservation
through improved boiler operations can be most significant.

Table 5-1

Summary of Industrial Boiler Size and Type Inventory

Capacity (MBtu/hr)	Furnace Design	1967 Boiler Population		Sales 1967-1974		Retired 1967-1974		1975 Boiler Population	
		No. of Units	Total Capacity 10^9 Btu/hr	No. of Units	Total Caapcity 10^9 Btu/hr	No. of Units	Total Capacity 10^9 Btu/hr	No. of Units	Total Capacity 10^9 Btu/hr
10-16	Watertube	7,300	91	375	5.2	176	2.4	7,499	93.8
16-100	Watertube	27,060	833	4,934	236.3	2,319	109.0	29,675	960.3
100-250	Watertube	4,015	658	1,157	180.3	845	131.6	4,327	706.7
250-500	Watertube	942	259	168	61.6	56	20.0	1,054	300.6
10-16	Firetube	9,970	126	6,615	85.1	1,190	15.3	15,215	195.8
16-30	Firetube	3,160	66	2,138	44.7	385	8.0	4,913	102.7
Totals		52,267	2,033	15,387	613.2	4,971	286.8	62,683	2,359.9

Source: KVB, Inc., Industrial Boiler User's Manual, p. 204.

MEANS FOR IMPROVED EFFICIENCY

Significant energy efficiency improvement potential exists for industrial boilers. The extent to which operational and equipment modifications will improve performance is determined by the type and condition of the boiler, which include the firing system and combustion controls, the fuel employed, the existing heat recovery equipment, and the general efficiency-related operating practices. However, in order to determine whether modifying operating practices or installing auxiliary equipment is needed, it would be necessary to conduct a complete engineering evaluation from the standpoint of economics, system capability, and practicality.

The areas which have the major potential for efficiency improvements include improving boiler operating cycles, improving maintenance improving combustion controls, installing heat recovery equipment, minimizing radiation losses, improving burner designs, minimizing power consumption of associated or auxiliary systems, and applying new boiler technologies.

Improving Operations

Boiler fuel consumption can be reduced through improved load management. Boiler efficiency varies, depending upon such factors as equipment age, boiler design, fuels used, and firing rate. Effective load mangement, then, would demand the use of the optimum operating range for each boiler in the system whenever practical and the maintenance of this load to minimize efficiency losses during load variations.

Facilities with multiple boilers should be managed to achieve optimum system performance. One way to accomplish this is by loading the most efficient boilers to the desired operating level first, moving down to the least efficient boiler. Conversely, the least efficient boilers should be removed from service first. Another variation is to schedule one boiler for a specified process, another for

space heating, and so on. A detailed cost analysis is necessary to
determine the net overall cost savings in multiple boiler systems.

The importance of operating at high load is illustrated in Fig.
5-2, which shows how the various efficiency losses change with variations in boiler firing rate. As indicated in the figure, the change
in excess oxygen with a change in load has a strong influence on the

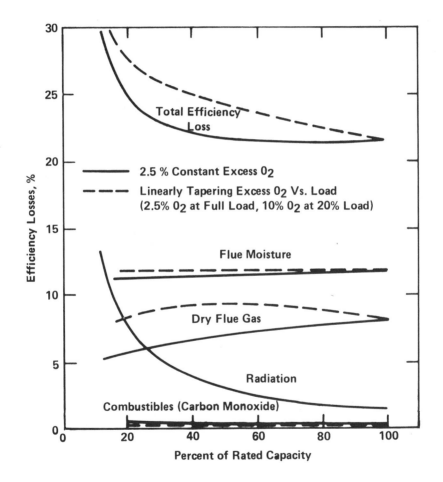

Figure 5-2. Variations in Boiler Efficiency Losses With Changes
in Boiler Firing Rate. Source: KVB, Inc., Industrial Boiler User's
Manual, p. 38.

eventual efficiency versus load profile. It is, therefore, advantageous to operate as close as possible to peak load for highest efficiencies, when there is a choice between partially loading several
boilers or operating fewer boilers at high loads.

Improving Combustion

Another source of efficiency loss in industrial boilers is caused
by incomplete combustion, or more air being supplied than is required
for burning the fuel. Combustion efficiency can be improved and fuel
saved when steps are taken to insure that only sufficient air needed
for burning the fuel completely and safely is supplied to the combustion chamber.

Controlling Excess Air

Excess air is required in all practical cases to assure complete
combustion, to allow for normal variations in the precision of combustion controls, and to insure satisfactory stack conditions for some
fuels. The optimum excess air level for the maximum boiler efficiency
occurs when the sum of the losses due to incomplete combustion and the
losses due to heat in the flue gases is at a minimum. The optimum
excess air level will vary with furnace design, type of burner, fuel,
and process variables, and can be determined by conducting tests with
different air-fuel ratios.

Efficiency gains obtained through reducing excess air are shown
in Fig. 5-3. A reduction in excess air is normally accompanied by
flue gas temperature reduction. The actual temperature reduction is
dependent on the initial stack temperature and on the excess air reduction. A reduction in excess air will have a greater effect when stack
temperatures are high. For example, a reduction in excess air of 10%
(i.e., from 20-10% or from 80-70% excess air) will produce a 0.94%
improvement in efficiency when stack temperatures are 600°F. However,
the same change in excess air, when stack temperatures are 300°F will
improve boiler efficiency only by 0.39%. These values are not significantly affected by the type of fuel.

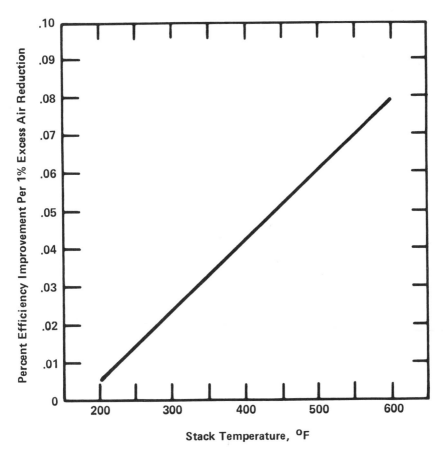

Figure 5-3. *Percent Efficiency Improvement for Every One Percent Reduction in Excess Air. Source: KVB, Inc., Industrial Boiler User's Manual, p. 122.*

To determine the impact of a 1% efficiency change, consider a boiler operating at 20,000 lb/hr steam flow for 7000 hours per year. The heat required to generate one pound of steam is approximately 1000 BTU divided by boiler efficiency as a decimal fraction. If the boiler efficiency were 80%, the annual fuel consumption would be

$$20,000 \text{ x } 7000 \text{ x } \frac{1000}{.80} = 175,000 \text{ MMBTU/yr.}$$

If fuel costs are $2.00 per MMBTU, the annual fuel bill would be $350,000.

If efficiency were improved to 81%, the fuel consumption would be

$$20,000 \times 7000 \times \frac{1000}{.81} = 172,840 \text{ MMBTU/yr.}$$

This is a savings of 2,160 MMBTU/yr or $4,320. If the steam flow were 150,000 lb/hr, and the efficiency increase were 3%, the annual savings would be $95,000. It is obvious that greater effort to improve boiler efficiency is necessary as output increases.

Figure 5-4 shows how the control of excess air can reduce fuel cost.

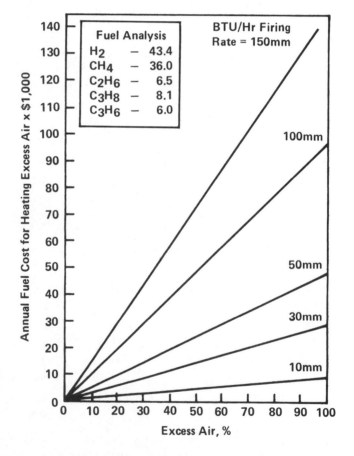

Figure 5-4. Fuel Cost VS Excess Air. Source: George Perry and James Monteaux, "Practical Analysis Can Cut Plant Energy Use, Trim Operating Costs," Oil and Gas Journal, 74 (February 23, 1976), 79.

This chart is based on a fuel cost of $2.00 per MMBTU and a stack temperature of 800°F. Data are plotted for firing rates of 10, 30, 50, 100, and 150 MMBTU per hour.

Combustion Control Systems. Combustion controls regulate the quantity of fuel and air flows in a boiler. In choosing the most effective type of control system, one must consider boiler·capacity, steam demands, expected performance levels, costs, pollution regulations, and safety. It has been shown to be economical to equip units as small as those that produce 25,000 pounds of steam per hour with fuel saving control systems, but caution should be exercised to avoid selecting a control specification that may be operationally impractical.

Various levels of sophistication in controlling excess air are available (4):

- A portable oxygen analyzer and a portable draft gauge represent the least costly system to implement. Periodic readings with these instruments will guide the operator in making adjustments for optimum operation with minimum excess air.

- A continuous oxygen analyzer with local readout plus a permanently mounted draft gauge with local readout represent the most common system. Readings, available continuously, enable the operator to monitor and adjust the excess air whenever required.

- Another system differs from the continuous oxygen analyzer by the addition of a remotely controlled pneumatic damper positioner. Draft and oxygen readouts are located in a control room, enabling one operator to remotely control a number of heaters.

- Automatic stack damper control is the next step in increased sophistication of combustion controls. The cost of this system is usually justifiable only on large heaters.

Excess air reduction of 20% is possible by periodic use of the first system. A further reduction of 10-15% excess air can be achieved with the second and third systems. The most suitable system must be determined by calculating the economics of each case.

Low Excess Air Burners. Another means for reducing excess air is the installation of low excess air (LEA) burners which were originally developed in Europe (5). These combustion systems employ

improved oil fuel atomization and controlled air fuel mixing to achieve
extremely low excess air (3-5%) operating levels, as well as reduce
pollution levels. Some restrictions have been encountered in the application of LEA burners to industrial boilers, however, especially
on a retrofit basis. These problems have been primarily related to
the compatibility of new flame shapes and heat release rates with existing furnace volume and design. LEA burners require a very sophisticated combustion control system, one that employs automatic oxygen
correction, to fully utilize its low excess air capability and to
prevent small excursions in excess air which might lead to high com-
bustibles and potentially unsafe conditions.

Other Equipment. Other equipment that can improve boiler efficiency include special oil atomization and viscosity control systems.
Proper pretreatment and atomization of the oil are vital to reduced
excess air operation.

Improving Maintenance

In many cases, substantial improvements in boiler operating efficiency can be achieved without requiring the purchase of new equipment or retrofit devices. Through proper maintenance, efficiency
losses can be minimized and can lead to the most efficient utilization
of existing steam generating equipment. Proper maintenance can also
have an important bearing on plant reliability, load carrying ability,
and safety.

Efficiency degradations can be traced directly to problems with
mechanical linkages controlling fuel and air flows, to malfunctioning
or poorly calibrated combustion analyzers, to inoperable or maladjusted
dampers or boiler instrumentation, to the degree of boiler tube clean-
liness, to boiler loss during blowdowns, and to other such factors.

A boiler tuneup is a very cost effective means of achieving efficient operation, saving fuel, and reducing operating costs. Adjustment and maintenance of fuel burning equipment and combustion controls
permit operation with the lowest practical excess air, and reduces
stack losses. These tuneups can be accomplished through plant

personnel, engineering consulting firms, boiler manufacturers, and service organizations.

Wall and Soot Blowers. Wall and soot blowers should be employed to remove carbon and slag deposits on heat transfer surfaces in boilers. Wall blowers remove slag deposits from the furnace walls of coal fired units, while soot burners remove fly ash and soot deposits from the boiler.

The properties of the fuel, the characteristics of the firing system, and the resulting temperature distribution in the boiler affect the type and accumulation rates of the deposits. These deposits can retard heat transfer and will eventually result in clogging a boiler. However, manufacturers claim that proper soot blowing can increase unit efficiency by up to 1% (6).

Insulation. Heat lost from the boiler jacket through its insulation is generally termed "radiation loss." The quantity of heat lost in this manner (BTU/hr) is fairly constant, even at different boiler firing rates; thus, it is an increasingly higher percentage of the total heat loss at the lower firing rates. This loss may be unavoidable to some extent, but deterioriated insulation and deteriorated furnace wall refractory will increase the loss at all loads.

Properly applied insulation can result in large savings in energy losses, depending on the type, thickness, and condition of the existing insulation. Chapter 2 provided information on savings effected by the installation of proper insulation thicknesses. But insulation provides other benefits besides reduced heat loss. These include controlled surface temperature for comfort and safety, fire protection, reduced noise, and increased structural strength.

Maintenance for Coal Fired Units. Efficiency improvements of up to 5% can be obtained by re-injecting fly ash from boiler hoppers or mechanical collectors into boilers. This process reduces carbon losses by reburning the coal. In addition, stoker grate surfaces must be maintained to provide good air distribution and to minimize shifting losses. Feeder devices must be kept in good repair to provide uniform fuel distribution. In pulverized coal units, the coal fineness should be checked regularly and the mills adjusted to provide the correct fineness. High carbon losses will result with poor grinding.

WASTE HEAT RECOVERY

The greatest potential for boiler efficiency improvement is the installation of waste heat recovery equipment to minimize stack gas losses and losses from blowdown water and expelled condensate. The majority of industrial boilers have very large flue gas losses because they operate at high excess air levels and at high stack gas temperatures. The furnace is operated at high excess air levels to insure complete combustion and safe operation, but this results in the unnecessary heating of tons of air that are subsequently dumped into the atmosphere. Waste heat energy losses in stack gases consist of the flue gas loss and the moisture loss due to latent and sensible heat in water vapor.

In examining efficiency losses, it is apparent that any reduction in the exit flue temperature will help minimize the flue gas losses. To recover all of the energy would require that the exit gas be cooled to ambient air temperature and the water contained in the flue condensed. However, the flue gas must remain above a certain temperature because of condensation on the heat transfer surfaces, which will deposit particulates and cause corrosion.

Stack gas waste heat recovery can be achieved through three primary means: the use of economizers, air preheaters, and turbulators.

Economizers

An economizer is an arrangement of feed water tubes located in an exhaust flue duct to absorb a portion of the heat energy that would otherwise be lost in the stack. It is a device which, using heat energy from the flue gas, preheats boiler feedwater, heats hot water for space heating, or provides heat for other process requirements.

The choice between an air preheater and an economizer for waste heat recovery is made on the basis of:

- Level of flue gas temperature

- Draft losses

- Operating steam pressure

- Investment and operating costs

- Maintenance requirements

In general, economizers are preferred over air preheaters in small units, i.e., in those which produce below 50,000 pounds of steam per hour. Economizers are also preferred over air preheaters because the capital investment is less, furnace heat absorption is lower, draft losses are lower, power requirements are less, and because there is no impact on NO_x emissions.

Air preheaters become competitive with economizers for larger units. On very large units, e.g., a unit with an operating pressure above 400 psig, a combination air preheater/economizer may be employed.

The overall boiler efficiency will increase approximately 2.5% for every 100^oF decrease in stack gas temperature or 2% for every 100^oF increase in combustion air temperature. Figure 5-5 can be used to estimate the efficiency improvement for a given reduction in flue gas temperature. An approximate gross annual fuel savings that can be achieved by installing an economizer can then be estimated by using the nomograph in Fig. 5-6.

Sample Problem: If a 16,000 lb/hr packaged boiler has an exit temperature of 600^oF, and if oil costs $2.00 per MMBTU, calculate the potential gross annual savings in fuel cost if an economizer is added and stack temperature can be reduced to 350^oF.

Solution: First, determine the approximate efficiency improvement by using Fig. 5-5.

$\eta = 6\%$

Then, using the nomograph in Fig. 5-6, see that the potential gross annual savings is $185,000.

Figure 5-7 provides an indication of the approximate equipment and installation costs of retrofitting an economizer into an existing

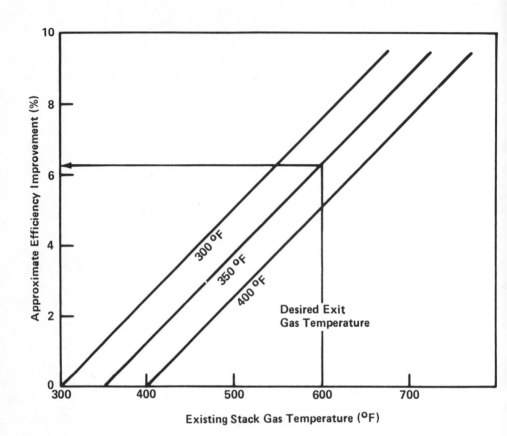

Figure 5-5. Approximate Efficiency Improvement with Decreased Flue Gas Temperatures. Source: KVB, Inc., Industrial Boiler User's Manual, Vol. 1, p. 36.

boiler. Of course, these costs will vary from application to application (7). Capital costs are influenced by such factors as:

- Desired performance

- Tube size and materials

- Arrangement of the economizer tubes

Installation costs vary from 30-100% of the capital costs, depending on:

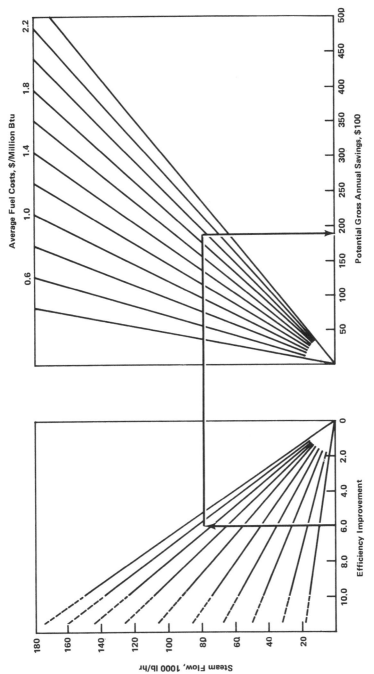

Figure 5-6. Potential Gross Savings in Annual Fuel Costs. Source: Sales Technotes from the Combustion Engineering Preheater Company, Wellsville, New York.

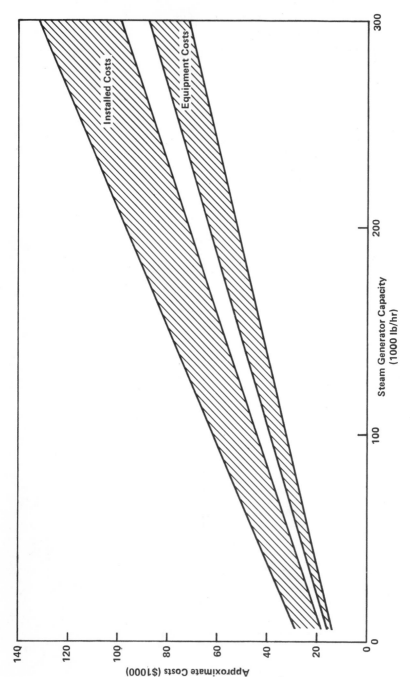

Figure 5-7. Approximate Economizer Equipment and Installed Costs. Source: KVB, Inc.,
Industrial Boiler User's Manual, p. 47.

- Feed water preheaters

- Plumbing and pumping alternatives

- Available space

- Duct modifications

- Air fan changes due to increased draft loss

The practical extent to which the feed water temperature can be increased by using an economizer is limited by flue gas temperature, economizer surface area, steam pressure, and corrosion characteristics of the flue gas.

Air Preheaters

An air preheater is employed to transfer heat from the hot flue gas to the incoming combustion air supply. Unit efficiency will increase approximately 2% for every $100^\circ F$ increase in combustion air supply. Figures 5-5 and 5-6 can be used to determine the annual fuel savings through the installation of air preheaters. However, application of air preheaters is limited by stack temperatures, corrosion considerations, maximum acceptable combustion air temperature, and associated NO_x emission levels.

Air preheating can be accomplished either through the use of recuperators, regenerators, or heat pipes. In recuperators, the heat energy is transferred directly from the flue gas to the combustion air. Regenerators transfer the heat energy from the flue gas to the combustion air through an intermediate heat storage medium. It is generally more common to use a regenerative unit in retrofit installations.

A new technological development for air preheating is the heat pipe. It offers a promising alternative to conventional heat transfer devices because it allows no cross contamination, has no external power or moving parts, and requires minimum maintenance. Its major disadvantage is a $600^\circ F$ maximum operating temperature, which would limit its application.

The cost of installing an air preheater varies from boiler to boiler. The approximate range of capital equipment and installed costs are given in Fig. 5-8. Capital costs vary, depending on the pressure drop that can be tolerated, air heater design, materials, and desired performance. Installed costs are much higher for preheaters than for economizers and can be as much as three to five times the capital costs. Significant annual costs are also incurred, including fixed costs, operating costs, and required maintenance costs. Fixed and operating costs may amount to 10-20% of the installed costs (8).

Turbulators

Turbulators are baffles which are installed into the secondary passes of a firetube boiler to increase flame turbulence and thereby increase convective heat transfer to the surrounding boiler water. These devices also allow the balancing of gas flow through the fire-tubes to achieve more effective utilization of existing heat transfer surfaces. However, a disadvantage is that turbulators cannot be used on coal fired units.

Figure 5-9 depicts a two pass Scotch Marine firetube boiler with and without turbulators. The improvement in boiler performance will be indicated by the decrease in stack gas temperature and the corresponding increase in steam generation. As described previously, a reduction of $100^{\circ}F$ in stack gas temperature will result in a 2.5% efficiency improvement.

Waste Water Heat Recovery

Waste water heat recovery involves the recovery of the waste heat energy contained in drum blowdown water and in expelled condensate. While not contributing directly to boiler efficiency, recovering this lost energy can result in very substantial fuel savings. For example, this heat energy can be effectively used to preheat boiler feed water, instead of just being discarded.

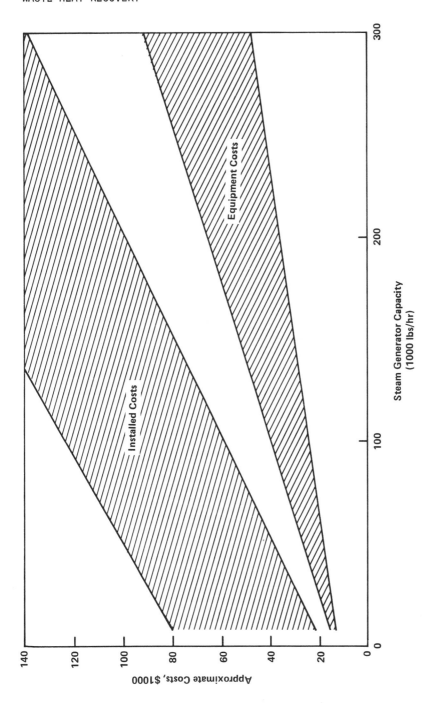

Figure 5-8. Approximate Air Preheater Equipment and Installed Costs.
Source: KVB, Inc., Industrial Boiler User's Manual, p. 37.

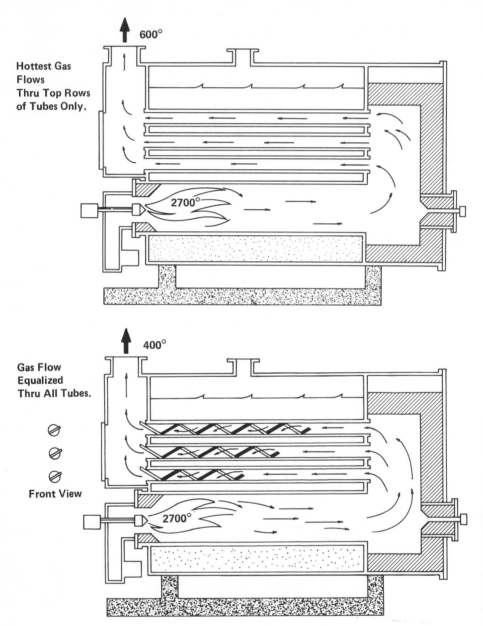

Figure 5-9. Two Pass Scotch Marine Firetube Boilers With and Without Turbulators. Source: Sales Literature for Fuel Efficiency, Inc., Newark, New Jersey.

Blowdown is a procedure which removes boiler water with high impurity concentrations that can affect steam quality and result in tube scale deposits. The amount of water discarded is dependent on boiler and make up water quality and can be as much as 5-10% of the total boiler steam flow. This waste water loss can occur from excessive drum water blowdown beyond what is required to maintain satisfactory dissolved solids concentration. This represents not only lost energy but also a waste of water and of the chemicals used in its treatment. The increasing cost of energy makes the recovery of these losses more economical.

A large portion of blowdown heat energy can be reclaimed by continuous blowdown extraction in which flashed steam from the blowdown water is recycled to the boiler feed water and heat energy is removed by heat exchangers from the remaining blowdown water. Use of frequent short blows, as opposed to infrequent lengthy blows, is preferred, since it reduces both treated water loss and sensible heat energy in the waste water.

Recovering heat energy normally lost in steam condensate can also reduce fuel consumption. Fuel savings will be realized by recycling as much condensate as possible at boiler operating temperature and pressure. Contamination and losses will limit the amount of condensate that can be recycled. The expected fuel savings are also dependent upon the existing intermittent blowdown patterns. Continuous blowdown from the upper drum can reduce the required intermittent bottom blowdown and allow the use of heat recovery devices, so that nearly all of the heat can be recovered. The potential savings is proportional to the amount of blowdown required, which is dependent on the percentage and purity of the makeup water.

Energy savings that can be achieved by recovering heat in blowdown water can be estimated by using Figs. 5-10 and 5-11. The percentage of blowdown will be fixed by the solids concentration of the makeup water and by the maximum allowable boiler water concentration.

Sample Problem: Determine the energy savings which can be achieved by a waste water heat recovery system in which:

- Boiler operating pressure is 400 psig

- Solids allowable in boiler water is 2,750 ppm

- Solids allowable in feed water is 250 ppm

Solution: Use Fig. 5-10 to determine percent blowdown required.

- Percent blowdown is 10%.

Use Fig. 5-11 to determine the fuel savings at 400 psig boiler operating pressure and 10% required blowdown.

- Fuel savings = 4%.

Actual energy savings depend on boiler operating pressure, flash
tank pressure, and on the use of the flashed steam. Preheating boiler
feed water with waste heat can amount to a 1% for every $10°F$ increase
in feed water temperature (9).

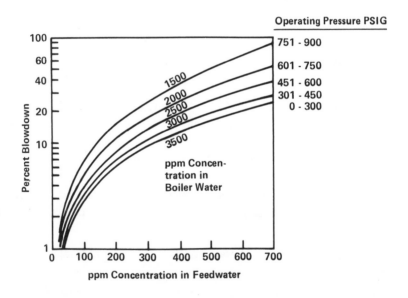

*Figure 5-10. Percent Blowdown Required to Maintain Predetermined
Boiler Water Dissolved Solids Concentration. Source: Otto De Lorenzi,
Combustion Engineering (New York: Combustion Engineering Co., 1947),
p. 21-18.*

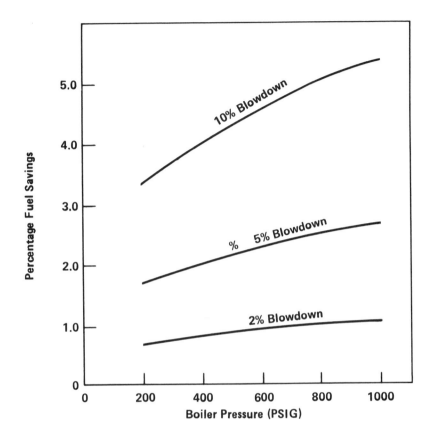

*Figure 5-11. Waste Heat Energy Recovery Potential Using Contin-
uous Blowdown System, Flash Steam Heat Recovery, and Heat Exchangers
on Expelled Blowdown Water. Source: KVB, Inc., Industrial Boiler
User's Manual, p. 71.*

FLUIDIZED BEDS

A promising new combustion method which can work for all kinds of
fuels, regardless of sulfur content, is the fluidized bed. Although
still in the early development stage, results obtained in Czechoslovakia have demonstrated its capability of burning, with 100% inter-
changeability, brown coal, fly dusts, mixtures of low grade solid fuels

with low grade inflammable liquid wastes from the chemical, pharmaceutical and paper industries, and solid and liquid fuels. This is a low temperature combustion process that uses limestone in a slowly moving bed. Much of the ash and sulfur is removed in the bed, greatly reducing the need for costly scrubbing systems.

Figure 5-12 shows a range of energy use and steam production, which is representative of about 60% of industrial plants. In this figure is a line for coal burned in a fluidized bed boiler. As can be seen, it depicts lower costs for generating steam than do other

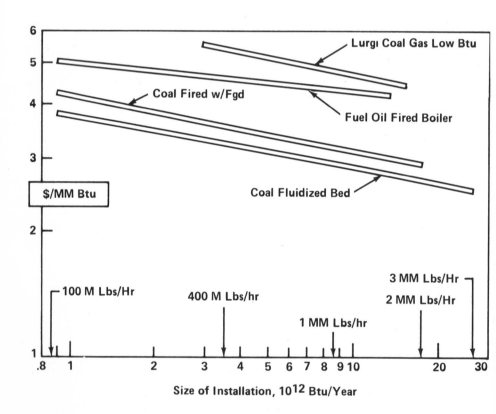

Figure 5-12. Steam Energy Costs VS Capacity (Smaller Facilities).
Source: R. W. Barnes, "Energy and Industrial Processes: A Look at the Future," Paper presented at the symposium on Advances in Energy Storage and Conversion, American Chemical Society National Meeting, San Francisco, California, September 1, 1976, p. 17.

boiler types. It is believed that there is a substantial potential for this new combustion technology in the intermediate to large industrial boilers.

The fluidized bed combustion technique is attractive for several reasons, not only from the economic viewpoint, but also from the environmental and operational aspects. It allows its users to burn coal in an environmentally acceptable way, whereas other techniques that control sulfur and nitrogen oxide emissions are not as satisfactory.

POTENTIAL FOR ENERGY CONSERVATION

The preceding sections discuss those techniques which could be used to enhance boiler operating efficiency. KVB, Inc., in a study done for the Federal Energy Administration, estimates the energy efficiency improvement potentials that exist for given boiler classes (10). Examples of the range of operating efficiencies for watertube boilers using various fuels are shown in Figs. 5-13, 5-14 and 5-15. On the basis of these estimates, Table 5-2 provides an indication of the potential for industrial boiler energy conservation. The baseline, η, represents the "as found" operating efficiencies. Efficiency improvements from actual or "as found" operating conditions is obtained through a boiler tune-up, which involves adjustment and maintenance of the combustion system and cleaning of boiler heat transfer surfaces.

The second group of estimates is based on tuned-up operating efficiencies. As shown in the table, there exists a 0.2-0.9% efficiency improvement potential between actual or "as found" and tuned-up conditions. In general, the larger units tend to have a greater improvement potential.

The maximum economically achievable efficiency level is the savings which could be attained by economically employing the appropriate available efficiency improvement equipment. Auxiliary equipment that can be added to improve efficiency include air preheaters, economizers, turbulators, improved combustion control systems, and improved burners.

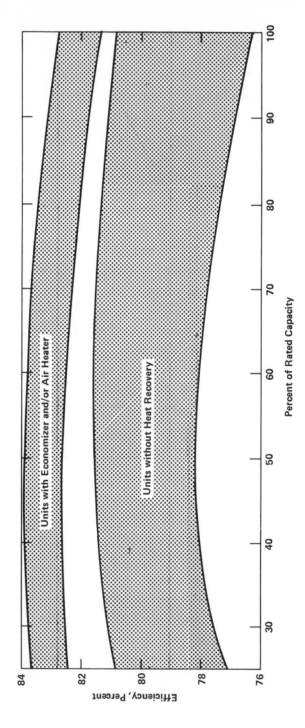

Figure 5-13. Typical Performance of Gas Fired Watertube Boiler

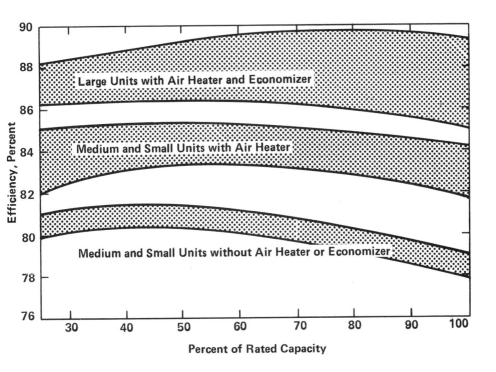

Figure 5-14. *Typical Performance of Oil Fired Watertube Boiler.*

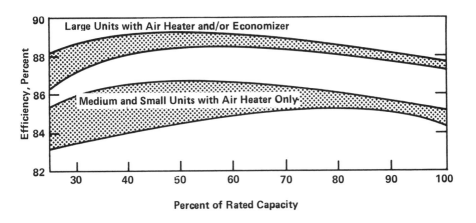

Figure 5-15. *Typical Performance of Pulverized Coal Fired Water-tube Boiler.*

Table 5-2

Industrial Boiler Energy Conservation Potential

10-16 k lb/hr

Fuel	Baseline η	Tuned Up Δη	Tuned Up % Fuel	Max. Econ. Attainable η	Max. Econ. Attainable Δη	Max. Econ. Attainable % Fuel	Maximum Attainable η	Maximum Attainable Δη	Maximum Attainable % Fuel
Gas	79.9	0.2	0.25	80.1	0.2	0.25	85.6	5.7	6.66
Oil	83.7	0.4	0.48	84.1	0.4	0.48	88.8	5.1	5.74
Coal-Stoker	81.0	0.3	0.37	81.3	0.3	0.37	86.4	5.3	6.13
Pulverized	83.2	0.6	0.72	83.8	0.6	0.72	89.5	6.3	7.04

100-250 k lb/hr

Fuel	Baseline η	Tuned Up Δη	Tuned Up % Fuel	Max. Econ. Attainable η	Max. Econ. Attainable Δη	Max. Econ. Attainable % Fuel	Maximum Attainable η	Maximum Attainable Δη	Maximum Attainable % Fuel
Gas	80.9	0.8	0.98	84.0	3.1	3.69	86.5	5.6	6.47
Oil	84.6	0.9	1.05	88.3	3.7	4.19	89.7	5.1	5.69
Coal-Stoker	81.8	0.2	0.24	85.5	3.7	4.33	87.3	5.5	6.30
Pulverized	86.1	0.4	0.46	88.8	2.7	3.04	90.4	4.3	4.76

16-100 k lb/hr

Fuel	Baseline η	Tuned Up Δη	Tuned Up % Fuel	Max. Econ. Attainable η	Max. Econ. Attainable Δη	Max. Econ. Attainable % Fuel	Maximum Attainable η	Maximum Attainable Δη	Maximum Attainable % Fuel
Gas	79.9	0.3	0.37	81.7	1.8	2.20	86.2	6.3	7.31
Oil	83.7	0.5	0.59	86.7	3.0	3.46	89.4	5.7	6.38
Coal-Stoker	81.2	0.7	0.86	83.9	2.7	3.22	87.0	5.8	6.67
Pulverized	83.3	0.6	0.72	86.8	2.5	2.88	90.1	6.8	7.55

250-500 k lb/hr

Fuel	Baseline η	Tuned Up Δη	Tuned Up % Fuel	Max. Econ. Attainable η	Max. Econ. Attainable Δη	Max. Econ. Attainable % Fuel	Maximum Attainable η	Maximum Attainable Δη	Maximum Attainable % Fuel
Gas	81.2	0.8	0.98	85.2	4.0	4.70	86.6	5.4	6.24
Oil	85.3	0.9	1.06	88.7	3.4	3.83	89.8	4.5	5.01
Coal-Stoker	82.5	0.2	0.24	85.8	3.3	3.85	87.4	4.9	5.61
Pulverized	86.3	0.4	0.46	89.1	2.8	3.14	90.5	4.2	4.64

Source: KVB, Inc., Industrial Boiler User's Manual, p. 226.

On the basis of the boiler efficiency improvement potentials in
Table 5-2 and of the boiler inventories described in Table 5-1, Table
5-3, which shows the national energy conservation potential, was developed. The data indicates that an additional 1.5-3.0% improvement
potential can be attained through using economically justified auxiliary equipment. Oil and stoker fired coal units tend to have the
greatest potential. The maximum efficiency improvements attainable
range from 4.0-7.0%. Fuel/size categories that have the highest ratio
of units with existing stack gas heat recovery equipment have the
lowest potential for attainable maximum efficiency improvement. This
potential generally increases for all fuel categories with decreasing
unit capacity. Gas and stoker units generally have the highest potential, with large pulverized units the smallest.

The maximum potential energy conservation, using current technology without consideration for cost for all boiler categories and fuels,
is 63 x 10^9 BTU/hr or 550 x 10^{12} BTU/yr. This represents approximately 6% of the current energy consumption in industrial boilers in
the U.S. The estimated potential energy conservation for boilers
equipped with efficiency improvement equipment considered to be cost
effective under 1976 fuel prices is 30 x 10^9 BTU/hr or 265 x 10^{12}
BTU/yr. If all boilers were tuned up from their actual or "as found"
operating efficiencies to a desirable optimum potential, this could
result in savings of 6 x 10^9 BTU/hr or 56 x 10^{12} BTU/yr.

As can be expected, the largest potential savings exist for
fuel/size categories corresponding to large total capacities and high
usage factors, even though these categories may not have the highest
efficiency improvement potential on an individual boiler basis. In-
terestingly enough, the total energy conservation estimates indicate
that only 48% of the attainable maximum energy savings potential can
be realistically achieved in the near term. To realize the full potential, a dramatic change in relative fuel price versus equipment costs
is necessary.

Table 5-3
National Energy Conservation Estimate
(10⁹ Btu/hr)

| Fuel | Rated Capacity Range (10^6 Btu/hr) | | | | | | | | | | | | | | |
| | 10-16 | | | 16-100 | | | 100-250 | | | 250-500 | | | Total | | |
	Tuned Up	Maximum Economical	Maximum Attainable	Tuned Up	Maximum Economical	Maximum Attainable	Tuned Up	Maximum Economical	Maximum Attainable	Tuned Up	Maximum Economical	Maximum Attainable	Tuned Up	Maximum Economical	Maximum Attainable
Gas	0.14	0.14	3.19	0.68	4.02	13.37	1.34	5.03	8.82	0.54	2.58	3.43	2.70	11.77	28.81
Oil	0.18	0.18	2.24	0.86	5.04	9.29	0.78	3.11	4.23	0.35	1.27	1.66	2.17	9.60	17.42
Coal-Stoker	0.04	0.04	0.75	0.70	2.61	5.41	0.14	2.45	3.22	0.04	0.58	0.85	0.92	5.68	10.23
Pulverized	0.02	0.02	0.89	0.12	0.46	1.21	0.27	1.78	2.79	0.14	0.98	1.45	0.55	3.24	6.34
Total	0.38	0.38	7.07	2.36	12.13	29.28	2.53	12.37	19.06	1.07	5.41	7.39	6.34	30.29	62.80

Source: KVB, Inc., Industrial Boiler User's Manual, p. 12.

NOTES AND REFERENCES

1. R. E. Thompson et al., "A Study to Assess the Potential for Energy Conservation Through Improved Industrial Boiler Efficiency," Report to the Federal Energy Administration, No. C-04-50085-00 (Washington, D.C.: Government Printing Office, October 1976).

2. G. A. Cato et al., "Field Testing: Application of Combustion Modifications to Control Pollutant Emissions from Industrial Boilers--Phase II," KVB, Inc. Report to the Environmental Protection Agency, No. EPA 600/2-76-086a (Washington, D.C.: Government Printing Office, April 1976).

3. S. G. Dukelow, "Charting Improved Boiler Efficiency," Factory, 7 (April 1974), p. 31.

4. George Perry and James Monteaux, "Practical Analysis Can Cut Plant Energy Use, Trim Operating Costs," Oil and Gas Journal, 74 (February 23, 1976), p. 80.

5. KVB, Inc., Industrial Boiler User's Manual, Vol. 2, Report No. FEA D-77/026 (Washington, D.C.: Federal Energy Administration, January 1977), p. 64.

6. Ibid., p. 76.

7   Ibid., p. 46.

8. Ibid., p. 35.

9. J. K. Salisbury, Kent's Mechanical Engineers Handbook (New York: John Wiley and Sons, Inc., 1950), pp. 7-36.

10. KVB, Inc., Industrial Boiler User's Manual.

Chapter 6

CO-GENERATION

The most common method of heating process streams in industrial processes is through the combustion of fuel. Heating may be direct, passing the process fluid through tubes inside a heater firebox, or indirect, using steam that is raised by combustion of fuel in a boiler. Since few industrial processes require the process stream to be heated above $1000^{\circ}F$ and since flame temperatures for normal combustion are in the region of $3000^{\circ}F$ or higher, there is a substantial waste of available energy between the flame and the process stream. Ideally, the combustion should be used to perform work of some kind at, for example, $3000^{\circ}F$, followed by a "cascade" of other heating requirements down to $300^{\circ}F$ or lower, at which time the flue gas would be exhausted to the atmosphere.

The use of process steam by industry accounts for some 16% of total U.S. energy consumption and for 40% of total industrial fuel and energy consumption. Whenever process steam is generated, there also exists an opportunity for generating electricity. In a central station fossil fuel fired powerplant, about one third escapes in the form of flue gases entering the atmosphere at temperatures of $300-600^{\circ}F$ and as steam condensate at temperatures of $70-100^{\circ}F$. If industry would generate steam and electricity together, the temperature of the "reject" heat from the power cycle would have to be raised to the level of the

steam required in the industrial processes. This would reduce the efficiency of generating electricity but would result in industrial process steam serving as the sink for the power cycle, which would decrease thermal discharges substantially and would raise the effective
efficiency of the industrial power plant generating steam and electricity together to 60-70%. Central station steam power plants today are,
at most, 35-40% efficient.

Electricity can also be generated in industry from the waste heat
of high temperature direct fossil fuel fired combustion processes
through the use of bottoming cycle engines. This waste heat recovery
potential is much smaller than the co-generation of electricity and
steam. However, it can provide an important resource that should not
be overlooked.

Only about 14% of industrial electricity consumption and 6% of
nationwide electricity consumption is generated inplant, with the remainder generated by electric utilities at a nationwide heat rate of
10,500 BTU per KWH. This can be compared to heat rates of 4000-
7000 BTU per KWH, which is obtainable from co-generation. Nydick
et al. have estimated that the nationwide potential for inplant generation of electricity varies with the topping cycle used. At 50 KWH of
electric power per million BTU of process steam, steam turbine topping
can produce about 35% of all the electricity generated nationwide at
a heat rate averaging 4500 BTU/KWH. Gas turbine topping can generate
nearly 200 KWH, and diesel topping 400 KWH per MMBTU of process steam.
Using the latter two systems, more electric power can be produced in
the U.S. at a heat rate in the vicinity of 5500 and 7000 BTU/KWH, respectively (1).

Approximately 30% of energy input into fossil fuel direct fired
heaters eventually leaves through the stack gases. It has been estimated that with bottoming cycles recovering this waste heat, about 6.7%
of the U.S. electricity can be potentially generated fuel free. In
combination, steam and gas turbine topping and steam and organic bottoming cycles have the potential of saving the U.S. the fuel equivalent
of almost five million barrels of oil per day, which is 15% of all the
fuels and 30% of all the petroleum consumed (2).

Only a fraction of the potential for inplant co-generation is currently being realized. The chemical, petroleum refining, and pulp and paper industries produce about two thirds of the electricity generated in-plant by all industry in the U.S. Other industries generate even less of their potential than do these three industries. In recent years, in-plant electric power generation has declined from 15% of the national electricity generation in 1950 to 5% in 1973. With the increased problems of higher energy costs, of reduced energy availability, and of the need for more efficient use of our national energy resources, the concept of co-generation must receive greater attention.

METHODS FOR CO-GENERATION

Whenever industrial process steam is required, there exists an opportunity to produce electric energy at small cost, in terms of fuel used. In a conventional electrical power plant, steam is produced and used to drive turbine generator units, after which it is condensed and returned to the boiler for further steam production. The loss of heat in the condenser constitutes between one half and two thirds of the total energy used in producing the steam. In a co-generation unit, the extraction of steam from one of the intermediate stages of the turbine or at its exit by using a higher pressure than in a conventional plant can result in eliminating the condenser or in reducing the proportion of heat lost in the condenser.

Four techniques for industrial co-generation or electricity production are feasible. Figure 6-1 shows a gas turbine topping cycle. In this method air is compressed in a compressor and is mixed with fuel in the combustor. The hot gases from the combustor are fed to a gas turbine which drives an electric generator. The hot exhaust gases from the gas turbine are fed to a waste heat recovery boiler which generates process steam at the desired temperature and pressure.

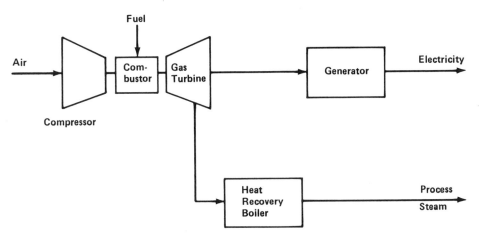

Figure 6-1. Gas Turbine Topping Cycle.

Figure 6-2 shows a combined process steam raising and electricity generating plant. In this method, steam is generated at a pressure greater than what is required for process steam and is then taken
down to the desired pressure through a back pressure turbine coupled
to a generator. The steam for electric power generation can vary from
10% of total steam use in an industrial steam plant that produces power
as a by-product to 70% or more for an electric utility that produces
steam as a by-product. Boiler pressures approach 1500 pounds per
square inch (psi) compared to 150-600 psi for all but the largest of
separate industrial process steam plants. The electrical heat rate
can be as low as 4500 BTU/KWH, as compared to over 10,000 BTU/KWH in
separate electric utility plants.

This type of plant has a number of advantages over a gas turbine
topping cycle. For one, a lower grade fuel, such as coal, can be used
for the steam topping cycle but not for gas turbine power. Also, for
lower pressures of process steam, steam turbines would probably be
preferable to gas turbines, although at higher pressures, power from

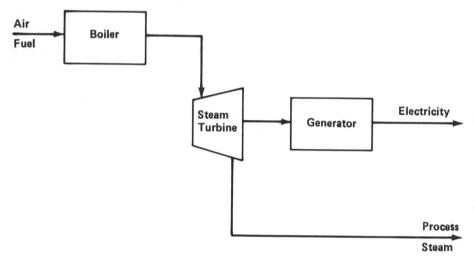

Figure 6-2.   Steam Turbine Topping Cycle.

steam turbines may be too small for good economy. On the other hand,
the gas turbine will generate the same power for process steam at
both high and low pressures.

Figure 6-3 shows a scheme for a combined gas turbine/steam cycle
plant. Combined cycle plants usually are comprised of a gas turbine
generator, a waste heat boiler, and a single steam turbine generator.
The compressor accepts inlet air from the atmosphere and compresses
it to 120-160 psig. The combustor burns the fuel in the proper air
mixture to provide hot gases for the gas turbine which generates electricity. The heat recovery boiler converts gas turbine exhaust gases
into steam energy. This boiler, which can be an unfired, supplementary
fired, or fully fired unit with an afterburner, provides steam to a
back pressure turbine for generating electricity and for obtaining low
pressure process steam.

This type of system may provide benefits in situations where the
process steam requirement is variable and the loss in power generation
capability resulting from following the steam demand requirements can
be made up with conventional power. In these cases, combined cycle

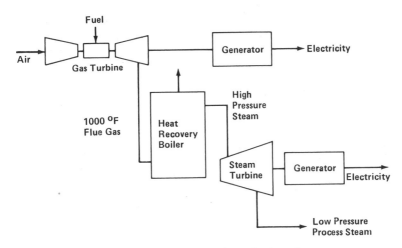

Figure 6-3.   Combined Cycle Topping.

plants are capable of power generation at approximately 8000 BTU/KWH
or at a 43% efficiency.

Combined cycle systems with unfired, supplementary, and fully
fired waste heat boilers have been supplied to industry, along with
back steam pressure turbines for obtaining low pressure process steam.
If the plant does not contain afterburners, the performance of the
system is similar to gas turbine topping.

In industrial processes, a significant amount of thermal energy
is released into the atmosphere at sufficiently high temperatures so
that, through the use of established technology, it can be further
utilized to produce useful work. As seen in Chapter 4, a significant
amount of this thermal energy is rejected in the 300-1000°F temperature
range. From a thermodynamic standpoint, the exploitation of this
wasted heat will result in higher overall plant efficiencies. The
increase in overall efficiency depends on how high a fraction of the
waste heat can be successfully turned into useful work. Figure 6-4
shows a scheme of a bottoming cycle plant. The waste gas from an industrial process heats a fluid which drives a turbine generator to produce electricity. The bottoming cycle working fluid can either be
water or an organic fluid which is used for lower temperature gases,

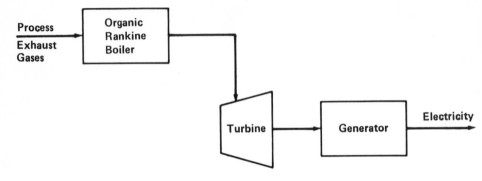

Figure 6-4.   Bottoming Cycle Plants.

depending on the temperature of the waste gas.   An attractive feature
of the bottoming cycle plant is the generation of electricity without
fuel consumption.

Figure 6-5 shows the maximum fraction of the waste heat output
that can be converted into power as a function of source gas temperature for steam and organic (Fluorinol-85) Rankine bottoming plants.
At the temperature range for most process furnace exhaust gases, the
Organic Rankine bottoming plants can supply significantly greater power
than can steam bottoming plants and at essentially the same installed
cost.

The performance of the Organic Rankine plant is shown in Fig. 6-6
in the form of the ratio of power outputs using Fluorinal-85 to steam
Rankine cycle bottoming plants as a function of source gas temperature.
The breakeven gas temperature is 1020°F.   For temperatures below about
550°F, organic working fluids other than Fluorinol-85 would provide
better performance.

IMPACT OF CO-GENERATION

The adoption of co-generation for satisfying industrial steam
and electricity requirements could have a significant impact on the

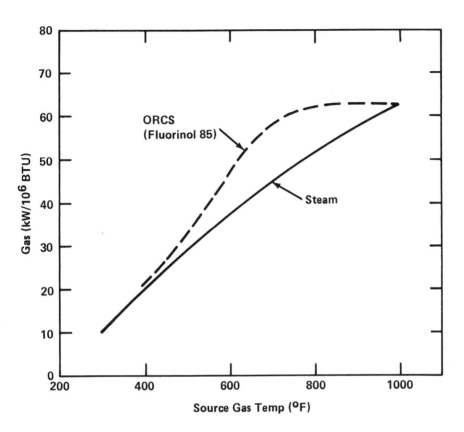

Assumptions: $T(Gas)_{in}$ — $T(Steam)_{out}$

Pinch Point = 40°F

Overall Steam
Turbine-Generator η = 68%

*Figure 6-5. Maximum Power Potential for Bottoming Plants.
Source: S.E. Nydick, J.P. Davis, J. Dunlay, R. Kukhuja, and S. Fam,
A Study of Inplant Electric Power Generation in the Chemical, Petro-
leum Refining and Pulp and Paper Industries, Report No. FEA/D-76/321
(Washington, D.C.: Government Printing Office, June 1976), p. 3-13.*

national problem of energy conservation. Its total adoption could accomplish a net fuel savings of about 15% of the fuel consumed by all
of industry or 30% of all fuel used by the electric utilities.

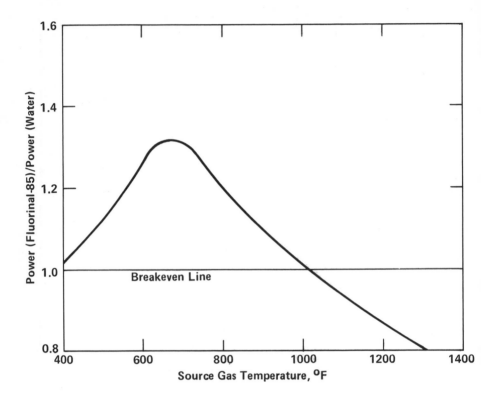

Figure 6-6. Ratio of Power Outputs from Fluorinol-85 and Steam Rankine Cycle Bottoming Plants As a Function of Source Gas Temperature. Source: Nydick et al., A Study of Inplant Electric Power Generation, p. 3-14.

Co-generation has a number of advantages over individual, separate electric utilities and over industrial steam and/or electrical generating plants.

The most significant impact of co-generation is that it substantially improves energy conversion efficiency. The heat rate in most co-generation units will be about 6000 BTU/KWH and, in the most efficient units, about 5000 BTU/KWH. Figure 6-7 indicates the relative efficiency of a good plant generating electricity only and of a co-generation plant where there is a good balance between heat and power requirements.

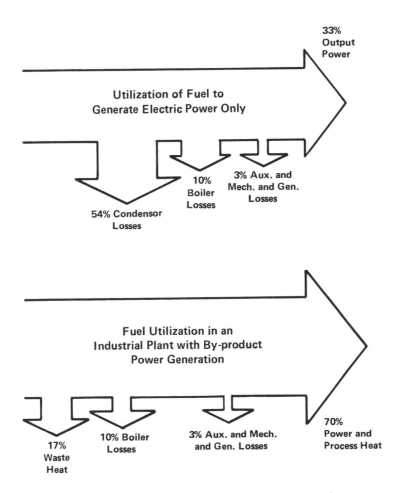

Figure 6-7.   Effectiveness of Fuel Utilization.

Potential for Increased Co-Generation

Table 6-1 shows the potential for fuel savings for the chemical, petroleum refining, and pulp and paper industries through increased co-generation. The fuel savings, based upon the maximum thermodynamic potential for inplant generation in 1974 is the fuel equivalent of about 550,000, 1,500,000, and 2,200,000 barrels of oil per day for steam turbine, gas turbine, and diesel topping, respectively. We see

Table 6-1

Comparison of Maximum Potentials for Fuel Savings by Inplant Generation or Electricity in the Chemical, Petroleum Refining, and Paper and Pulp Industry

Region	Topping of Process Steam			Topping of Process Heat				Maximum Potential[1]		
	Steam Turbine	Gas Turbine	Diesel Topping	Steam Turbine	Gas Turbine	Diesel Topping	Bottoming of Waste Heat	Steam Turbine	Gas Turbine	Diesel Topping
Chemical (10^{12} Btu/yr)	324.7	109.4	1882.4	34.0	114.0	170.1	157.1	515.8	1377.7	2199.4
(Bbl oil/day)	153,000	517,000	890,000	16,000	54,000	80,000	70,000	244,000	650,000	1,040,000
Petroleum (10^{12} Btu/yr)	159.1	676.9	995.9	107.7	262.6	392.3	185.5	452.2	1125.0	1574.0
Refining (Bbl oil/day)	75,000	320,000	470,000	51,000	124,000	185,000	88,000	214,000	552,000	743,000
Paper and (10^{12} Btu/yr)	186.5	415.2	793.4	–	–	–	–	231.7	465.5	891.8
Pulp (Bbl oil/day)	88,000	200,000	381,000	–	–	–	–	110,000	220,000	421,000
Total (10^{12} Btu/yr)	670.3	2404.3	3671.5	141.7	376.6	362.3	342.6	1154.6	3123.3	4576.3
(Bbl oil/day)	316,000	1,140,000	1,740,000	67,000	178,000	266,000	162,000	546,000	1,500,000	2,170,000

[1]Maximum Potential (Bbl oil/day) – topping of process steam and heat and bottoming of waste heat.

Source: Nydick, et al., A Study of Inplant Electric Power Generation, p. 6-25.

that the chemical industry has the greatest fuel savings potential.
The fuel savings potential in the pulp and paper industry is less than
in the other two industries, because it already produces more inplant
power, and little, if any, potential exists in the pulp and paper industry for power generation from topping of process heat and from bottoming of waste heat.

We do know, however, that if the present trends for inplant generation were to continue, the percent of self generated power used by
industry would remain essentially at today's levels and would decline
as a percentage of total utility generation because of the greater
growth rate of utility consumption, despite higher fuel costs and rates
of return which are attractive in comparison to average industry returns.

Improved Capital and Operating Cost Economics

The development of a single large co-generation plant, rather
than of a number of smaller separate steam generating plants, creates
capital and operating economies per BTU, i.e., a single large plant
does not require as much backup equipment as do several smaller plants.
In addition, the replacement of a large number of small package boilers
with a smaller number of large boilers would have the following results. Over the long run and taking boiler life into account, economies of scale should reduce capital requirements. Better design, better control systems, and better generation should raise boiler efficiency from a typical 70% to around 85%. It would also be possible
to generate more industrial steam from coal and nuclear fuels, and less
from the oil and natural gas which package boilers typically require.

Increase in fossil fuel costs will affect power costs twice as
much for an electric power plant as for an industrial co-generation
unit. Figure 6-8 shows the impact of higher future fuel costs and
the increased savings which can accrue for a co-generation plant.
The savings are based on 8400 hours per year operation at average load
and at a heat rate of 4,720 BTU/KWH, assuming that increased utility
fuel costs are passed through as purchased power costs. In addition,

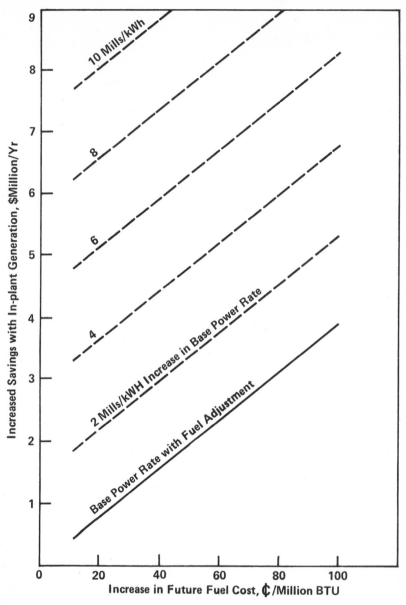

Figure 6-8. *Increased Annual Savings for Inplant Generation Resulting from Higher Fuel and Power Costs in Future. Source: W. G. Wilson, "In-Plant Generation: Profitable Today, More Profitable Tomorrow," TAPPI Manuscript, 1975. Basis: 1) 8400 Hrs/Yr Operation at Average Load. FCP=4720 BTU/kWh; 2) Increased Utility Fuel Costs Passed Through in Purchased Power Costs; 3) Dashed Lines Indicate Effect of Increases in Base Purchased Power Rates for Reasons Other than Fuel Effect.*

the industry investment in co-generation units will be in terms of
today's dollars, whereas purchased power costs will reflect future
utility costs that will include additional generation and transmission
investment costs to meet demands that have been doubling about every
ten years and higher interest costs for new capital as well as for re-
financing existing debt obligations.

On the other hand, industry ownership of inplant generation facilities would require the transfer of investment requirements from utilities to the industrial sector. It has been estimated that investment
requirements range from \$8.5-19 billion for industry self-sufficiency
in electrical requirements and involves considerably greater amounts
if excess power is exported (3). Table 6-2 provides a comparison of
inplant versus central station investment between 1975 and 1985.

Environmental Considerations

The major environmental impacts and trade-offs of increased co-
generation will depend on the specific locations and techniques utilized. Reduced use of energy nationally should lead to reduced environmental effects. However, because of differences in types of fuel used,
in types of fuel burning equipment, and in locations of fuel burning
to generate electricity, the exact reduction in environmental effects
is not known. Co-generation results in more dispersed emissions because the emissions occur at individual industrial plants, as compared
to a central generating station. This could be an important consideration for industrial plants located in air quality basins that are sensitive to increased emissions.

Similar impacts will be noticed for water pollution. Increased
co-generation will reduce water needs for central power stations and
reduce the thermal pollution. However, it will increase water requirements at industrial plants. The total increase in water requirement
will be substantially less than the decrease at central power stations,
but the change in location of the requirements will have an impact on
existing water supply systems.

A potential adverse impact is the land requirement. For existing industrial plants, particularly for those in urban areas, the

CO-GENERATION

Table 6-2

Comparison of Inplant Versus Central Station Investment
Between 1975-1985

Industry Ownership of Inplant Generation

	Incentive[1]	Inplant Option					Central Station Option (No Inplant)
		Generation		Investment			
		By Inplant	By Central Station	By Industry	Central Station (By Utility)	Total	Investment By Utility
A. Industry Self Sufficiency		kWh/yr		10⁹ Dollars			
Steam Turbine	1	115	190	8.5	21.7	30.2	34.7
	2	173	132	16.7	15.0	31.7	34.7
	3	196	109	18.8	12.4	31.2	34.7
Gas Turbine	1	305	0	8.1	0	8.1	34.7
	2	305	0	8.1	0	8.1	34.7
	3	305	0	8.1	0	8.1	34.7
Diesel	1	305	0	11.7	0	11.7	34.7
	2	305	0	12.1	0	12.1	34.7
	3	305	0	12.2	0	12.2	34.7
B. Maximum Power							
Steam Turbine	1	115	0	8.5	0	8.5	13.1
	2	173	0	16.7	0	16.7	19.7
	3	196	0	19.4	0	19.4	22.3
Gas Turbine	1	485	0	15.2	0	15.2	55.3
	2	728	0	23.5	0	23.5	83.0
	3	730	0	23.7	0	23.7	83.2
Diesel	1	610	0	28.5	0	28.5	69.5
	2	1089	0	54.4	0	54.4	124.1
	3	1190	0	57.9	0	57.9	135.7
C. Baseload Option (22% of 1985 Utility Generation)							
Steam Turbine	1	115	633	8.5	72.1	80.6	85.2
	2	173	575	16.7	65.6	82.3	85.2
	3	196	552	18.8	62.9	81.7	85.2
Gas Turbine	1	485	763	15.2	30.0	45.2	85.2
	2	725	23	23.3	2.6	25.9	85.2
	3	730	18	23.5	2.1	25.6	85.2
Diesel	1	610	138	28.5	15.7	44.2	85.2
	2	748	0	36.4	0	36.4	85.2
	3	748	0	35.6	0	35.6	85.2

[1] Incentives   1 – 50% tax rate, 10% investment tax credit, 0% financing
2 - 50% tax rate, 10% investment tax credit, 50% financing
3 - 50% tax rate, 50% investment tax credit (Industry) or 25%
investment tax credit (Utilities), 0% financing

Source: Nydick, et al., A Study of Inplant Electric Power Generation, p. 7-5.

land needed for co-generation may not be available. For new plants,
land use will not be a constraint, since it can be considered in the
planning phase. However, the land requirement for central power stations will be reduced.

Other Impacts

Another major industrial impact would be the additional labor
and maintenance requirements and skills which may be foreign to industry. Also, under present rate regulations, industry would be severely
penalized if it could not be completely self-sufficient. If it had to
purchase utility power, especially for emergency purposes or outages
during peak periods, severe penalties may be invoked, or power may not
be made available. This would create severe reliability problems.
With industrial ownership, the inplant generation facilities would be
subject to industrial tax regulations that are significantly different
from utility regulations.

Because of their larger size, co-generation plants can more economically use coal fired boiler plants. This technique provides an
opportunity to shift from the relatively scarce industrial fuels, such
as oil and gas, to the more abundant and domestically secure coal.

BARRIERS TO CO-GENERATION

Although there are many advantages to co-generation, there are a
number of factors which mitigate against the planning and construction
of these plants. These barriers can be designated technical, economic,
legal/regulatory, and institutional.

Technical Barriers

The major technical barrier to co-generation is load mismatch.
Even though industrial plants requiring both steam and electricity
might desire to use co-generation, their demand may not be feasible

in terms of the ratio of steam to electrical energy required, of steam
pressure requirements, of daily, weekly, and seasonal cycling characteristics, and of reliability requirements in terms of percent allowable down time.

There are also constraints due to improving technologies. Improved co-generation equipment that are cost effective must be developed. There is also a great need for a cost effective backpressure
turbine system which is applicable to industrial needs. Because of
the high reliability requirements, an equipment or technological improvement having energy efficiency advantages cannot be used until it
is thoroughly proven in practice.

The potential retrofit market is also limited by site specific
constraints. These include factors such as the lack of space, the remaining life of the existing plant, and the location of the users from
the co-generation plant. Typically, potential users should be within
five to seven miles of the plant for cost effective transmission of
the steam.

Economic Barriers

Industry operates on a competitive basis; thus, profits must be
earned that are large enough to attract additional capital for expansion. In many cases, product oriented investments can produce larger
dollar volumes of earnings than can inplant generation or other energy
conservation strategies. To ensure earnings in their primary product
oriented business area, industry will invest in these ventures, leaving
little capital for inplant generation. Many industries have been
plagued by severe shortages of equity capital as a result of depressed
stock market conditions and poor profits, requiring them to finance
investments by increasing debt. Again, they are limited in the amount
of funds they can obtain because of an already high debt to equity ratio resulting from past borrowing for product oriented ventures. In
addition, the cost of borrowed funds increases as the debt to equity
ratio of the company increases. Thus, funds used to finance ancillary

investments, such as co-generation, are more expensive than funds used
for financing product oriented ventures.

In the majority of cases, co-generation will cause an excess of
electricity produced. When electric utilities purchase electric power
from industrial plants (if they consider purchasing it at all), they
may set an artificially low purchase price based only on their own
equivalent fuel savings. This treatment will reduce the economic attractiveness of inplant generation for industrial companies.

Another barrier is that the cost of industrially generated power
may be greater than power purchased from the utility. Because of the
incremental electric power pricing structures of many utilities, companies often find that they cannot produce electrical power at a full
cost competitive with the incremental (declining block price structures) costs of electricity from the utility. Still another indirect
factor that increases the cost of industrial electric power generation
is the huge fee charged by utilities for backup power to a company's
on-site generating system. This can effectively raise the overall
annual cost of having an on-site generating system.

When a steam supplying utility is required to operate an additional boiler in order to provide required industrial steam reliability, its steam supply costs will be higher. An example would be an
electric utility setting artifically high prices for steam supplied to
outside customers. This occurs because the utilities allocate inor-
dinately high costs to the steam supply operations versus the main
electrical generation operation, as a result of pressure to keep their
electrical generation costs down.

All of these potential barriers affect the return on investment
that a co-generation plant project would yield and can, therefore,
cause the investment to fall below a given industrial company's investment cut-off rate (hurdle rate).

Legal/Regulatory Barriers

Extensive use of the inplant electricity generation concept will
depend in part on the role played by federal and state regulatory

bodies. Cravath, Swaine, and Moore prepared the landmark document
concerning regulatory issues in the Dow Chemical Company study of
industrial energy centers (4). The analysis investigated a large number of issues that would arise in judgments of the Federal Power Commission (now the Federal Energy Regulatory Commission) and the Securi-
ties and Exchange Commission, and to a lesser extent, it covered the
likely problems in environmental and antitrust matters. The extent
of jurisdiction of regulatory bodies was judged dependent in large
measure on the form of the corporation under which the co-generation
is to take place.

    Some of the potential barriers include:

- Possible public utilities commission or FERC regulation of
  industrial companies tying into the electrical grid. Based
  on discussions with industrial companies planning co-generation facilities, there is a clear concern about co-generation
  falling under public utilities commissions or FERC regulations because these companies would be electricity suppliers
  tying in to the utility grid.

- Possible public utilities commission regulation of utility
  steam supply. For electric utilities in co-generation,
  there is a distinct advantage if the steam supply is not
  regulated by the public utility commissions. Freedom from
  regulation makes price negotiations between potential industrial users and utilities easier.

- Threat of future allocation of coal or other fuel supplies.
  Industrial users must have guaranteed, long term supplies of
  coal or other fuels. The threat of fuel allocation in any
  future energy shortage reduces the attractiveness of co-
  generation investments, although alternative investments
  are not always less risky.

- Potential antitrust constraints. There is a possibility that
  the companies involved in the corporate ownership of co-
  generation units will be subject to antitrust restrictions.
  However, the potential restraint is not as significant as it
  would be for a cooperative effort in the marketing activities
  of the·same companies.

- Financial difficulties. The financing of cooperative ventures
  will inevitably be affected by private law considerations.
  This may involve charter or by-law restrictions peculiar to one
  corporation or common to an industry.

• Regulatory constraints. The present regulatory postures of
the FERC and the state public utilities commission and the
existing rate structures, based on a fair and just return on
the valuation of utility property and on distributed costs,
are not conducive to utility investment in co-generation.

Institutional Barriers

Power generation is the primary product of utilities. These
utilities should be anxious to invest in co-generation, since in many
regions of the country co-generation could result in returns equal to
or greater than that obtainable from central station power plants.
The utilities, however, are to a large extent negatively disposed to
co-generation. They want to maintain their place in the industrial
market for electric power and are reluctant to compete with industry
owned co-generation.

There are other potential institutional barriers to co-generation, which relate mainly to electric utilities:

• Utility franchise position. Through its charter, a utility
has a monopoly for electric transmission in its geographic
area. Industrial companies planning co-generation plants
indicate that some utilities are extremely reluctant to allow
a tie-in to their grid or to share the electric generation
within their area.

• Utility plant flexibility constraints. Normally, an electric
utility can find convenient opportunities to shut down any
one of its plants, because it has several plants supplying
the electric grid. However, for utility co-generation units
supplying steam to industrial users, it is more difficult to
shut the plant down, since the users may have limited alternative steam supplies.

• Uncertainty in providing long term utility steam. Electric
utilities do not have to worry about a long term decrease in
electric power demand. However, if they supply steam to a
relatively limited number of industrial users, there is a
clear risk that steam demand will be terminated if a user
closes down a plant, changes his product line, or takes other
action affecting steam demand.

Other institutional barriers include restricted availability
of special equipment (e.g., turbines), manpower constraints, and
capital constraints.

POLICIES FOR ENHANCED CO-GENERATION

Technical and operational barriers to increased co-generation are not regarded as forbidding. Technologies are available and are in use in industry today. Economic and institutional barriers, on the other hand, may cause significant impact on the implementation of co-generation systems. As for institutional barriers, some sort of governmental action seems warranted.

Five different governmental roles can be identified:

1. There is intergovernmental cooperation and persuasion. Frequently, there are many actions the federal government can take with respect to state agencies, such as public utilities commissions or local governmental bodies, to clarify policy and, in general, to show support for certain activities. The government's role may involve creating awareness of certain opportunities and of educational activities designed to motivate companies toward certain ends.

2. The government can participate more directly by providing technical assistance, either directly with certain agency staffs or through contractors, working with private companies and public institutions in support of project planning or implementation.

3. When technological development requirements exist, the federal government can provide research and development funding support or directly fund a demonstration installation or plant.

4. The federal government can act, either legislatively or administratively, to remove or ease non-financial barriers. This has some similarity to the intergovernmental cooperation role but is directed to specific regulatory or environmental controls that impact on a particular development. This role may involve easing electric utility institutional barriers to facilitate industrial plant tie-in to the electrical grid system, rationalizing utility steam and electric power generation costs with industrial generation costs, and removing potential antitrust constraints.

5. The government can provide financial incentives to stimulate certain private industry actions. These incentives include tax features, loan guarantees, and governmental financing supports.

In a study for the Federal Energy Administration, ThermoElectron Corporation performed an economic analysis to determine the impact of loan guarantees on achieving the potential for co-generation in three industries--chemical, petroleum refining, and pulp and paper (5). Table 6-3 summarizes both the economic potentials for inplant generation with and without federally guaranteed loans and the industrial and utility ownership of facilities in 1975.

The study estimates that with federally guaranteed loans to industry and leasing companies of up to 50% of the required investment, about 60-70% of the potential for inplant generation could be realized, resulting in a substantial increase in power generation over what is achieved today at costs less than or competitive with utility power. With steam turbine topping, central station power would still be required in these three industries, but inplant generated power would increase to 67% of the requirements, up from 30% in 1974-1975. With gas turbine topping, export of power by the three industries could amount to about 13% of the electricity generated by utilities in 1985, increasing to 24% with diesel topping. Fuel savings in 1985 could amount to the equivalent of about 500,000, 1,600,000, and 1,900,000 barrels of oil per day for steam turbine, gas turbine, and diesel topping systems, respectively. Utility ownership of inplant generating capacity can result in even greater savings of 760,000, 1,900,000, and 2,500,000 barrels of oil per day for steam turbine, gas turbine, and diesel topping systems respectively. The optimum ownership of inplant generation facilities will be predicated on local considerations and attitudes.

Additional policy issues that need to be explored include questions concerning the manner in which public utilities and public utility commissions should allocate costs between electricity and discharge heat; issues concerning wheeling excess electricity from small private systems over public utility transmission and distribution systems to other private users and/or to utility customers; and problems concerning the reliability of composite systems, relative to the reliability of single and interconnected electric public utility systems.

Table 6-3

Summary of Maximum Thermodynamic and Economic Potentials
And Fuel Savings for Inplant Generation
Of Electrical Power in 1985

	Economic Potential for Inplant Generation			Economic Potential for Export of Power			Fuel Savings		
	Steam Turbine	Gas Turbine	Diesel	Steam Turbine	Gas Turbine	Diesel	Steam Turbine	Gas Turbine	Diesel
	Percent of Utility Generation			Percent of Utility Generation			Bbl/Oil Equivalent/Day		
1.0 Industry or Leasing Company Ownership									
Without Economic Incentives	4.0	15.0	18.0	(5.0)	6.0	10.0	270,000	1,100,000	1,000,000
With Federally Guaranteed Loans (50% of Investment)	6.0	22.0	33.0	(3.0)	13.0	24.0	500,000	1,600,000	1,900,000
2.0 Utility Ownership									
Without Economic Incentives	7.0	23.0	38.0	(2.0)	14.0	29.0	600,000	1,800,000	2,100,000
With Federally Guaranteed Loans (50% of Investment)	8.0	24.0	44.0	(1.0)	15.0	35.0	760,000	1,900,000	2,500,000
3.0 Maximum Thermodynamic Potential	10.0	35.0	70.0	1.0	25.0	60.0	800,000	2,200,000	3,300,000

( ) –Purchase of Power

Source: Nydick et al., A Study of Inplant Electric Power Generation, p. 1-9.

NOTES AND REFERENCES

1. S. E. Nydick, J. P. Davis, J. Dunlay, R. Sukhuja, and S. Fam, A Study of Inplant Electric Power Generation in the Chemical, Petroleum Refining and Pulp and Paper Industries, Report No. FEA/D-76/321 (Washington, D.C.: Government Printing Office, June 1976), p. 21.

2. Ibid., p. 2-3.

3. Ibid., p. 7-4.

4. Dow Chemical Company, Energy Industrial Center Study, Report prepared for National Science Foundation (Midland, Michigan: n.p., June 1975).

5. Nydick, et al., A Study of Inplant Electric Power Generation.

Chapter 7

IRON AND STEEL INDUSTRY

The United States is the world's largest producer of raw steel, manufacturing about 150 million tons in 1974. The industry comprises approximately 400 steel production and fabrication plants in thirty-seven states. About 130 of these plants produce raw steel, and the remainder are steel rerolling, finishing, and fabrication plants. The major concentration of the industry's operations occurs in the Ohio River Valley and in the states bordering the Great Lakes. Ten major corporations account for about 80% of U.S. steel production. The largest of them, U.S. Steel Corporation, produces one quarter of the national total, while the smallest in this group produces 2.1%. See Table 7-1.

Virtually all U.S. raw steel production is derived via one of three processes:

• Coke oven - blast furnace - oxygen converter process

• Coke oven - blast furnace - open hearth process

• Scrap - electric furnace process

The raw steel is then rolled to final shapes, such as structural steel, sheets, bar, pipe, wire, etc., in a rolling mill operation.

The relative contributions of the different steelmaking processes have been changing since the early 1960s. See Fig. 7-1. Prior to

Table 7-1

Major Corporate Steel Producers
(1973)

Company	Raw Steel Production (million tons)	Total National Production (%)
United States Steel Corporation	34.97	23.2
Bethlehem Steel Corporation	23.70	15.8
National Steel Corporation	11.32	7.5
Republic Steel Corporation	11.29	7.5
Armco Steel Corporation	9.46	6.3
Inland Steel Company	8.16	5.4
Jones and Laughlin Steel Corp.	7.99	5.3
Youngstown Sheet and Tube Co.	5.85	3.9
Wheeling Pitssburgh Steel Corp.	4.41	2.9
Kaiser Steel Corporation	3.17	2.1
All others	30.10	20.1
Total U. S. Production	150.42	100.0

Source: Steel Industry Financial Analysis for 1973, Iron Age.

this period, the open-hearth furnace was the dominant steelmaking process. The open-hearth has since declined in importance, and no new furnaces are being built. In fact, none have been built for several years. Total operating annual capacity for open hearths is now less than 40 million tons of raw steel, and at least a quarter of this capacity is slated for replacement over the next few years. In any case, by 1985 the open hearth process will no longer be a significant factor in U.S. steel production.

The open hearth process has been taken over largely by two other competing steelmaking processes--the oxygen converter and the electric arc furnace. These processes have definite cost and operating advantages. For instance, the quality of the steel from the oxygen converter is similar to that from an open hearth, but it is produced about ten times faster. Also oxygen converter heats can be tapped on a predictable, periodic basis. In 1974, oxygen converters--basic oxygen furnaces (BOF) and bottom blow processes (Q-BOP)--were responsible for about 56% of U.S. raw steel production, with a total installed capacity of approximately 90 million tons. Its share of domestic raw steel

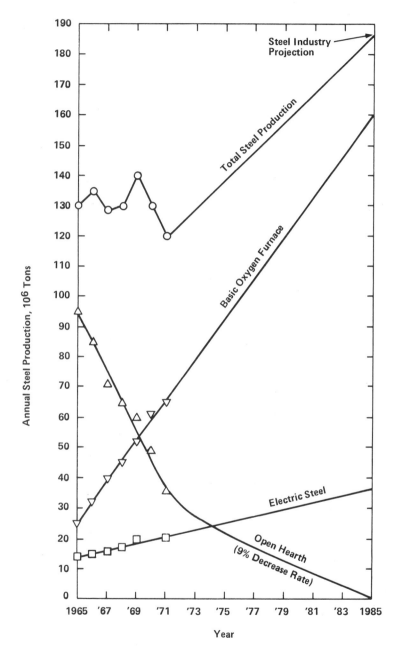

Figure 7-1. Projected Annual Steel Production to 1985. Source: D.H. Larson, "A Projection of the Energy Demand by the Iron and Steel Industry," Paper presented at the Energy Conservation Through Effective Utilization Conference, Henniker, New Hampshire, August 19-24, 1973, p. E-18.

production is projected to equal about 125 million tons in 1980 and
about 160 million tons in 1985.

The electric arc furnace has also witnessed a surge in capacity
over the past decade. In 1974, it comprised about 20% of the domestic raw steel output and is expected to capture as much as 30 million
tons in 1980, rising to about 35 million tons by 1985. This pattern
of capacity growth is spurred on by:

- The more readily available scrap at economically attractive
  prices.

- The development of high power arc furnace technology.

- The establishment of the economic viability of the "mini mill"
  concept.

- The commercialization of iron ore direct reduction processes.

THE STEELMAKING PROCESS

The major process steps in iron and steel making are depicted
in Fig. 7-2. The initial steps are the agglomeration of the iron
ore, coke-making, and the production of limestone, which is fed into
the blast furnace that produces molten pig iron. The output of the
blast furnace is transported to the open hearth furnace or to the
basic oxygen furnace, either of which refines the iron and produces
raw steel. In the case of the electric furnace, scrap is charged
into it to manufacture steel.

The output of these furnaces is then either continuously cast
into slabs, blooms, and billets or cast into ingots. Because of
the time delays between casting and rolling, the metal becomes
solidified. The solidified product is then heated in a reheat furnace rolled, heat treated, and sometimes coated to form a wide
variety of products, such as steel plates, rails, structural materials, rods and bars, pipes and tubes, hot rolled sheets and strips,
wires and wire products, galvanized products, and other plated products.

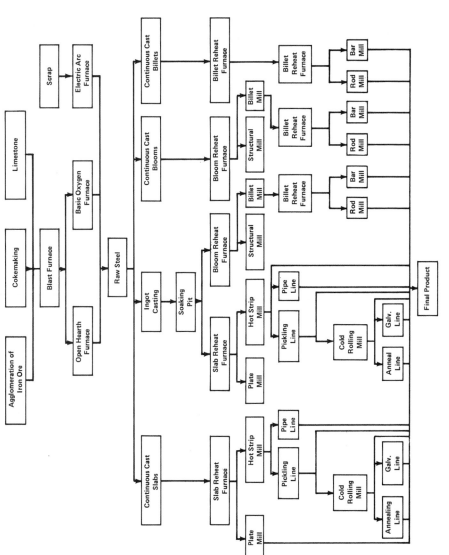

Figure 7-2. Steelmaking Operation.

ENERGY UTILIZATION PATTERNS

The iron and steel industry is a major energy consumer in the industrial sector. The industry uses about 6% of all the energy consumed in the U.S., or about 17% of the total industrial energy requirements, which amounted to 3.6×10^{15} BTU in 1973 (1).

The steel industry has had a long history of increasing its output while decreasing its consumption of energy per ton of steel produced (see Fig. 7-3). In 1950 it required about 37 million BTU per

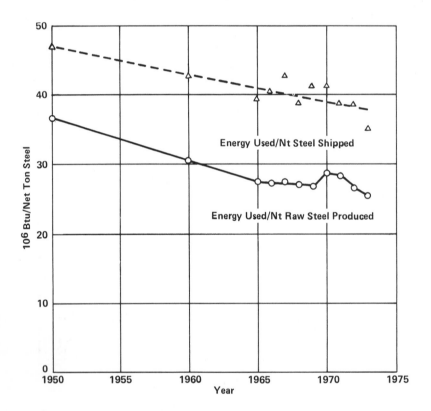

Figure 7-3. Trend in Consumption of Total Energy in U.S. Steel Plants. Source: Battelle Columbus Laboratories, Potential for Energy Conservation in the Steel Industry (Columbus, Ohio: Battelle, May 1975), p. III-20.

ton of steel produced, and in 1973 it required about 26 million BTU
per ton. In terms of steel shipped in 1950, the apparent consumption
of energy was about 47 million BTU per ton; in 1973 it was about 27%
lower, at about 35 million BTU.

In terms of the type of fuel used in the steel industry, Fig. 7-4
provides a historical summary of the energy consumption per ton of raw
steel. The consumption of various forms of energy for production of
steel in 1973 is shown in Table 7-2. As is readily evident, when compared to other manufacturing industries, the steel industry uses a
large fraction of its total energy in the form of coal. The technological changes that allowed the reduction in total energy requirements

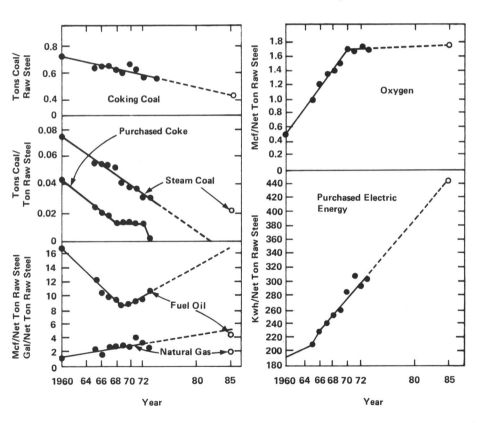

*Figure 7-4. Historic Consumption of Energy Per Ton of Raw Steel.
Source: AISI Annual Statistical Reports.*

Table 7-2

Steel Industry Energy Use
(1973)

Energy Source	Amount	% of Total (BTU Basis)
Coal	90.4×10^6 tons	64.5
Fuel Oil	2000×10^6 gal	8.4
Natural Gas	725×10^9 cu ft	20.6
Purchased Power	43.5×10^9 kWh	4.2
Other	94×10^{12} BTU	2.6

Raw Steel Production = About 151×10^6 tons

Source: Environmental Protection Agency, Environmental Considerations of Selected
         Energy Conservation Manufacturing Process Options, Vol. II: Industry
         Priority Report, EPA 600/7-76-034b (Washington, D.C.: Government
         Printing Office, December 1976), p. 65.

for making a ton of steel were accompanied by a 2.5 fold increase in
the use of purchased electricity. The total energy obtained from oil
and natural gas has changed very little from 1950-1973, with gas being
slightly more intensively used.

Table 7-3 provides a summary of 1975 and of future projections
of energy consumption characteristics for each of the phases in the
steelmaking process. As is evident, the major consumption of energy
occurs in reducing iron ore to remove its oxygen content. This process step in the blast furnace sequence accounts for about two thirds
of the energy consumption per ton of raw steel.

The two major sources of recycled fuel gas within the steel manufacturing complex are: (1) coke oven gas, a sulfur containing, medium
BTU fuel gas produced in the manufacture of coke and (2) blast furnace
gas, a low sulfur, low BTU value fuel.

Table 7-3

Energy Consumption Characteristics

(All in 10^6 BTU/Ton of Finished Steel)

Fuel Type / Process	1975 (Base)			1980			1985		
	Oil/Gas	Coal[1]	Elec.[2]	Oil/Gas	Coal[1]	Elec.	Oil/Gas	Coal	Elec.
Ore Mining & Processing	0.13	0	0.15	0.220	0	0.340	0	0	0
Coke Production	0	15.44	0.02	0.065	14.36	0.030	0.062	13.71	0.03
Sinter	0	0.10	0.04	0.020	0	0.120	0.017	0	0.10
Blast Furnace	0.61	0	0.03	0.790	0	0.360	0.440	0	0.16
BOF	0	0	0.05	0.105	0	0.170	0.080	0	0.12
Q-BOP	0	0	0	0.010	0	0.009	0.091	0	0.08
Elec. Arc	~0	0	0.34	0.012	0	0.910	0.010	0	0.73
Open Hearth	0.65	0	0.05	0.630	0	0.070	0.170	0	0.04
Steel Finishing	3.11	0	0.63	2.540	0	0.550	2.090	0	0.46
Co-Generation	1.41	0.70	-0.26	0.427	0.57	-0.380	0	2.91	-0.55
Conversion	0	0	0	-1.070[3]	0	-0.260[4]	-1.02[3]	0	0.26[4]
Sum	5.93	16.24	1.05	3.750	14.93	2.440	1.940	16.62	1.44
		Σ 23.22			Σ 21.120				20.00

[1]Based on 26.6 x 10^6 Btu/ton (26 Heating value + 0.6 mining & transport).
[2]Based on 10,500 Btu per kwh.
[3]Conversion of captured off gases to useable fuel.
[4]Electricity for generation of oxygen included here.

Source: F. T. Sparrow and T. F. Dougherty, Energy Conservation in the Iron and Steel Industry (Texas: University of Houston, April 1977), p. 10.

IRON ORE PREPARATION

High grade iron ores are being rapidly depleted worldwide and appears to be nearing exhaustion. About 96% of the iron ore mined in the U.S. in 1974 was from open pit mines, the balance from underground mines (2). By 1972, the U.S. steel industry was dependent on low grade magnetic taconite for 80% of its requirements, and the average iron content of domestic ores was down to 34% (3).

The commercially important iron ores are magnetite, hematite, and limonite, each of which has impurities in its natural state. All of the ore is purified or beneficiated to reduce smelting costs. These techniques include crushing, screening, and separation by gravity, by flotation, or by magnetic means. The beneficiated ores are known as "concentrates," which are agglomerated into coarse particles for the subsequent processing steps.

One method of agglomeration is sintering. Sintering converts a mixture of fine ores into granular lumps which are suited for the blast furnace at a temperature of 2400-2700°F. The energy requirements for the sintering process is about 1.6 million BTU/ton (4).

Another agglomeration method is pelletizing, which consists of adding a binder to the ore and forming pellets by means of heating in a rotating drum or disk at 2400°F. Fuel requirements are about 0.6 million BTU/ton (5). See Table 7-4 for the energy requirements for sintering and pelletizing.

The production of sintered ore has been decreasing steadily for the past decade. On the other hand, pelletizing has shown a steady increase. Figure 7-5 indicates the trends toward less sintering and more pelletizing, projected to 1985. Pelletizing almost exclusively uses natural gas as heating energy, but fuel oil can be substituted for this application. Historically, sintering uses 50% natural gas or oil and 50% inplant gases. With the energy requirements fixed at 1.6 MMBTU/ton for sintering and at 0.6 MMBTU/ton for pelletizing, the total projected energy requirements for ore treatment can be calculated. As shown in Fig. 7-6, the change in treatment method almost stabilizes energy consumption between 1971 and 1985, while production grows from 150-220 million tons per year.

New technologies are being developed for beneficiation of iron ores, in order to achieve both more economical processing of low grade ores and higher iron contents in blast furnace charges. Such technologies should result in energy conservation and include (6):

• Cationic flotation of silica for more efficient removal of silica from low grade ores.

Table 7-4

Sintering of Concentrates and Roasting of Magnetic Pellets

Method Used	Average Heat Consumption of Sinter of Pellets 1000 Btu/ton	Average Productivity of Standard Plants, tons/day	Sintering Plants
Grill Sintering	1800	1400-1700 for 53-cu-m conveyor	Extaca, Oliver Iron Mining Division, U.S. Steel Corporation United States
Grate Sintering	1200/1300	Approximately 3000 for Greenawalt machine with 8 grates, i.e., 20-30 tons/cu-m-day	Domnarvet, Stora Kopparberg, Sweden
Rotary Furnace Sintering	1800	1450 per furnace 106 m long with 850-cu-m capacity	Extaca, Oliver Iron Mining Division, U.S. Steel Corporation, United States
Pellet Roasting in Shaft Furnace	500/600	Approximately 1000 per furnace	Erie Mining Plant, United States
	240/360	Approximately 650 per double furnace	Malmberget plant, Sweden
	360	Approximately 120	Segre mine, France
Grill Roasting Pellets	700	2500/3000 per 94-cu-m conveyor; i.e., 25-30 tons/cu-m-day	Reserve Mining, United States (6 primary machines)
Grate-Kiln Pellet	600	Estimated output: 1 plant: 3000 1 plant: 4500	Plans under way for construction

*Estimated

Source: United Nations, Economic Aspects of Iron Ore Preparation

- Floculation of iron oxide prior to flotation to minimize loss of iron during flotation.

- Conversion of low grade non-magnetic iron ore to magnetic iron ore by roasting, so as to allow magnetic beneficiation.

- Prereduction of iron ores. The objective is to raise the iron content of low grade ores so that the smelting required in the blast furnace is minimal, with consequent reduction in coke requirements, and very high iron content agglomeration that can be refined directly is produced.

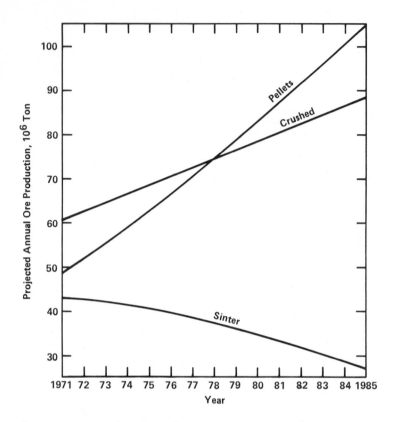

Figure 7-5. Projection of Ore Treatment Trends. Source: Larson, "A Projection of the Energy Demand by the Iron and Steel Industry," p. E-8.

COKEMAKING

Coke is the primary fuel used in the blast furnace for the reduction of iron ore into pig iron. The coke oven is a long, narrow
chamber lined with refractory brick. Coke plants consist of a series
of many ovens, each of which operates in rotation to produce a continuous supply of coke oven gas. Crushed coal is fed into the oven,
which is then sealed. The exterior of the oven is heated by combustion of previously produced coke oven gas, causing the coal to fuse

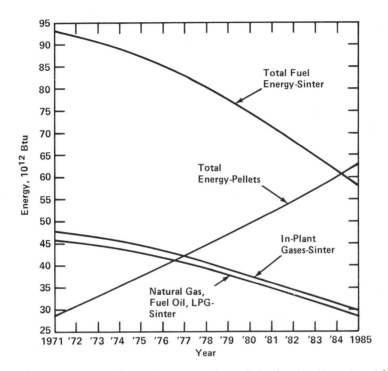

Figure 7-6. Projected Consumption of Fuels for Iron Ore Sinter-ing and Pelletizing. Source: Larson, "A Projection of the Energy Demand by the Iron and Steel Industry," p. E-9.

at the walls, which are at about $2000°F$. Coking proceeds gradually towards the center of the coal charge, while volatile products are re-moved for recovery. The coking is completed in about seventeen hours, at which time a door at the end of the oven is opened and a ram pushes the hot coke into a quenching car. The car takes the coke to a quench-ing station, where it is sprayed with water. The coke is finally al-lowed to cool and, before being fed into the blast furnace, is screened.

Two techniques for improvement of energy efficiency in the coke-making process are reasonably practicable: preheating of coking coal and dry coke quenching.

Preheating of Coking Coal

Coal is generally fed to the coke ovens at ambient temperature with a moisture content of 6-12%. One means of achieving energy conservation is to preheat the coal with waste heat from the plant. Preheated coal is charged at temperatures high enough to assure that the coal is dry.

The principal advantages cited for preheating of coking coals are:

• Increased productivity from coke ovens.

• Improved quality of coke.

• Minimization of air pollution problems at the coke ovens.

A number of authorities have discussed the preheating of coal for improving energy efficiency and increased productivity (7).

There are two basic types of coal preheater systems: the Rosin designed two stage flash pneumatic dryer preheater and the Cerchar designed fluidized bed preheater. Pater and Webster, in their analysis of the two types of preheaters, discuss various methods of handling and charging hot coal to the coke ovens (8). The Rosin type preheater has slightly lower thermal requirements than the Cerchar type, but the reported difference is not significant because the absolute thermal requirements will depend on the moisture content of the incoming coal and on the preheat temperature.

There are several potential problems associated with the preheating and handling of coal. In particular, these are the hazards of handling hot coal dust in suspension, the substantially higher rate of gas evolution (about twice the rate), the higher rate of tar emission, and the higher carryover of solids into the by-product system during charging. The safety aspect of any preheating unit requires careful design and good housekeeping practice.

A 1971 Battelle report presented an estimate of 0.10 MMBTU per ton of coke as potential energy savings associated with preheating of coal for coke ovens (9). Since then, a number of preheated coking plants have indicated net energy savings which are comparable to Battelle's estimates. Bruce and Staneforth observe an overall fuel

savings of 5% or about 0.24 MMBTU per ton of coke (10). Beck et al.
estimate a net reduction in heat consumption of about 0.13 MMBTU per
ton of coke (11). Preheating of coking coals, then, has a potential
for energy conservation of 0.1-0.3 MMBTU per ton of coke.

Dry Coke Quenching

An alternative to wet quenching is a dry system that quenches
the incandescent coke pushed from the coke oven. Dry quenching is not
a new concept, having been developed by the Sulzer Brothers in the
1920s. At one time, over seventy coke plants in Europe were equipped
with dry coke quenching units, but most of these were closed when low
cost natural gas became readily available.

The Soviet Union began using dry coke quenching in the early 1960s
and has now made it mandatory for all new coking facilities. As of
1973, about forty-seven Russian dry quenching facilities were in operation, with about ten newly-installed Russian facilities anticipated
per year (12). Although no dry quenching units are presently operating
in the U.S. or Japan, several dry quench construction projects have
recently been announced by Japanese steel companies (13).

Quenching plants in the Soviet Union are comprised of independent
tower boiler blocks. Each block includes a cooling tower, a waste
heat boiler, dust collectors, and a gas blower, as shown in Fig. 7-7.
The incandescent coke, which is between $1900^{o}F$ and $2000^{o}F$, is pushed
from the coke oven into a bucket. An electric locomotive transports
the bucket to the cooling tower, and a hoist lifts the bucket from the
locomotive to the charging hole. As the hoist approaches the hole,
it automatically opens, and a hopper is placed over the prechamber
so that the bucket and the red hot coke enters the prechamber. During
this time, the pressure at the charging hole is between 0.02 and 0.03.
After the prechamber is charged, the coke guide hopper is rotated,
the charging hole closes automatically, and the bucket is returned
to the locomotive. After forty to fifty minutes, the coke drops
through the prechamber and falls into the cooling zone. As it does so,
circulating gases cool the coke to between $400^{o}F$ and $500^{o}F$ (14).

As the gases rise in the cooling chamber, heat is transferred from the hot coke to the circulating gases. After the gases are heated to between $1,380^{\circ}$ and $1,470^{\circ}F$, they pass from the cooling chamber into dust dropout chambers, where coarse particulates are removed. The gases then pass from the dust dropout chamber into the waste heat boiler, where they are cooled to between $350^{\circ}F$ and $390^{\circ}F$, and high pressure steam is raised.

The use of the coke's sensible heat via the dry quenching process can yield 900-1000 pounds of high pressure steam per ton of coke quenched (15). Introducing the steam enthalpy (typically, 1200-1400 BTU per pound, depending on the steam characteristics), the energy savings associated with the dry quenching process ranges from 1.1-1.3 MMBTU per ton of coke cooled. Thus, dry quenching can provide an effective energy savings of 3.8% in the coking operation.

Dry quenching also has the potential for additional energy savings over the wet quenching method as a result of lowered air and water pollution problems and through lowered coke rates per ton of blast furnace hot metal. The dry quenching operation essentially occurs within a totally enclosed system, and the gas borne particulates are captured within the system.

Comparisons of wet quenched and dry quenched coke in blast furnace trials have shown that with the dry method, the amount of coke required per ton of hot metal is lowered, and the blast furnace productivity is increased. Coke reduction of 2-3% and furnace productivity increases of similar percentages have been reported for the Soviet blast furnace trials with dry quenched coke (16). This represents an energy savings of about 0.38 MMBTU per ton of hot metal.

On the other hand, the physical installation of the dry quenching system is more complex, increasing the capital cost significantly. The difference between a dry coke quenching station and a wet coke quenching station is about $7 million for an annual production of one million tons of coke (17). It also appears that a standby coke quenching system needs to be available to ensure reliable operations.

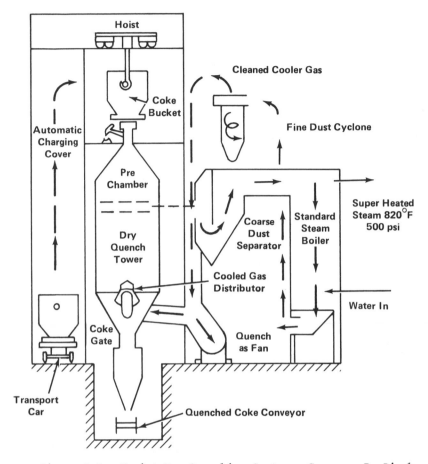

Figure 7-7. Soviet Dry Quenching System. Source: B. Linsky et al., "Dry Coke Quenching, Air Pollution and Energy: A Status Report," Journal of Air Pollution Control Association, 25 (September 1975), 918-24.

Considering the reasonably large energy savings potential and other attributes of the dry quenching process, it seems inevitable that this process will someday find usage in the U.S. industry. The date of the introduction of this technology in the U.S. will probably be affected directly by the cost and availability of steam coal and other conventional fuels. The U.S. iron and steel industry will probably not adopt dry coke quenching on any significant scale during the next fifteen years.

Formed Coke

Formed coke refers to a shaped carbonized or partially carbonized product prepared from coals that need not meet the restricted
specifications of metallurgical coking coals. Interest in formed
coke has intensified in the past decade because of the decreasing
supply and increasing cost of metallurgical coking coal and because
of increasingly stringent air pollution standards, which are potentially easier to meet using formed coke than using conventional
cokemaking processes.

Formed coke can be prepared in a variety of ways. The particular
method selected depends on the coals available and on the end use of
the product. But regardless of the method selected, three stages
exist in the manufacture of formed coke:

• Thermal pretreatment of the coal

• Forming, shaping, briquetting or pelletizing

• Thermal post treatment of the briquet

The differences among the various processes currently under consideration for manufacturing formed coke arise from the different types of
coal employed and from the details of the thermal treatment and shaping
operations. With respect to raw materials, some processes require low
volatile coal, some require high volatile coal, and some processes
accept any type of coal.

Extensive and costly blast furnace trials have been performed
to prove the formed coke process. Long lead times are required between
the conception of a process for making formed coke and the acceptance
of the product for replacement of conventional coke in the blast furnace. The Saporhnikov process, for example, after fifteen years work,
has so far been limited to 50% replacement of conventional coke (18).
Other processes that have good success in blast furnace trials include
those developed by FMC Corporation (19), by Bergbauforschung (Germany)
(20), and by ICEM (Rumania) (21), and a process developed jointly by
Keiham Retan (Japan) and Didier-Kellogg (Germany) (22).

It is not clear whether the production and use of formed coke in steelmaking provides a significant potential for energy conservation. Intuitively, one would not anticipate overall energy savings, because formed coke processing begins with a lower grade coal than does conventional cokemaking. In addition, it requires energy in the briquetting operation. On the other hand, some of the energy losses encountered in conventional batch processes, such as heat lost to the atmosphere during the pushing stage, may be diminished in the continuous production of formed coke.

BLAST FURNACES

Essentially all iron ore is reduced to metallic iron in blast furnaces. The blast furnace consists of a vertical vessel about 100

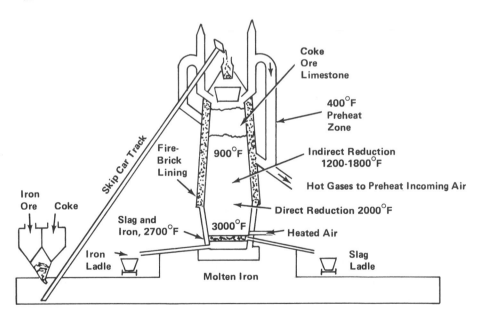

Figure 7-8. Blast Furnace Schematic. Source: Gordian Associates, The Data Base: The Potential for Energy Conservation in Nine Selected Industries, Vol. 6: Steel, (Washington, D.C.: Federal Energy Administration, 1975), p. 124.

feet high, lined with refractory brick. See Fig. 7-8. Iron ore,
coke, and fluxes are charged into the top, and as they move slowly
down, blasts of hot air, frequently enriched with oxygen, are injected
into the bottom of the furnace through holes, or "tuyeres." Temperatures in the blast furnace rise to over $3000^\circ F$ in the combustion zone.
The hot gases from this zone travel up the furnace and heat the descending "burden."

The smelting process occurs in several stages. The ore, coke,
and flux are preheated at the top of the blast furnace by the hot exit
gases. As the charge descends, a partial reduction begins to occur
at temperatures around $1200^\circ F$. As the burden continues to descend, the
temperature rises to around $2000^\circ F$, at which time direct reduction occurs. In this reaction, the incandescent coke and iron oxides react to
produce metallic iron and carbon monoxide gas. The final reaction occurs in the hearth of the furnace, where temperatures finally rise to
about $3000^\circ F$. In the hearth, the iron and slag become molten and are
periodically removed by tapping the furnace. The pig iron produced
from the blast furnace requires further processing to make steel.

The primary source of energy in blast furnace operations is coke.
During the last twenty-five years, the amount of coke required to
produce one ton of iron has decreased steadily. In the past ten years,
the single most important reason for the decline in coke consumption
has been the prereduction of the ore in the pelletizing processes.
Figure 7-9 shows the reduction in coke rate and the increase in production due to increased prereduction of iron ore.

In 1950, 0.92 tons of coke was required to produce one ton of
pig iron. By 1971, coke consumption had decreased approximately 32%
to 0.625 tons of coke per ton of pig iron, as shown in Fig. 7-10. The
steel industry predicts that coke requirements will decrease to only
0.5 tons per ton of pig iron by 1985. This decrease is based on the
increased use of fuel injection and higher blast air temperatures, on
oxygen injection, on higher top pressure, and on external desulfurization, all of which reduce coke requirements.

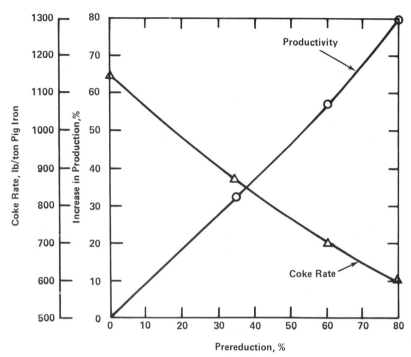

Figure 7-9. Effect of Prereduction on Coke Rate and Blast Furnace Productivity. Source: R. B. Rosenberg, D. H. Larson, W. W. Waterman, and F. Ekman, "Energy Use Patterns in the Metallurgical Industries," Efficient Use of Fuels in the Metallurgical Industry (Chicago, Illinois: IGT, 1974), p. 26.

High Blast Temperatures

An operating change that contributes to decreasing the coke consumption rate is an increase in blast air temperature. Prior to 1950, blast air temperatures were usually below 1200°F, but current blast temperatures are between 1300°F and 1400°F.

There are two conditions which govern the upper blast temperature value without blast modification. They are (1) a high combustion zone temperature, above which the furnace will get sticky and hang, and (2) a minimum quantity of reducing gas per ton of hot metal, below which the quality of the metal cannot be maintained (23). As blast

temperature is increased at constant wind, coke rate decreases and
productivity increases. Table 7-5 shows the effect of combining minimum moisture with high blast temperature. As noted, coke savings is
moderate; however, productivity increases are substantial.

Table 7-6 shows the effect of high blast temperature with minimum natural gas injection on blast furnace operation. As noted, both
coke rate and production rate increases are substantial. Table 7-7
demonstrates the effect of blast temperature and of oil rates which

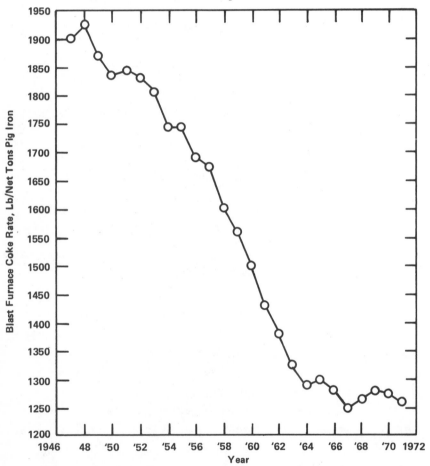

*Figure 7-10. Average Blast Furnace Coke Rate. Source: Rosen-
berg et al., "Energy Use Patterns in the Metallurgical Industries,"
p. 29.*

Table 7-5

Effect of High Blast Temperature With Minimum Blast Moisture

	Base	Minimum Moisture
Wind rate, scfm	1469	1478
Blast temperature, °F	1610	2078
Blast moisture, gr/ft^3	7	20
Coke rate, lb/THM	1262	1207
Production rate, tpd	32.9	37.5

Source: P. L. Woolf, "Improved Blast Furnace Operation," Efficient Use of Fuels in the Metallurgical Industries (Chicago, Illinois: IGT, 1974), p. 271.

Table 7-6

Effect of High Blast Temperature With Minimum Natural Gas Injection

	Base	Minimum Gas
Wind rate, scfm	1394	1395
Blast temperature, °F	1700	2240
Blast moisture, gr/ft^3	7	7
Natural gas, pct blast air	–	1.8
Natural gas, ft^3/THM	–	953
Coke rate, lb/THM	1263	1027
Production rate, tpd	30.1	37.8

Source: Woolf, "Improved Blast Furnace Operation," p. 271.

Table 7-7

Effect of High Blast Temperature With
Fuel Oil Injection

	Base	Minimum Oil	High Oil	Maximum Oil
Wind rate, scfm	1344	1359	1397	1454
Blast temperature, °F	1810	2255	2015	2025
Blast moisture, gr/ft^3	7	7	7	7
Oil rate, lb/min	—	2.3	5.9	7.0
Oil rate, lb/THM	—	90	263	307
Coke rate, lb/THM	1270	1023	965	918
Production rate, tpd	31.6	37.2	32.1	32.7

Source: Woolf, "Improved Blast Furnace Operation," p. 271.

range from the minimum, permitting the high blast temperatures, to the maximum, with which metal quality could be maintained. As oil rate increases coke rate is reduced, but productivity, after an initial surge, begins to decline markedly. It appears that the minimum oil rate is the optimum selection.

In tests, run by the Bureau of Mines, that use pellets, fluxed sinter, and combinations of sinter and ore burdens, the following conclusions regarding average coke savings for injections and increased blast temperature were reached (24):

- Increase of 100°F in blast temperature saves thirty-six pounds of coke per ton of hot metal.

- Increase of one grain per cubic foot of blast in blast moisture requires an additional nine pounds of coke per ton.

- Heavy oil will replace coke on an equal weight for weight basis.

- The addition of 1000 cubic feet of natural gas will save forty pounds of coke.

Hydrocarbon Injection

Hydrocarbon injection is used with more frequency today, because it increases blast furnace production by increasing the proportion of hydrogen available, which reduces iron oxides faster than conventional carbon monoxide derived from coke. Injection of either natural gas or oil lowers the coke rate and increases furnace productivity. But while these two injectants are most widely used and preferred, their restricted availability and increasing costs can be expected to limit their future usage.

Based on theoretical considerations, coal, with its high ratio of carbon to hydrogen, appears as one of the more desirable auxiliary fuels for injection. As a result of the need for reasonably elaborate coal grinding and handling systems for injection, few companies have utilized the coal injection practice. Coal has been injected through the tuyeres of a total of five American, two French, one British, and at least one Soviet blast furnaces (25).

An approximation of the energy savings associated with the injection of a given auxiliary fuel is simply the difference between the energy content of the replaced coke and the added fuel. The energy savings for the injection of prescribed amounts of a given injectant can be calculated and is summarized in Table 7-8. As indicated in the table, all the injected auxiliary fuels lower the energy that is required to produce a ton of hot metal, relative to its production with no injection. Of the selected injectants, coking coal appears to offer a relatively large energy savings per ton of hot metal (THM).

The effects of coal injection on furnace productivity have not been well documented. However, based on Bureau of Mines experimental blast furnace studies, a productivity increase of about 3% could be expected to result from coal injection (26).

High Top Pressure

Energy efficiency improvements in blast furnaces can also be attained through the use of high top pressure blast furnaces. Increasing the top pressure of the blast furnace permits the blowing of more

Table 7-8

Energy Savings for Selected Blast Furnace Injectants

Injectant	Fuel Energy Value, Million Btu/ton	Replacement Ratio	Tons of Injectant THM	Energy Savings Million Btu/THM
None (base case)	31.5	–	–	–
Coal	26.0	1.1	0.125	1.02
Tar	33.6	1.4(b)	0.075	0.79
Oil	38.4	1.4(c)	0.075	0.43
Natural Gas	45.6	1.5(c)	0.05	0.08

(a)   Energy saving is measured relative to base case of no injection.
(b)   The replacement ratio for tar injection has been reported to vary from 1.0 to 1.98. Most
       values are between 1.3 and 1.5.
(c)   The coke replacement ratio varies with the amount of injectant. These are typical values
       for the given units of injectant/THM.

Source: Battelle, Potential for Energy Conservation in the Steel Industry, p. V–35.

wind without fluidizing the burden or greatly increasing flue dust.
Another benefit is the increase in the retention time of the gases,
which promotes improved shaft efficiency and coke savings. Figures
7-11 and 7-12 show the effect of top pressure on coke rate and productivity, respectively. Note that the crossing of the 10 and 20 psig
top pressure curves result in lower coke rate at 2100 cfm and 10 psig
than at the same wind rate with twenty pounds top pressure. This out-
come suggests that there is an optimum top pressure for a given wind
rate. Also, very high top pressure will be beneficial only at extremely high wind rates, relative to furnace size. The savings in
coke rate by increasing top pressure from two to ten pounds could be
as much as 100 pounds, assuming the wind rate at the low pressure operation is near the practical maximum.

Associated with high top pressures are specially designed furnace tops which, in addition to acting as pressure chokes, have the
purpose of distributing the burden in a uniform manner. Burden distribution has been found to be an important parameter for efficient
furnace operation. At the high wind rates associated with high top
pressures and modern furnaces, appreciable energy can be recovered

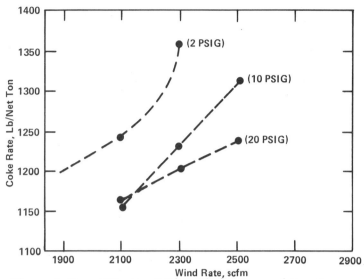

Figure 7-11. Effects of High Wind Rates and Top Pressure on Coke Rate With a Pellet Burden. Source: Woolf, "Improved Blast Furnace Operation," p. 268.

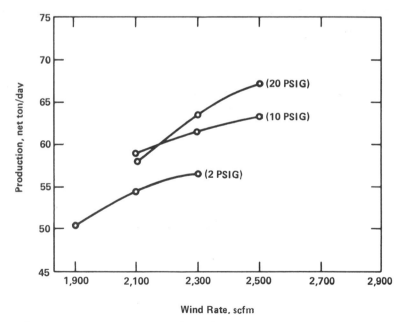

Figure 7-12. Effects of High Wind Rates and Top Pressure on Production with a Pellet Burden. Source: Woolf, "Improved Blast Furnace Operation," p. 268.

from the air stream by using expansion turbines. In this way, much
of the extra energy used to raise the pressure can be recovered.

Kawasaki Steel has installed a gas expansion turbine generator
on one of its Mizushima blast furnaces (27). The generator utilizes
the pressure difference between the inlet gas entering the turbine and
the outlet gas leaving to generate electricity. The power output is
6500 KW, with an estimated equivalent energy production of about
200,000 BTU per ton of hot metal. Such a facility is only useful
when the blast furnace is operated at a top pressure of at least about
40 psi. Blast furnaces in the U.S. do not operate under such high top
pressures, but this concept should be considered when older furnaces
are modernized or when new large furnaces are constructed.

Blast Oxygen

The use of oxygen-rich air as blast in the blast furnace has been
suggested. An oxygen or oxygen-rich air blast will have significant
effects on the thermochemistry and operation of the blast furnace.
Higher temperatures will be produced at the same blast parameters.
The blast furnace gas with oxygen blast will have a higher heating
value than with air blast. Assuming that the carbon monoxide/carbon
dioxide ratio remains constant, the heating value of the resultant gas
with oxygen blast would be about 250 BTU per cubic foot. The total
amount of heat in the blast furnace gas would be the same under these
conditions, but the mass flow rate with oxygen blast would be about
half that for air blast. The ratio of carbon dioxide/carbon monoxide
will probably increase, producing more heat in the furnace but less
heating value in the blast furnace gas.

The energy saved by using oxygen blast is estimated to be
3.35×10^6 BTU/ton (28). The economic advantages are difficult to
determine, however. On one hand, since the gas flow rate will be less,
the volume of the blast furnace and the size of the stoves and compressors could be reduced. On the other hand, because of the high
cost of manufacturing the oxygen, the use of oxygen blast will probably be uneconomical.

External Desulfurization

External desulfurization refers to those processes and techniques which are used to lower the sulfur content of hot metal external to the blast furnace. In conventional practice, blast furnace sulfur content is completely controlled by adding limestone to form a sulfur bearing slag and by limiting the sulfur content of the metallurgical coke.

External desulfurization is viewed as a means for producing lower sulfur content steels and as a method for increasing blast furnace productivity. The use of external desulfurization could allow the replacement of low sulfur content fuels with high sulfur equivalents, thereby extending the range of cokes and auxiliary fuels which would be used in the blast furnace, external desulfurization could provide an energy savings of about 0.9 million BTU per ton of hot metal. This savings would lower the amount of energy required to make raw steel and steel mill products.

The benefits of lowering the blast furnace slag volume include a coke saving of about twenty to thirty pounds per ton of hot metal and a production increase of 3-4% per 100 pounds of decrease in slag volume (30). Thus, 0.2-0.3 pound of coke is replaced per pound of slag volume decrease. Brunger notes that a study of world blast furnace practice suggests that an average replacement factor is 0.16 ton of coke per ton of slag volume (31).

Several external desulfurization materials and systems are currently used in the U.S. These include the plunging of bells containing magnesium impregnated coke (referred to as Mag-Coke) and the lance injecting of lime and/or calcium carbide. Most of the major U.S. steel producers are now testing and/or using external desulfurization in combination with various blast furnace operations. But desulfurizing agents and injection techniques are still undergoing improvement and development, and there is no preferred or standard industry procedure.

CONTINUOUS CASTING OF STEEL

Most of the steel produced in the industry is cast into ingots, allowed to cool, and then reheated for rolling into a product. Continuous casting has the potential for saving fuel because it eliminates the ingot stage of the steelmaking process as well as increases yields by about 10% over the ingot casting route.

There are three types of continuous casting machines, as shown in Fig. 7-13 (32). In all three machines, liquid steel is poured into a tundish from a ladle. Slide gates allow the liquid steel to proceed into the mold and cooling chamber. The liquid steel, as it moves through the mold, forms a thin solid skin. Thermal contractions cause the solid skin to move away from the mold, and because of the low thermal diffusivity of liquid steel, solidification of the center core in the mold is very slow.

The casting machines differ both in the shape of the mold and in the manner in which the solidified slab or billet is changed from the vertical to the horizontal position. The vertical casting machine requires at least seventy feet from the tundish to the floor to complete the solidification process. In the vertical caster, the solidified steel is placed in the horizontal position by precutting lengths and rotating the lengths 90° or by a series of rollers guiding the hot soft steel to the horizontal position. The curved mold machines require about twenty feet of head room between the tundish and the floor, and the mold and cooling chamber is shaped in a big arc. Curved mold machines are advantageous in that the metal strand is bent when the steel is in the liquid state, rather than in a partially solidified state as occurs in bending. The result is that strands in curved molds are less susceptible to surface and corner cracks caused by a combination of bending and thermal stresses.

The temperature of the liquid steel must be precisely controlled in the ladle and tundish to insure fluidity and a minimum amount of fuel use. Usually, the tapping temperature in the steelmaking furnace is slightly higher for a continuous cast than for an ingot cast

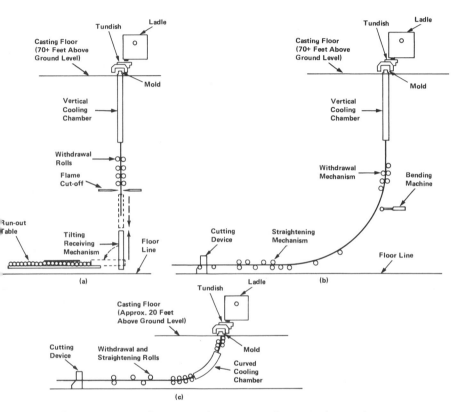

Figure 7-13. Three Continuous Casting Configurations. Source: Sander Nydick and John Dunlay, Recommendations for Future Government Sponsored Research and Development in the Paper and Steel Industries, Report No. 4212-7-77 (Waltham, Massachusetts: ThermoElectron Corporation, August 1976), p. 2-40.

heat. A major concern is preventing the billet or slab from sticking to the mold as it solidifies, so that ruptures and skin cracks are avoided.

Continuous casting installations in the U.S. are divided almost equally between curved and straight mold machines. Since 1971, manufacturers appear to have a slight preference for curved mold slab machines and straight mold billet machines. Straight mold machines, both for billet and slab casting, tend to have larger capacity than curved mold machines.

Continuous casting has a substantial potential for saving fuel in steel processing. In addition to fuel saved in steel processing, electricity consumption can be reduced because of the potential higher yield of continuous casting, compared with ingot casting. ThermoElectron has estimated a total saving of about 1.4×10^6 BTU/ton of raw steel. This is about 5.5% of all fuels consumed in steelmaking (33).

Although continuous casting seems to have many advantages, there are a number of limitations which have inhibited its installation in the U.S.:

• Some grades of high alloy steels cannot be continuous cast easily, e.g., silicon steels.

• Continuous casting facilities require close control of metal chemistry, temperature, and operating procedures.

• Continuous casting results in loss of flexibility of product mix.

• The process requires a longer period of time for breaking in equipment and personnel.

• Existing plants have ingot casting facilities which do not need to be replaced or would be expensive to replace.

A recent commentary notes that 109 of the 158 steel plants in the U.S. do not currently have continuous casting facilities (34). The largest number of continuous casters in the U.S. have been billet casters installed in mini mills.

Although the rated U.S. continuous casting capacity relative to the total national rated capacity for raw steel is low at about 11%, even more significant is the fact that U.S. annual steel production by continous casting in 1973 was only 5.5% of the national output. This indicates that only about one half of the U.S. continuous casting rated capacity is being used. In contrast, in 1973, the West German steel and the Japanese steel industries continuously cast about 17 and 21% of their raw steel, respectively (35).

Continuous casting will, however, be on the uprise in the future. Projections shown in Fig. 7-14 estimate that this process will produce up to 20% of all finished products by 1985. The major facilities based on continuous casting will be mini mills. New integrated steelmaking capacity will probably use continuous casting, even though there are few present plans for new capacity to be installed.

SOAKING PITS

Most of the ingots produced by the steel industry are heated to rolling temperature in soaking pits. Fuels used in soaking pits include blast furnace gas and coke oven gas, straight natural gas, and residual oil. Problems in fuel utilization are associated with variations in flame length and with heat release patterns that occur in changing from one fuel to another. This is particularly true in changing from gaseous fuel to residual oil firing. Actual mill fuel requirements per ton for various fuels, air preheat temperatures, and the amounts of cold steel charged is given in Table 7-9, which shows that the fuel requirement varies from 545,000 to 2×10^6 BTU/ton. The average estimate of overall fuel requirements for soaking pit operations is 1.0×10^6 BTU/ton.

The efficiency of soaking pits is a subject of attention at most steel companies. The energy used can vary from 300,000 to 3×10^6 BTU per ingot ton (36). The wide variation in soaking pit energy requirements can be attributed to a number of factors (37):

- Basic furnace design (type, shape, size, insulation used, etc.)

- Size and type of steel

- Temperature of the steel at the time it is charged to the furnace.

- Air leakage through openings, cracks in refractories, etc.

- Mill delays requiring longer periods in the furnace

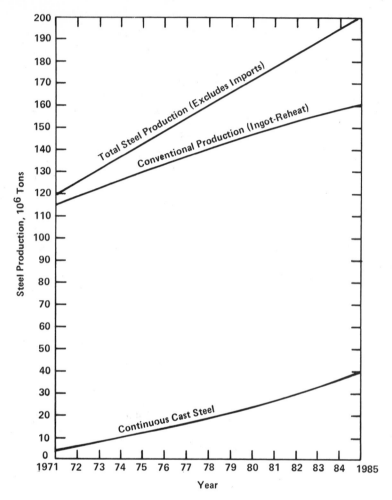

Figure 7-14. Projected Growth of Steel Production. Source: Larson, "A Projection of the Energy Demand by the Iron and Steel Industry," p. E-21.

- Variation from the ideal of fuel/air ratio at the furnace

- Combustion conditions throughout the furnace

- Production rates

- Hearth coverage

Table 7-9

Soaking Pit Fuel Requirements

Type Recuperator	Air Temperature, °F	Fuel	% Cold Ingots	Average Hot Ingot Temp, °F	Fuel Requirement 1000 Btu/ton
Tile	1200-1400	Mixed, coke oven and natural gas	10 or Less	1600	600
None	Ambient	Mixed, coke oven and natural gas	10 or Less	1600	850
Tile and Metallic	900-1200	Mixed, coke oven and natural air	7-10	1600	1000
None	Ambient	Natural gas	15	1400	2000
Ceramic	1500	Residual oil	6	1600	545
Ceramic	1500	Coke oven gas	6	1600	580
Ceramic	1500	Blast furnace gas and coke oven gas	6	1600	800
None	Ambient	Coke oven gas	6	1600	945
None	Ambient	Natural gas	6	1600	950
None	Ambient	Natural gas	10	1600	1270
None	Ambient	Natural gas	13	1600	1450
Metallic-Stack	1000	Oil or mixed gas	100	Ambient	1450
Metallic	975 air 890 gas	Blast furnace gas enriched with coke oven gas	6-7	1600	580 to 730
Metallic	975 air 890 gas	Oil	100	Ambient	1440

Source: D. H. Larson, M. Fejer, and J. Nesbitt, "Improving Efficiency in Reheating, Forging, Annealing and Melting," Paper presented at Industrial Efficiency Seminar, AGA Marketing Conference, Atlanta, GA. March 5-7, 1975. p 15.

• Condition and capability of the waste heat recovery device

• Furnace refractory and insulation condition

Large quantities of energy will be lost if heat is not recovered from the large volumes of high temperature gases exhausting from soaking pits. Soaking pit furnaces should be equipped with some form of heat recovery device, such as waste heat boilers, recuperators, or regenerators.

Mill operating practices can also have a major effect on the fuel economy of soaking pits. Delays in sending ingots from the BOF to the soaking pit can have major fuel impacts. Figure 7-15 shows

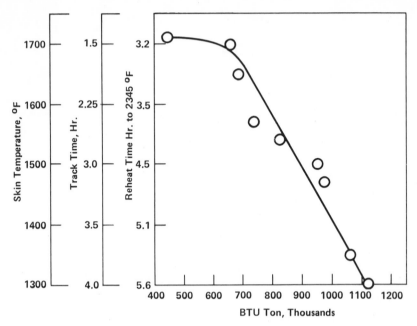

Figure 7-15. Relationship Between Ingot Skin Temperature and Holding Time and Loss of Heat. Source: D.H. Larson, M. Fejer, and J. Nesbitt, "Improving Efficiency in Reheating, Forging, Annealing and Melting," Paper presented at Industrial Efficiency Seminar, AGA Marketing Conference, Atlanta, Georgia, March 5-7, 1975, p. 15.

the relationship between track time, skin temperature, and amount of heat required to bring the ingot back to a proper rolling temperature. An ingot held for as little as four hours of holding time after being poured requires about 670,000 BTU more heat per ton of steel than if it were charged one and a half hours after being poured.

REHEAT FURNACES

Slab, bloom, and billet reheat furnaces are required for heating the semi finished shapes to a temperature at which they can be hot rolled to the desired size and shape. For this, there is a wide variety of furnace designs in use, such as the continuous, semi continuous pusher type in which the steel shapes are intermittently pushed

horizontally through the furnace. See Fig. 7-16. The steel is supported on a system of water cooled pipes so that it can be heated from
both sides in the preheating and heating zones of the furnace. The
steel is in continuous contact with skid rails that are attached to
the tops of the skid pipes.

Energy requirements for reheating vary quite widely, depending
on furnace type. Typical heating requirements for six types are given
in Table 7-10. These energy values can vary from furnace to furnace
for each type, depending on such factors as insulation, production
down time, and loading of billets or slab.

Practically alloof the slabs, billets, and blooms reheated in
the industry start from ambient conditions. In typical practice the
shapes are cooled to allow inspection and conditioning of the surface
in order to remove defects and to make the steel more suitable for
subsequent processing.

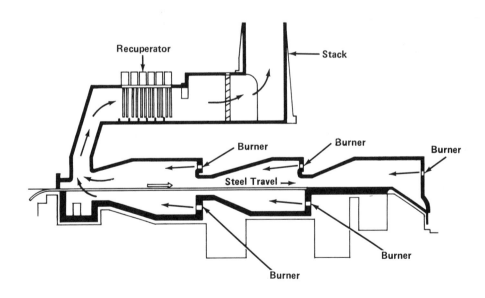

*Figure 7-16. Profile of Five Zone Reheating Furnace with Over-
head Metal Recuperator.*

Table 7-10

Reheat Furnace Heating Requirements

Type of Preheat Furnace	Typical Energy Use, Million Btu Per Ton of Steel Preheated
Three-zone slab, pusher	2.1
Billet, pusher	1.7
Billet, side charge	1.6
Rotary-hearth	3.0
Five-zone slab, pusher	2.3
Five-zone slab, walking beam	1.8 – 3.2

Source: Battelle, Potential for Energy Conservation in the Steel Industry, p. IV-16.

Comparisons between the actual and theoretical values of net fuel consumption show that the thermal efficiencies of the reheat furnaces are in the order of 25-30%. The input energy not recovered averages about 2.0×10^6 BTU per ton (38). This energy leaves the furnace system via the hot waste gases and water in the skid pipe system, via the loss of hot gases and radiation from various openings in the furnace, and via heat conduction through the walls and roof of the furnace.

The major losses are due to the hot exhaust gases and to the skid pipe cooling water. The heat lost to the cooling water in the skid pipe system is an area in which major savings of energy is possible. Present practice is to completely insulate the furnace, run the furnace for fifteen to eighteen months, and then reinsulate. The average insulation during this period will run in the range of 50%. If nothing more than scheduled maintenance of insulation is done every six months, insulation coverage will increase from 50 to almost 80%. Figure 7-17 provides the impact of insulation coverage on fuel consumption. Note that at the 50% figure, the monthly average fuel rate is 3.2×10^6 BTU/ton and will be reduced to approximately 2.8×10^6 BTU/ton at the 80% level. This is a savings of 400,000 BTU/ton, or approximately 13% (39).

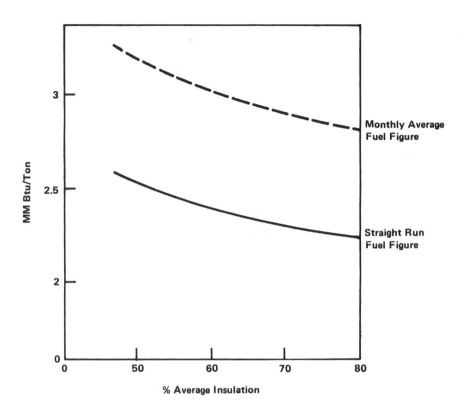

Figure 7-17. Effect of Insulation Coverage on Fuel Consumption.
Source: Reprinted from the August 1974 issue of IRON AND STEEL EN-
GINEER.

Recovery of energy in the waste gas stream can also be accomplished by operating a recuperator system for preheating the combustion air or by using a waste heat boiler to generate steam. Recuperator systems have been in use on reheat furnaces for many years. Figure 7-18 shows that a flue gas temperature of 2600°F, increasing the
preheat temperature from 700°F to 1000°F, reduces fuel consumption by
about 12%. A preheat temperature of 1500°F would save about 12% more
energy than a temperature of 1000°F, and it would save 28% more than
at 700°F, if the flue gas temperature were 2600°F. Metallic recuperators are not capable of withstanding such high temperatures, and,

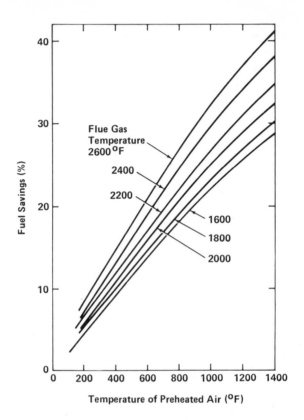

Figure 7-18. Fuel Saving From Preheating Air. Source: Nydick and Dunlay, Recommendations for Future Government Sponsored Research and Development, p. 2-50.

unfortunately, ceramic recuperators are not available commercially at this time.

Preheat temperatures of 900°F for the combustion air can be achieved with metallic recuperators. Typical heat balances show that fuel savings of 30% or more can be achieved. In most cases, the preheat temperatures on existing furnaces with metallic recuperators are in the range of 600-800°F. These temperatures are representative of industrial practices that give reasonable recuperator life.

A new development in reheat furnaces is the monobeam furnace.
The mechanism in contact with the slab is made of ceramic, and the
remaining mechanism can be covered with more insulation than either
a walking beam or a pusher furnace. The monobeam furnace is fired
from bottom and top in a lengthwise path, either parallel or counterflow to the slab movement. The developers of this furnace estimate
10-15% fuel savings (40). Others estimate 6% savings (41).

Table 7-11 provides estimates of the potential industrywide load
reduction resulting from the various means of improving energy utilization in slab, bloom, and billet heating. Note that the percentage
improvements on a specific furnace design will necessarily differ
from the industrywide average.

HEAT TREATING

The amount of fuel used in heat treating and working of steel
varies, depending on the individual operations. Heat treating of
plates, bars, and structural shapes require large amounts of gas,
whereas sheet, strip, and wire production will use lesser amounts.
Production of pipe is another large consumer of gas. Energy consumption per ton of product for various steel working and heat treating
operations are summarized in Table 7-12. The thermal efficiencies of
these systems are relatively low, averaging from 8-40% (42).

Strip steel and other cold rolled products constitute about one
third of the total U.S. steel shipments. In the cold rolling process,
the grain structure of the metal is elongated in the direction of the
rolling, which results in a highly stressed, hard product that is
difficult to work. Annealing is the process of heating the cold
rolled product to a temperature of 1100-1300°F for soak times of two
to twenty hours to obtain recrystallization of the metal and to reduce
its hardness. Table 7-13 provides a summary of the typical fuel requirements for batch and continuous annealing. Most annealing furnaces are not equipped with heat recovery devices, and it is not possible to estimate accurately how many may be retrofitted. There

Table 7-11

Potential Industrywide Load Reductions
for Various Means of Improving Utilization in
Slab, Bloom, and Billet Furnaces

Improvement	% Savings	Load Percentage	Potential Load Reduction, 10^9 CF/yr*
Roof-Firing With Preheated Air	15	0.10	0.37
Control Improvements			
Programmed Firing Control	10	0.50	1.24
Improved Ratio Controls	4.35	0.50	0.55
Full Computer Control	25	0.25	1.6
Application of Existing Recuperator Designs to Furnaces Rated at Less Than 100 Tons/hr	15	0.15	0.56
Skid Rail Insulation	7.5	0.70	1.3
Roof Insulation	1.4	0.75	0.26
Improved Door Design	3.6	0.50	0.44

*Load reduction numbers are not additive. Effective of combinations must be calculated in separate heat balances.

Source: J. D. Nesbitt, "Increased Efficiency of Gas Utilization in Industrial Processing," Institute of Gas Technology, Chicago, Illinois, p. E-89.

is, however, a major potential for improvement of energy efficiency by 20-30% in this area.

Designs for batch type coil annealers range from the single stand annealer to large multistand annealers. Single stand annealers have a great amount of flexibility for handling a variety of jobs in coping with the cyclic nature of steel production. The multistand annealer, on the other hand, offers the advantages of high quality production with economic benefits, in terms of reduced space requirements, reduced material handling, improved fuel economy, and reduced maintenance.

In the past, indirect heating with alloy radiant tubes have been used on single and multistand annealers. The trend in radiant tube design and firing practice in the U.S. has been to conserve alloy by

Table 7-12

Fuel Use for Working and Heat Treating of Steel

Process	Natural Gas Use, 1000 CF/ton
Soaking Pits	1.5
Reheat Furnaces	2.5
Heat Treatment of Plates and Bars	
Carbon Steel	1.8
Alloy Steel	2.45
Stainless Steel	2.5
Heavy Structural Shapes	
Alloy Steel	2.45
Rolled Steel Car Wheels	
Carbon and Alloy	2.45
Sheet and Strip	
Low Carbon Sheet and Strip	1.6
Black Plate	1.0
Electrical Sheet and Strip	1.7
Stainless Steel Sheet and Strip	1.7
Steel Wire	
Low Carbon	1.0
Medium Carbon and Alloy	1.8
High Carbon	1.8
Stainless Steel	2.5
Tubular Products	
Skelp Heating for Buttweld Pipe	2.5
Seamless Pipe (Heat for Hot Working)	2.5
Alloy Pipe	2.45
Stainless Steel Pipe	2.5

Source: Rosenberg et. al., "Energy Use Patterns in the
Metallurgical Industries," p. 41.

use of larger diameter tubes with high firing rates, which result in
very large differentials between alloy temperature and flue gas temperature. Few attempts have been made to recover waste heat by recuperation or other means. In Europe and Japan, high fuel costs encourage the use of recuperators on both radiant tube and direct fired
furnaces.

Table 7-13

Typical Fuel Requirements for Batch and Continuous Annealing

Type of Annealer	Firing Method	Air Preheat Temperature, °F	Annealing - Temperature, °F	Fuel Required,* 1000 Btu/ton
Single- or Multistand	Radiant Tube	Cold	1250	1150-1250
Single- or Multistand	Direct Fired	Cold	1250	900-1000
Single- or Multistand	Direct Fired (Recuperative)	750-900	1250	700-750
Single Stand (Open Coil)	Radiant Tube	Cold	1250	1400
Continuous	Radiant Tube	Cold	1300	925
Continuous	Radiant Tube (Recuperative)	600-650	1300	750

*Fuel required for atmosphere–gas preparation not included.

Source: Larson, Fejer, and Nesbitt, "Improving Efficiency in Reheating, Forging, Annealing and Melting." p. 30.

The demand for large production of steel strip has led to the development of continuous annealing lines, particularly for tin plate. The typical production rate for a modern continuous annealer line is 40-50 tons per hour at a line speed of 1500-2000 feet per minute. A strip may be 0.01 inch thick and 24-36 inches wide. A continuous annealing line is depicted in Fig. 7-19. The heating furnace is provided with radiant tubes that are located between the strip passes and are mounted one half on one sidewall and one half on the opposite sidewall to obtain the most uniform temperature profile.

Both batch and continuous annealing furnaces are heated with gaseous fuels. Fuel conservation measures can be applied with regard to design of new equipment, upgrading of existing equipment, operating practices and maintenance practices. Batch annealing furnaces should be designed for or converted to direct firing from radiant firing. There is ample evidence that not only can the fuel required be reduced, heating uniformity, particularly for multistand covers, will be improved.

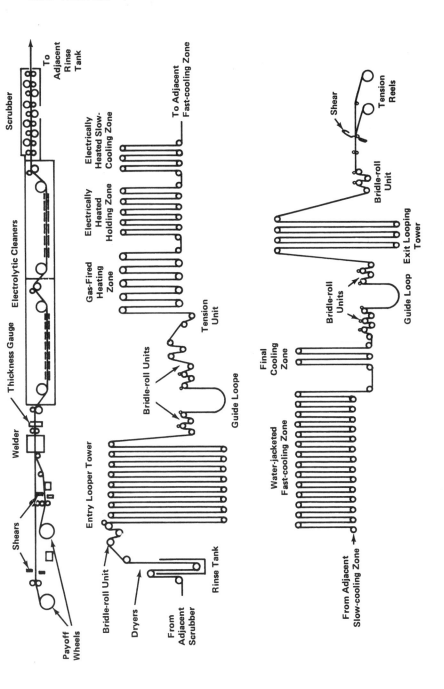

Figure 7-19. Continuous Annealing Line. Source: Larson, Fejer, and Nesbitt, "Improving Efficiency," p. 28.

A conversion from radiant tube firing to direct firing will bring about a 20% fuel reduction. A further conversion from cold air to hot air will improve fuel efficiency by another 20%. Thus, a program to convert radiant tube fired furnaces to recuperative direct firing will bring about a 35% reduction in fuel consumption (43). The actual savings can amount to 400,000-500,000 BTU/ton. The actual energy consumption will vary, depending on the type of product, the coil size, load factors, and the effectiveness of the base fan, base distribution, and convector plate. As much as 10-15% increase in fuel consumption can be required as a result of poor loading practices.

Figure 7-20 shows a recuperative direct fired single stack furnace. The recuperator units are mounted in multiple stub stacks, and the burners can be either radiant, flat flame, or tangential. The projected savings for the hot air system results from close attention to maximum input ratings, to lower waste gas temperatures, and to the value of the sensible heat in the preheated air.

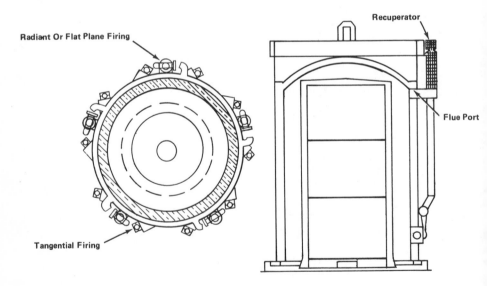

Figure 7-20. Recuperative Direct Fired Single Cover Annealing Furnace. Source: Reprinted from the August 1974 issue of IRON AND STEEL ENGINEER.

Another means for improving the energy efficiency of annealing
furnaces is to use a recirculating system. Figure 7-21 shows a recirculating furnace with a batch or car type arrangement. It has a pressurized hot gas feed system, which allows feed pressures in the range
of 4-6 inches, and positive control over distribution along the furnace length. Gases are collected from the furnace, fed through a hot
fan into a recirculating gas heater, and returned to the furnace. The
gas heater is designed with the return gas as a coolant to the combustion chamber. Temperature drop in the hot gas feed system is minimal,
and the actual heat loss from the hot gas feed system can be held to
2-3% of the total input. An important advantage of a recirculating
type furnace is the considerable flexibility in the type of fuel used,

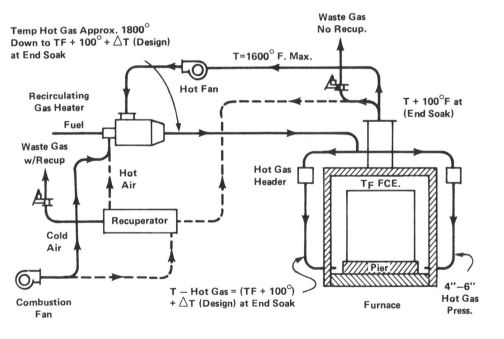

*Figure 7-21. Recirculating Furnace. Reprinted from the August
1974 issue of IRON AND STEEL ENGINEER.*

compared to a direct fired arrangement. Heavy oil, light oil, rich gaseous fuels, or low BTU gas can be used.

Figure 7-21 also shows an alternate recuperation system which is reasonable only in the higher temperature heat treating ranges. This system uses high pressure injection, unlike direct fired systems. An added advantage of this arrangement is the increased recirculation of gases within the chamber itself. The injection velocities will be substantially higher than that of the direct fired burner approach, and the resultant forced recirculation within the furnace chamber will be substantially increased. This should shorten cycle times for a given uniformity requirement.

Figure 7-22 compares fuel consumption between the recirculating concept which fires in ratio with a typical constant air direct fired furnace for a specific cycle on a car type application. A significant difference in fuel consumption is developed during the cutback section of the cycle and during the soak. If the charge were removed after ten hours, allowing a five hour soak, the recirculating furnace would use 41% less fuel than the direct fired constant air furnace. If the soak cycle were ten hours instead of five, the recirculating furnace would use only 51% of the fuel required for the constant direct fired furnace. Thus, this concept, if applied to heat treating and annealing, can cut fuel usage in half (44).

The preceding discussion indicates that major energy savings can be attained through improved heat treating operations. The conversion of batch type annealing furnaces from radiant tube heating to direct firing can lower fuel consumption per ton by 20-35% (45). Using a recirculating system could cut the fuel usage in half. In some situations it is possible to substitute low density ceramic fiber refractories for the high density brick or rammed refractories which are normally used; ceramic fiber refractories lower heat losses through furnace walls and rolls so that less energy is required to heat them when the furnace is started up. The change to fiber refractories in annealing furnaces can lead to fuel savings of as much as 33% (46). In the various other finishing operations, the existing technology is

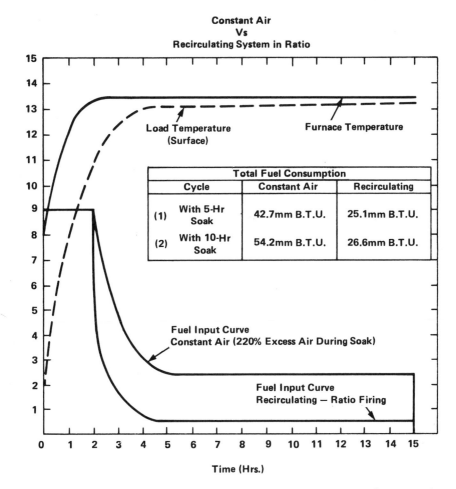

Figure 7-22. Comparative Heat Cycles With Constant Air VS Recirculating System in Ratio. Source: Reprinted from the August 1974
issue of IRON AND STEEL ENGINEER.

not very amenable to changes which will provide any large energy savings. Minor improvements, however, can be achieved under ongoing
plantwide energy conservation programs.

OFF GAS USE

Large volumes of low BTU gas (about 250 BTU/ft^3) are generated during the oxygen injection phase of basic oxygen furnace (BOF) operation. This gas, approximately $2300^\circ F$ in temperature and containing up to 90% carbon monoxide, is formed by burning carbon out of pig iron with pure oxygen. Its heating value is about 0.75×10^6 BTU/ton of raw steel (47), making the total available energy represent about 2-3% of the total energy required in raw steel production.

In the U.S., BOF exhaust gases are collected under a hood. In conventional practice, since there is no provision to prevent air from entering, these hot gases combust spontaneously in the gas collecting hood, are cooled, scrubbed, and then released to the atmosphere. In the past, because of low fuel costs, heat recovery was not economical. However, because fuel costs have always been relatively high in Europe and Japan, the off gases from BOFs are collected and burned in heat recovery boilers for process steam. Use of the off gases also results in better pollution control.

The Japanese have designed a closed hood off gas collection system to eliminate pollution resulting from oxygen blowing (48). See Fig. 7-23. Nitrogen is used to dilute and purge the gases in the system, in order to prevent combustion and explosion of the off gases and to seal against air entering back into the system. Another BOF gas recovery system has been developed in France at the Institute de Recherches de la Siderurgie Francaise (IRSID) and at the Compagnie des Ateliers et Forges de la Loire (CAFL) (49). The primary difference between the two systems is that the Japanese system utilizes nitrogen as a purge gas, whereas the French system utilizes a nitrogen CO_2 mixture which separates the combustible gases from the air aspirated during the non blowing and tapping period.

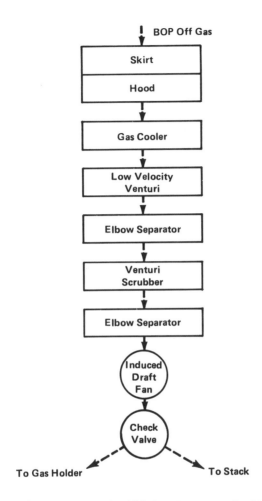

Figure 7-23. OG Off Gas Recovery System.

The Japanese OG gas collection hoods are currently installed in three BOF's in the U.S.--Armco, Inland, and U.S. Steel--and one is under construction in Canada. In all the U.S. installations the gases are being flared, but provisions have been made for the installation of gas collection and storage equipment. Worldwide, about 100 BOFs, largely in the U.S., Europe, and Japan, have closed hood off gas collection systems (50).

Gas recovery systems have been reported to collect 1900 cubic feet of gas per ton of raw steel having a low sulfur content and a caloric value of about 250 BTU/ft^3 or 470,000 BTU/ton of steel. In addition to the collection of the calorific value of the off gases for use elsewhere, the heat value of BOF off gases can be used to generate steam, to preheat scrap for the BOF, and to contribute additional heat for the steelmaking process.

Steam Raising

The off gas burning that takes place in the hoods are steam or water cooled. Since in most steel plants steam is a useful commodity, some of the heat from burning off gas should be recovered to produce steam. This can be done by heat transfer to the water cooled hood or by radiant heat transfer. When such steam is generated, the energy recovered from the off gas is estimated at 150,000-250,000 BTU per ton of steel, comprising only about 20-30% of the total sensible heat and fuel value contained in the BOF off gas (51).

A more effective means of using the BOF off gas heat would be to install waste heat boilers over the BOF so that the heat can be extracted more effectively. With boiler efficiency at 80%, it would be possible to recover over 500,000 BTU per ton of steel. But at the present price of energy, it may not be economical in the U.S. to improve steam generating efficiency by installing waste heat boilers to increase what is presently attainable with water cooled or steam cooled hoods.

Preheating Scrap

BOF furnaces are typically charged with about 25-30% scrap. Since increased scrap utilization would result in significant energy savings, various experiments for preheating scrap, both inside and outside the BOF vessel, have been attempted.

A potential source of energy for preheating scrap is the BOF off gas. One method for utilizing off gas would be to collect it with a closed hood system and use it as a fuel to preheat the scrap in a

separate unit prior to charging the scrap into the BOF. But because
of the high cost of closed hood off gas collection, scrap preheating
by this means is not likely to be widely practiced in the near future.

Another method for utilizing the BOF gas to preheat scrap is depicted in Fig. 7-24. The feasibility of preheating scrap using this
method was investigated by the Bureau of Mines in a pilot plant (52).
In this study, off gases generated during the oxygen blowing of a
molten pig iron and scrap charge were passed through a static bed of
shredded auto scrap. Scrap quantities varying from 22-40% of the BOF
metallic charge were preheated. Average scrap bed temperatures of
1,650-1,150°F were achieved during the blows, which provided up to 44%
of the energy required for melting. The total heat recovered in the
preheated scrap ranged from 90,000-12,000 BTU per ton. These studies
are currently being extended to investigate heat transfer efficiency,
the effect of scrap density, and dust removal. If these experiments
prove successful, it is likely that larger scale demonstrations will
be conducted. At this time, scrap preheating appears to hold the most
promise for utilizing BOF off gas energy.

ENERGY CONSERVATION POTENTIAL

Until the advent of the energy crisis, steel companies concentrated their efforts on decreasing their cost of fuel and energy per
ton of steel to accrue higher profits and to assure long term supply
sources of fuels and energy. Profit was and will continue to be the
driving force that results in decreases in energy consumption per ton
of steel. As all types of fuel and energy increase in price, the pressures upon the steel companies will be to continually minimize their
consumption of energy.

As a result of technological advances, the steel industry has been
experiencing changes in each of the major steps of the steel producing
operations. Improvements in metallurgical processes, product properties, and production economics have been especially pronounced over

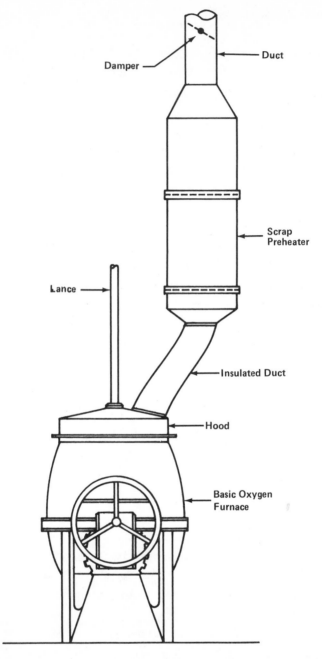

Figure 7-24. BOF Preheater System. Source: W.M. Mahan and B.D. Daellenbach, "Thermal Energy Recovery by Basic Oxygen Furnace Off Gas Preheating of Scrap," *Efficient Use of Fuels in the Metallurgical Industries* (Chicago, Illinois: Institute of Gas Technology, 1974), p. 463.

the past decade and have led to major changes in the operation of steel complexes.

Since World War II, advances in iron ore beneficiation have made possible the exploitation of lower grades of ore. Blast furnace production has increased, using faster blowing rates with high top pressure. Coke oven throughputs are becoming larger, and a wider variety of coals can now be blended to make quality metallurgical coke. Newer steelmaking technologies are supplanting the older technologies. With significant quantities of local scrap being available and with a direct reduction of iron ore, steel making with electric furnaces has also been growing quite rapidly. Continuous casting of billets and, more recently, of slabs are gaining widespread acceptance in new mills over conventional ingot casting, and hot strip mills seem to be growing continuously larger.

The steel industry is relatively efficient thermally, when compared with other manufacturing industries in the same general category. For examples of high thermal "efficiencies" reported in the Battelle report for some of the processes involved in iron and steelmaking, see Table 7-13. The report concludes that in comparing the actual use of available energy to theoretical minimum available energy, the steel industry is about 30% effective. The theoretical minimum energy required for steel production is 7.52×10^6 BTU/ton. The steel industry presently uses approximately 26×10^6 BTU/ton produced. Other conclusions from that study include the following observations (53):

- The coke and ironmaking processes are the largest energy consuming processes and are thermally very efficient.

- The practical limits to improved thermal efficiencies are strongly dependent upon the economics of plant modification.

- Although the theoretical minimum available energy requirement cannot be achieved in any real process, there appears to be an opportunity for significant increases in the effectiveness of fuel utilization.

Opportunities for energy conservation are also indicated by comparing U.S. energy usage with the average usage for steelmaking

in West Germany and Japan, who in 1973 produced steel mill products
at a respective 13-24% below average U.S. energy consumption (54).
The West German coke rate is low at 0.49 tons of coke per ton of hot
metal, compared to 0.60 in the U.S. and 0.58 in the UK. Although the
average size of West German furnaces does not differ significantly
from those in the U.S., German plants, on the average, are newer and
more modern than U.S. plants as a result of rebuilding after World War
II. Germany also uses continuous casting for about 17% of its steel,
compared with 6% in the U.S. These differences account for much of
the higher efficiency of West German steelmaking operations, relative
to U.S. operations.

The Japanese Steel industry is a model for improving the present
state of the art of energy conservation in steelmaking. Japan's 1973
average coke rate of 0.44 tons of coke per ton of hot metal was the
lowest in the world. This was accomplished by a relatively high
preparation of materials before charging into blast furnaces which are
the largest and most modern in the world. Japan continuously casted
about 21% of its steel in 1973, compared to about 6% using the same
method in the U.S. Because Japan must import most of its steelmaking
energy, it imports raw materials in the highest grades and in the optimum mix.

In answering the leading question, "Why are the overseas plants
more efficient in the use of energy?", Edgar Speer, Chairman of U.S.
Steel Corporation, replies, "The high cost of money is preventing
U.S. Steel Company from using technological capability of constructing a new steel mill that would cut manhours by at least 50% and energy 35-40% (55)." It should also be understood that overseas steel
industries have been working intensively at energy conservation for
at least a decade longer than in the U.S. This is due to the fact
that energy for steelmaking was scarcer and higher priced than in the
U.S. Also, the other countries have plants and equipment that are
newer than those of the U.S. The newer plants are partially the result
of the rebuilding which followed demolition of old plants during World
War II and partially the result of active involvement of the foreign
governments in the operations of the steel companies.

Notwithstanding the above arguments, much can be done to improve plant operations and toward the installation of improved technologies for significant conservation of energy. The steel industry is made up of some older facilities, a few new ones, and many in-between. Energy consumption in the newer facilities is typically lower than 30 x 10^6 BTU per ton, but the older ones can range from 40-45 x 10^6 BTU per ton. One could say that the answer to improving energy efficiency in the steel industry is to replace all of the old plants with new plants. This will improve energy efficiency by one third, but this is much simpler said than done.

Table 7-14 provides a summary of the actions which have the major impacts on improving energy efficiency and of the potential energy savings which can accrue from them. Based on the estimated energy conservation potential in Table 7-15, summing of the individual potentials suggests that the total energy conservation, if all of the new technology measures were instituted concurrently, would be 7.08 x 10^6 BTU per ton of raw steel. The energy savings would be 25% of the present energy consumption.

Table 7-14

Thermal Efficiencies in Steelmaking

	Thermal Efficiency, %*
Cokemaking	95.6
Blast furnace	67.3
Blast stoves	76.0
BOF steelmaking	87.3
Electric-arc furnace	83.0

*(Useful energy output/energy input) x 100.

Source: E. H. Hall, Evaluation of the
Theoretical Potential for Energy
Conservation in Seven Basic
Industries (Columbus, Ohio:
Battelle Columbus Laboratory,
July 1975), pp. 81, 83, 84, 86, 89.

Table 7-15
Summary of Energy Conservation Potential

Conservation Practices or Techniques	Potential Savings 10^6 BTU/ton of raw steel	Potential Savings* 10^{12} BTU/year
Increased use of continuous casting	1.4	280
Increased utilization of BOF off gas	0.5	100
Preheating coal to coking ovens	0.2	40
Injection of hydrocarbons into blast furnaces	0.7	140
Internal desulfurization of blast furnace metal	0.9	180
High top pressure blast furnace	0.2	40
Increased efficiency of soaking pits	0.4	80
Insulation of skid pipes	0.4	80
Increased efficiency of reheat furnaces	0.6	120
Increased efficiency of annealing furnaces	0.5	100
Co-generation of electricity	0.9	180

* Calculated on the basis of 200×10^6 net tons of raw steel produced.

Adding the energy conservation potentials yields a numerical total that is too high to be obtainable in actual practice in the short run. Combinations of individual savings generally result in a total savings that is lower than the sum of the individual items. Every savings of 1 million BTU makes the saving of the next million BTU more difficult. It is likely that the steel industry will make energy savings by the replacement of old equipment and unit operations with improved systems involving existing technology.

NOTES AND REFERENCES

1. Environmental Protection Agency, Environmental Consideration of Selected Energy Conservation Manufacturing Process Options, Vol. III, Iron and Steel, EPA 600/7-76-034c (Washington, D.C.: Government Printing Office, December 1976), p. 14.

2. U.S. Department of Interior, Minerals Yearbook, 1974 (Washington, D.C.: Government Printing Office, 1976), p. 653.

3. J. C. Brevard et al., Energy Expenditures Associated with the Production and Recycle of Metals, ORNL-MIT 132 (Oak Ridge, Tennessee: Oak Ridge National Laboratory, November 1972), p. 20.

4. R. B. Rosenberg, D. H. Larson, W. W. Waterman, and F. Ekman, "Energy Use Patterns in the Metallurgical Industries," Efficient Use of Fuels in the Metallurgical Industries (Chicago, Illinois: Institute of Gas Technology, 1974), p. 23.

5. Ibid., p. 23.

6. U.S. Department of Interior, Minerals Yearbook, 1974, pp. 596-98.

7. See K. Beck et al., "A New Technique for Preheating Coking Coal Blends for Carbonization in Slot Type Recovery Ovens," AIME Proceedings of Ironmaking Conference, Vol. 31 (New York: The Metallurgical Society of AIME, 1972), pp. 185-92; D. Marting and R. Davis, "Coaltec System for Preheating and Pipeline Charging of Coal to Coke Ovens," AIME Proceedings of Ironmaking Conference, Vol. 31, pp. 174-84; L. Aldermon and R. Chambers, "Preheating and Charging Coal to Coking Ovens," AIME Proceedings of Ironmaking Conference, Vol. 31, pp. 193-200; and R. Marcellini and J. Geoffroy, "Development of a New Process to Preheat Coal Blends Used for Coking," AIME Proceedings of Ironmaking Conference, Vol. 31, pp. 166-73.

8. V. Pater and J. Webster, "Methods of Charging Preheated Coal," Developments in Iron Making Practice (London: The Iron and Steel Institute, 1973), pp. 53-62.

9. Battelle Columbus Laboratory, Energy Use by the Steel Industry in North America, Final report to Energy in Steelmaking Research Group (Columbus, Ohio: Battelle, July 30, 1971), pp. IV-B-15.

10. J. Bruce and W. Staneforth, "Some Aspects of Experience on the Brookhouse Project," Developments in Ironmaking Practice, pp. 63-77.

11. Beck, et al., "A New Technique for Preheating Coking Coal," AIME Proceedings, p. 185.

12. Proceedings of Symposium on Soviet Dry Quenching, Sponsored by Patent Management, Inc., and VO Licensintorg, Washington, D.C., 1973.

13. "Latest Energy Recovery Technology: New Coke Processing Method--CDQ," Tekko Kai Ho, Tokyo, July 11, 1974.

14.   B. Linsky et al., "Dry Coke Quenching, Air Pollution and Energy:
      A Status Report," Journal of Air Pollution Control Association,
      25 (September 1975), pp. 918-24.

15.   Battelle Columbus Laboratory, Potential for Energy Conservation
      in the Steel Industry (Columbus, Ohio: Battelle, May 30, 1975),
      p. V-16.

16.   Ibid., p. V-17.

17.   Environmental Protection Agency, Environmental Considerations,
      Vol. III, Iron and Steel, p. 9.

18.   D. A. Tsikarev, "Experimental Blast Furnace Smelting Using
      Molded Coke," Koks Khim, 4 (1974), pp. 58-59.

19.   J. K. Holgate and P. H. Pinchbeck, "Use of Formed Coke: BSC
      Experience 1971/1972," Journal of Iron and Steel Institute, 211
      August 1973), pp. 547-66..

20.   Ibid., p. 547.

21.   I. Barbu and I. Stefanescu, "Use of ICEM Formed Coke in the
      Blast Furnace and in Other Applications," Journal of the Iron
      and Steel Institute, 211 (October 1973), pp. 685-88.

22.   K. Sugasawa et al., Production of Formed Coke and Trials in
      Blast Furnace (Japan: Mitsubishi Research, October 1974). See
      also D. Wagener, "Discussion," AIME Proceedings of 31st Iron-
      making Conference, Vol. 31 (New York: The Metallurgical Socie-
      ty of AIME, 1972), pp. 297-98.

23.   P. L. Woolf, "Improved Blast Furnace Operation," Efficient Use
      of Fuels in the Metallurgical Industries, p. 269.

24.   Ibid., p. 270.

25.   Ibid., p. 269. See also L. Cochs, "Why Injection?--Some Facts
      and Figures," Blast Furnace Injection: Proceedings of the Sym-
      posium on Blast Furnace Injection (Sidney: The Australian
      Institute of Mining and Metallurgy, 1972).

26.   Woolf, "Improved Blast Furnace Operation," p. 269.

27.   "New Kawasaki Unit Uses Furnace Gas to Generate Power," Ameri-
      can Metal Market, 81 (December 24, 1974), p. 3.

28.   Sander Nydick and John Dunlay, Recommendations for Future
      Government Sponsored Research and Development in the Paper and
      Steel Industries, Report No. 4212-7-77 (Waltham, Massachusetts:
      ThermoElectron Corporation, August 1976), p. 2-85.

29.   J. C. Agawal and J. F. Elliott, "High Sulfur Coke for Blast
      Furnace Use," AIME Proceedings of Ironmaking Conference, Vol.
      30, 1971, pp. 50-67.

30.   Woolf, "Improved Blast Furnace Operation," p. 268.

31.   R. Brunger, "Sulfur Control by Furnace and External Means,"
      AIME Proceedings of Ironmaking Conference, Vol. 31, 1972, pp.
      169-87.

32.   Nydick and Dunlay, Recommendations for Future Government Spon-
      sored Research and Development, p. 2-39.

33.   Ibid., p. 2-38.

34.   "Continuous Casting--More for Less," Commentary by Institute
      for Iron and Steel Studies, No. III-3, Washington, D.C., March
      1974.

35.   Battelle, Potential for Energy Conservation in the Steel Indus-
      try, p. V-70.

36.   Battelle, Energy Use by the Steel Industry in North America.

37.   Battelle, Potential for Energy Conservation in the Steel Indus-
      try, p. IV-12.

38.   Ibid., p. IV-27.

39.   James Hovis, "Energy Conservation--A Must," Iron and Steel
      Engineer, 51 (August 1974), p. 54. Reprinted from August 1974
      issue of IRON AND STEEL ENGINEER.

40.   Loftus Engineering Corp., Pittsburg, Pennsylvania.

41.   Energy Conservation in the Steel Industry (Washington, D.C.:
      American Iron and Steel Institute, 1976), p. 111.

42.   Rosenberg et al., "Energy Use Patterns in the Metallurgical
      Industries," Efficient Use of Fuels in the Metallurgical Indus-
      tries, p. 42.

43.   Hovis, "Energy Conservation--A Must," Iron and Steel Engineer,
      p. 56. Reprinted from August 1974 issue of IRON AND STEEL
      ENGINEER.

44.   Ibid., p. 57.

45.   H. B. Helm, "Converting Batch Type Annealing Furnace from Radiant
      Tube to Direct Firing," Iron and Steel Engineer, 48 (August 1971),
      pp. 80-84. Reprinted from August 1971 issue of IRON AND STEEL
      ENGINEER.

46.   R. C. Olson, "Using Ceramic Fiber Refractories in Heat Treating
      Furnaces," Metal Progress, 103 (April 1973), p. 85.   (c) Ameri-
      can Society for Metals.

47.   Nydick and Dunlay, Recommendations for Future Government Spon-
      sored Research and Development, p. 2-33.

48.   H. P. Boyce, "The OG Gas Clearing System," Operation of Large
      BOFs (London: The Iron and Steel Institute, 1972), pp. 1-9.

49.   A Maubon, "Technical and Economical Considerations of the
      IRSID/CAFL Oxygen Converter Gas Recovery System," Iron and
      Steel Engineer, 50 (September 1973), pp. 87-97.   Reprinted
      from the September 1973 issue of IRON AND STEEL ENGINEER.

50.   J. K. Stone, "Worldwide Roundup of Basic Oxygen Steelmaking,"
      Iron and Steel Institute, 214 (April 1976), p. 311.

51.   Battelle, Potential for Energy Conservation in the Steel Indus-
      try, p. 45.

52.   W. M. Mahan and B. D. Daellenbach, "Thermal Energy Recovery by
      Basic Oxygen Furnace Off Gas Preheating of Scrap," Efficient
      Use of Fuels in the Metallurgical Industries, pp. 457-66.

53.   E. H. Hall et al., Evaluation of the Theoretical Potential for
      Energy Conservation in Seven Basic Industries (Columbus, Ohio:
      Battelle Columbus Laboratory, January 1975).

54.   Battelle, Potential for Energy Conservation in the Steel Indus-
      try, pp. I-17.

55.   Mark Meadows, "Speer Sees Growth Stymied," American Metal Market,
      81 (September 18, 1974), p. 1.

Chapter 8

THE ALUMINUM INDUSTRY

Aluminum was commercially introduced at the beginning of this century. Since then, it has exhibited a pattern of very rapid growth because of the unique characteristics of strength, light weight, and versatility. Figure 8-1 contrasts the growing rate in the production of aluminum with those of other light metals. During the years 1940 to 1970, the total U.S. output of aluminum increased at an annual rate of 5.7%. Production growth is projected to continue through the year 2000, albeit at a lower rate of 3.8%.

U.S. aluminum consumption has risen gradually over the years, with the exception of a period during the 1940's when wartime requirements for aircraft production caused a sharp rise in the rate of production. Figure 8-2 shows the distribution of aluminum markets for 1960 and 1972. The building and construction industries are the largest end users of aluminum (26.5% of the market in 1972), with transportation (18.5%), packaging (15.2%), electrical users (12.7%), and consumer goods (9.2%) also playing major roles. Of these markets, packaging and transportation are the fastest growing.

The aluminum industry is comprised of two basic operations: (1) the production of alumina from bauxite by the Bayer process and (2) the reduction of alumina to aluminum metal by the Hall-Heroult electrolytic reduction process. These two operations are performed in separate locations within the U.S.

255

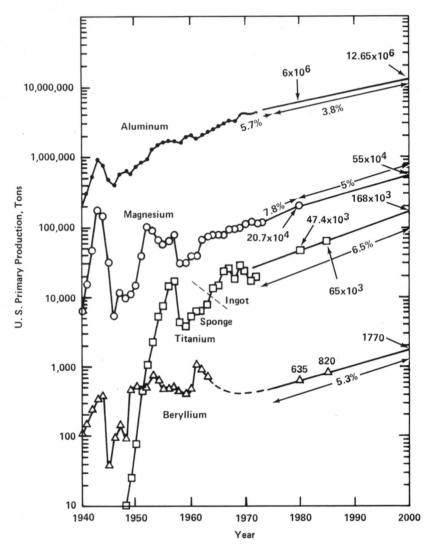

Figure 8-1. U.S. Production of Light Metals. Source: P. R. Atkins and C. N. Cochran, "Future Energy Needs in the U.S. Light Metals Industry," Efficient Use of Fuels in the Metallurgical Industries (Chicago, Illinois: Institute of Gas Technology, 1974), p. 735.

There are nine alumina production plants in the U.S. Table 8-1 provides a summary of the locations and capacities of the plants.

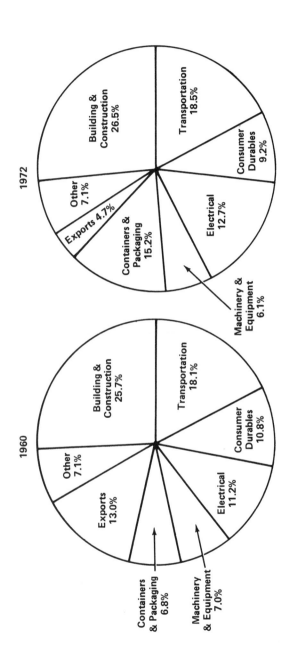

Figure 8-2. Aluminum Shipments (Distribution by Markets). Source: Atkins and Cochran, "Future Energy Needs," Efficient Use of Fuels in the Metallurgical Industries, p. 736.

Table 8-1

U.S. Alumina Plants

Location and Capacity — 1975

	Short Ton/Yr
Aluminum Company of America	
Mobile, AL	1,025,000
Bauxite, AR	375,000
Point Comfort, TX	1,350,000
Martin Marietta	
St. Croix, VI	370,000
Kaiser Aluminum and Chemical	
Baton Rouge, LA	1,025,000
Gramercy, LA	800,000
Reynolds Metals Company	
Hurricane Creek, AR	840,000
Corpus Christi, TX	1,385,000
Ormet Corporation	
Burnside, LA	600,000
	7,770,000

Source: Environmental Protection Agency, Environmental
          Considerations of Selected Energy Conserving
          Manufacturing Process Options. Vol. III,
          Alumina/Aluminum Industry Report (Cincinnati,
          Ohio: Environmental Protection Agency,
          December 1976), p. 92.

Total aluminum production capacity is estimated at 7.7 million short
tons, with individual plant capacities ranging from 370,000 to 1.38
million short tons per year. The U.S. alumina plants are small in
comparison with new foreign installations which have capacities of one
million to two million tons per year. Also, most of the U.S. alumina
plants are relatively old, having been built between 1940 and the earl
1950s. The only new plant, built in the 1960s, is at St. Croix. The
alumina plants are located on the Gulf Coast because of the availabili-
ity of natural gas, or in Arkansas, close to the only domestic bauxite

deposits. A few of the alumina plants are located near the aluminum smelters they serve. Since a new alumina plant involves a major investment in mining and production facilities, plants are not usually designed and constructed until firm long term commitments have been obtained for marketing the alumina.

There are thirty-one alumina plants in the U.S.; these are operated by twelve companies. See Table 8-2. These plants are located primarily along the Mississippi and Ohio Rivers, in the Pacific Northwest, in Texas, and in upper New York state. The location of the smelters is determined by the accessibility to river systems for the transportation of alumina and by the availability of low cost power. Table 8-3 shows the locations and ages of the various smelters, as well as the different smelter technologies used. Soderberg and prebake smelters were built in the 1940s and 1950s, but in the last fifteen years, only prebake smelters have been constructed. Since 1970, eleven new facilities have been built.

Total aluminum capacity in 1975 is estimated at five million short tons; individual plant capacities range from 36,000-285,000 short tons per year. The industry is basically integrated: operations include the location and mining of ores, purification of ores, production of the aluminum metal, and the fabrication of the metal into products.

The industry is comprised of both the large integrated producers who cover most of the above areas and the medium and small companies that are involved in only one portion, such as castings or extrusions, or produce only one product.

PRODUCTION OF ALUMINUM

The production of primary aluminum involves the mining, handling, and drying of bauxite at the mine site, transporting the bauxite to the alumina plant, refining the bauxite into alumina, and smelting the alumina into aluminum. Auxiliary operations include the manufacture of soda, ash, lime, and cathode and anode.

Table 8-2

U.S. Aluminum Plants

Location and Capacity — 1975

	Short Tons/Year
Aluminum Company of America	
Alcoa, TN	270,000
Badin, NC	120,000
Evansville, IN	280,000
Massena, NY	140,000
Point Comfort, TX	180,000
Rockdale, TX	285,000
Vancouver, WA	115,000
Wenatchee, WA	190,000
Anaconda Aluminum	
Columbia Falls, MT	180,000
Sebree, KY	120,000
Consolidated Aluminum	
New Johnsonville, TN	141,000
Lake Charles, LA	36,000
Martin Marietta	
The Dalles, OR	90,000
Goldendale, WA	115,000
Eastalco	
Frederick, MD	174,000
Intalco	
Bellingham, WA	260,000
Kaiser Aluminum and Chemical	
Chalmette, LA	260,000
Mead, WA	220,000
Ravenswood, WV	163,000
Tacoma, WA	81,000
Ormet	
Hannibal, OH	260,000
Noranda	
New Madrid, MO	70,000
National Southwire Aluminum	
Hawesville, KY	180,000
Revere Copper and Brass	
Scottsboro, AL	114,000
Reynolds Metals Company	
Arkadelphia, AR	68,000
Corpus Christi, TX	114,000
Jones Mills, AR	125,000
Listerhill, AL	202,000
Longview, WA	210,000
Massena, NY	126,000
Troutdale, OR	130,000
	5,019,000

Source: Environmental Protection Agency, *Environmental
Considerations*, Vol. VIII, *Alumina/Aluminum*, p. 93.

Table 8-3

U.S. Aluminum Smelters — Age and Technology

Company & Location of Smelter	Age	Smelter Technology
Aluminum Company of America		
Alcoa, TN	pre-1946	PB
Badin, NC	pre-1946	PB
Evansville, IN	1960	PB
Massena, NY	pre-1946 & 1958	PB
Point Comfort, TX	1950	VSS
Rockdale, TX	1952	PB
Vancouver, WA	pre-1946	PB
Wenatchee, WA	1952	PB
Anaconda Aluminum		
Columbia Falls, MT	1955	VSS
Sebree, KY	1973	—
Consolidated Aluminum		
New Johnsonville, TN	1971	PB
Lake Charles, LA	1971	—
Martin Marietta		
The Dalles, OR	1958	VSS
Goldendale, WA	1971	—
Castalco		
Frederick, MD	1970	PB
Intalco		
Bellingham, WA	1966	PB
Kaiser Aluminum		
Chalmette, LA	1957	HSS
Mead, WA	pre-1946	PB
Ravenswood, WV	1957	PB
Tacoma, WA	pre-1946	HSS
Ormet		
Hannibal, OH	1958	PB
Noranda		
New Madrid, MO	1971	PB
Revere Copper & Brass		
Scottsboro, AL	1970	PB
Southwire		
Hawesville, KY	1969	PB
Reynolds Metals Company		
Arkadelphia, AR	1952	HSS
Corpus Christi, TX	1952	HSS
Jones Mills, AR	pre-1946	PB
Listerhill, AL	pre-1946	HSS
Longview, WA	pre-1946	HSS
Massena, NY	1959	HSS
Troutdale, OR	1959	PB

PB = Prebaked
HSS = Horizontal Soderberg System
VSS = Vertical Soderberg System

Source: Environmental Protection Agency, *Environmental Considerations*, Vol. VIII, *Alumina/Aluminum*, pp. 95-96.

The only domestic source of bauxite, the major raw material in
the production of alumina, is in Arkansas. The U.S. industry has always relied largely on imports from the Caribbean, northern South
America, and Australia for most of its supply of bauxite and alumina.
Most of the bauxite lies near the surface of the earth, so it is surface mined. It is then crushed, ground, and kiln dried to remove
excess moisture.

Alumina Production

All production of alumina for the U.S. market use the Bayer process, with some process variations to account for differences in ore
quality. Figure 8-3 provides a process flow sheet for this old and
well developed process. Introduced in 1888, it is applicable only
to bauxite as raw material. This process has the ability to treat
both trihydrate and monohydrate bauxites, although the cost for treating monohydrate bauxite is much higher.

In the process, finely ground bauxite is chemically digested with
a caustic soda solution that is produced in the alumina plant by reacting soda ash with lime. This is done under pressure and at elevate
temperatures. The alumina hydrate in the bauxite dissolves as sodium
aluminate. The insoluble components of the bauxite, primarily the
oxides of iron, silica, and titanium, are removed by thickening and
filtration. The separated solids, known as red mud, are discarded.
Silica is a particularly undesirable impurity in bauxite, since it
is readily dissolved in the caustic liquor.

Following digestion, which requires approximately an hour, the
caustic slurry is cooled to 250°F in a series of flash tanks. The
steam flashed off during cooling is used to preheat the new fresh
bauxite-caustic mixture prior to its entry into the digesters. After
cooling, the red mud is removed from the caustic slurry in the thickeners and goes to a mud washer filter, where the mud is washed with
water to recover sodium hydroxide, which goes back to the thickener
and into the main liquor process stream.

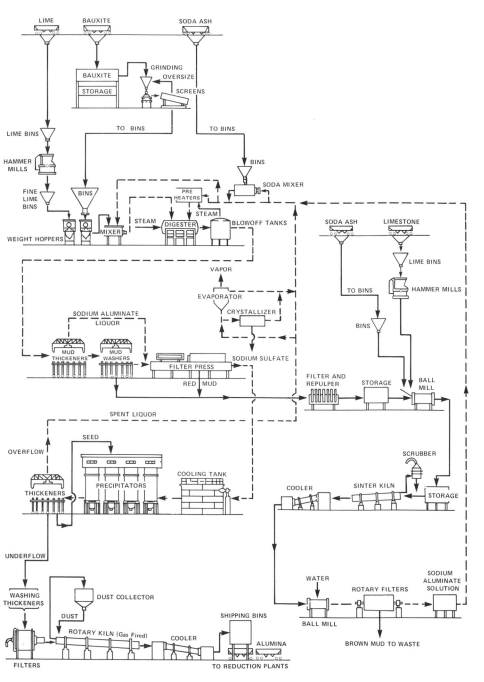

Figure 8-3. Bayer Process Flowsheet. Source: *Pre-Investment Data for the Aluminum Industry, Studies in Economics of Industry*, No. 2 (New York: United Nations, 1966).

263

The sodium aluminate solution goes to a clarifying filter, and
this clarified "green liquor" then goes to a precipitator. The liquor is cooled to 120-140°F in a heat exchanger and placed into large
precipitation vessels that are seeded with alumina trihydrate crystals
and mildly agitated to precipitate dissolved alumina trihydrate. Approximately half of the alumina is precipitated during a 34-36 hour
period. The resulting trihydrate is separated by settling and filtration. Spent liquor goes on to a spent liquor treatment operation,
and the washed alumina hydrate goes to calcination.

Spent liquor is a caustic solution containing about 50% of the
sodium aluminate originally present before precipitation. This is recycled to the process for reuse. The caustic content of this recycled
liquor is increased to the desired level for digestion by evaporation
of excess water and addition of makeup caustic. This latter step can
be accomplished by adding caustic soda directly, but more typically,
it is accomplished by the addition of lime and soda ash. The loss of
caustic in the spent liquor is approximately equivalent to 90% of the
silica content of the bauxite. Typical makeup requirements per ton
of aluminum produced are 100-200 pounds of soda ash and about the same
amount of calcined lime.

In most U.S. Bayer alumina plants, the alumina hydrate is calcined in rotary kilns which operate at about 2100°F to remove moisture and water of hydration. The resulting alumina is called pot feed
alumina, which is the raw material for the production of aluminum
metal via the Hall-Heroult electrolytic reduction process.

Aluminum Reduction

The alumina is reduced to aluminum metal using the Hall-Heroult
process. See Fig. 8-4. The alumina is dissolved in a molten cryolite
(sodium-aluminum fluoride) electrolyte. Cryolite, which melts at about
1800°F, dissolves alumina up to as much as 10-20% of its weight, with
a resultant decrease in melting point of the solution. The electrolytic reduction cell ("pot") breaks alumina down into its components,
aluminum and oxygen. Electrolysis is carried out in carbon lined boxe

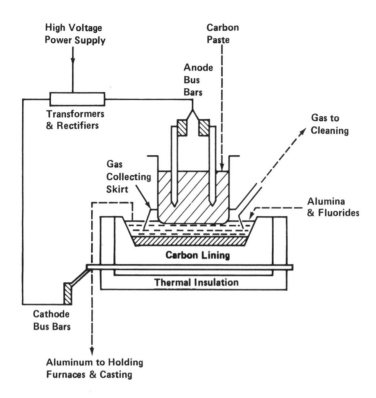

Figure 8-4. Alumina Reduction Cell Soderberg System. Source: Pre-Investment Data for the Aluminum Industry, Studies in Economics of Industry No. 2 (New York: United Nations, 1966).

into which a carbon anode projects. High ampere, low voltage direct current is applied, with the box serving as the cathode and the rods as anodes. Upon electrolysis, the alumina is decomposed as the aluminum metal is deposited in molten state at the cathode and the oxygen is deposited at the anode. The liquid aluminum, being heavier than the cryolite, sinks to the bottom of the vessel and is removed by tapping. The carbon anode is oxydized by the electrolysis to carbon dioxide.

Aluminum Processing

After smelting, some of the refined metal is cast into ingots.
However, a fairly large proportion of the liquid metal is also delivered to on-site casting houses within a given plant and to outside
users. These liquid metal transfers are routinely made within a ten
mile radius, and occasionally, for even longer distances. C. L. Brooks
has estimated that the present energy for melting might be 25-40%
greater in the absence of such liquid metal deliveries (1).

At the casting sites, the molten primary aluminum is charged into
reverberatory type holding furnaces, along with scrap and alloying
additions. A substantial amount of scrap is recycled in the melting
operation, an average of 55 tons of scrap being added for each 43 tons
of primary metal and two tons of alloying additions to form a melt of
a hundred tons (2). See Fig. 8-5, which points out that from a 100
ton melt only 53 tons of shipments result. Initial processing of cast
ingots is usually conducted at 800-900°F in either a hot rolling
(blooming) or extrusion operation. For some products, intermediate
conditioning and continued hot rolling is necessary. Most mill products are cold rolled or drawn to finish dimensions.

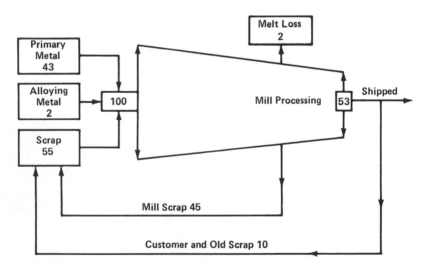

*Figure 8-5. Aluminum Industry Flow Diagrams. Source: C. L.
Brooks, "Energy Conservation in the Aluminum Industry," Efficient
Use of Fuels in the Metallurgical Industries, p. 710.*

ENERGY CONSUMPTION IN THE INDUSTRY

Aggregate energy usage within the aluminum industry has been increasing as total production has increased. See Table 8-4. However, a review of the historical data shows that there has been a steady decrease in the energy used per unit of output in aluminum manufacture. The major energy consumption in aluminum production is in the reduction process, when alumina is reduced to aluminum metal. As is also evident, both the alumina production process and the manufacture of the carbon anodes are energy intensive.

Table 8-5 provides a summary of data for the individual operations in aluminum manufacture. The major energy consumption in aluminum production is in the reduction process, when alumina is reduced to aluminum metal. As is also evident, both the alumina production process and the manufacture of the carbon anodes are energy intensive.

Table 8-6 provides a summary of the energy use distribution by type of energy for each of the components of aluminum manufacture. As can be seen, electricity is the largest component of energy use, followed by natural gas. Energy use patterns also vary, depending on the manufacturing process used.

Because aluminum production plants are large consumers of power, the industry has historically located its reduction plants near sources of low cost (hydroelectric power) or sources of low cost fuel. The industry has recognized the importance of reducing electrical energy consumption and through the years has made some gains in reducing electricity consumption. These gains are the results of:

- Replacing the higher power consuming, older Solderberg pot lines with the more efficient prebaked anode reduction cells.

- Using larger cells that permit reduction of heat losses per ton of aluminum.

- Making minor improvements in efficiency through reduction in electrical resistance losses in the anode and cathode electrical connectors.

Table 8-4

Energy Use in the Aluminum Industry

		1971[2] Estimates	1967	1962	1958	1954	1947[3]
Production							
1	Primary metal production (10[9]) lbs.........	7.850	6.539	4.236	3.131	2.921	1.144
2	From scrap (domestic) (10[9]) lbs............	2.100	1.756	1.164	.708	.626	.690
3	(imports)(10[9]) lbs	.113	.055	.012	.018	.015	.028
4	Total scrap (10[9]) lbs....................	2.213	1.811	1.176	.726	.641	.718
5	Total domestic production.................	10.063	8.350	5.412	3.857	3.562	1.862
GEC based data							
6	Heat rate for electricity GEC						
6	Heat rate for electricity GEC	10,500	10,432	10,588	1,1,085	12,180	15,600
7	Electricity purchased kwh (10[9])............	45.3	41.957[1]	26.883	16.170	17.239	9.331
8	Row 6 × row 7 = BTU(10[12])	475.7	437.69	283.83	179.24	209.97	145.56
9	Other fuels purchased BTU (10[12])	160.0	151.90	141.00	124.20	111.50	11.00*
10	Total energy to SIC 3334 BTU (10[12])	635.7	589.59	424.83	303.44	321.47	156.56*
UE based data							
11	Electricity generated kwh (10[9])............	18.1	11.648[1]	9.387	8.968	9.044	1.070*
12	Total electricity row 7 + 11	63.4	53.605	36.270	25.138	26.283	10.401
13	Row 9 × (3412 BTU/kwh) UE (10[12])	154.6	143.16	91.725	55.172	58.82	31.84
14	Total row 13 + 9 UE = BTU (10[12])	314.6	295.06	232.725	179.372	170.32	42.84*
15	Row 14/row 1...........................	40.076	45,123	54,940	57,289	58,309	37,447*
Ratios							
16	Row 10/row 1 BTU/lb	80.98	90,165	100,290	96,914	110,055	136,853
17	Row 10/row 5 BTU/lb	63.172	70.609	78.498	78.672	90.250	84.081
18	Row 12/row 1 kv:h/lb	8.08	8.198[1]	8.029	9.00	9.092	
Capacity							
19	Published U.S. capacity-thous. sh. tons......	4,666	3,321	2,489	2,194	1,413	630
20	Ratio (Row 1/2000)/line 19	.84	.98	.85	.71	1.03	.91

[1] Data for 1967 are considered to be understated. We have obtained data covering roughly 80% of 1967 production, and we have estimates of the minimum datum for an added 6% of production. These imply that the balance of the industry was producing at under 6kwh/lb. metal in smelting, or roughly 6.5 kwh/lb. total usage in the relevant establishments. Such a low figure we view as highly improbable.
[2] Adjusted to match as closely as possible the establishments included in the *Census of Manufacturers*.
[3] Data for 1947 are suspect and ought to be viewed as unreliable. Asterisks mark the anomalous data.

These improvements were accomplished within the last decade, when energy costs were rather stable and there was not the incentive to reduce energy consumption that there is today. With a higher incentive for reducing energy consumption and, in particular, electricity consumption, there are real possibilities for efficiency improving technologies being implemented.

Table 8-5

Table 8-5

Energy Consumption in Aluminum Production

	Unit	Units Per Net Ton of Aluminum	10^6 Btu Per Unit	10^6 Btu Per Net Ton of Aluminum
Mining				
Drilling	kwh	1.0	0.0105	0.01
Drill bits, drilling machines	lb	Negligible		
Explosives	lb	0.80	0.030	0.02
Subtotal				0.03
Shovel loading				
Electrical energy	kwh	10.25	0.0105	0.11
Materials, repair, and maintenance	Btu	0.03×10^6		0.03
Subtotal				0.14
Truck transportation				
Diesel fuel oil	gal	0.70	0.139	0.10
Truck materials, tires, and repair	Btu	0.02×10^6		0.02
Subtotal				0.12
Crushing, washing, and screening				
Crushing and screening electrical energy	kwh	12.5	0.0105	0.13
Pumping electrical energy	kwh	6.4	0.0105	0.07
Machinery wear and service energy	Btu	0.02×10^6		0.02
Subtotal				0.22
Drying	Btu	1.90×10^6		
Transportation	net ton-mile	9,500.0	0.00025	2.38
Bayer processing				
Crushing and grinding				
Electrical energy	kwh	31.43	0.0105	0.33
Lime	net ton	0.10	8.5	0.85
Subtotal				1.18
Digestion				
Steam	lb	12,143.0	0.0014	17.00
Caustic Soda	net ton	0.15	30.00	4.50
Subtotal				21.50
Clarification				
Electrical energy	kwh	30.48	0.0105	0.32
Starch	—	—	—	0.00
Cooling				
Electrical energy	kwh	5.71	0.0105	0.06
Precipitation-filtration				
Electrical energy	kwh	66.67	0.0105	0.70
Evaporation				
Steam	lb	6,829.0	0.0014	9.56
Spent liquor recovery				
Electrical energy	kwh	69.52	0.0105	0.73
Net steam usage	lb	593.0	0.0014	0.83
Subtotal				1.56
Calcination				
Natural gas	ft^3	7,720.0	0.001	7.72
Carbon anode manufacture				
Raw petroleum coke	net ton	0.425	30.0	12.75
Coke transportation (500 miles by rail)	net ton-mile	212.5	0.00067	0.14
Calcining				
Hydrocarbon fuels	Btu	—	—	1.0
Electrical energy	kwh	20.0	0.0105	0.21
Crushing and grinding				
Electrical energy	kwh	5.0	0.0105	0.05
Pitch binder	gal	28.44	0.16	4.55
Pitch transportation (400 miles by rail)	net ton-mile	52.4	0.00067	0.04
Natural gas for baking	ft^3	2,094.0	0.001	2.09
Subtotal				20.83
Carbon cathode manufacture				
Anthracite	net ton	0.02	25.94	0.52
Anthracite transportation (500 miles by rail)	net ton-mile	10.0	0.00067	0.07
Electrical energy for calcining	kwh	40.0	0.0105	0.42
Crushing and grinding				
Electrical energy	kwh	0.2	0.0105	0.00
Pitch binder	gal	0.74	0.16	0.12
Pitch transportation (500 miles by rail)	net ton-mile	0.17	0.00067	0.00
Electrical energy for baking	kwh	8.0	0.0105	0.08
Subtotal				1.21
Reduction				
Makeup cryolite (Na_3AlF_6)	net ton	0.035	155.0	5.44
Cryolite transportation (300 miles by rail)	net ton-mile	10.5	0.00067	0.01
Makeup aluminum fluoride	net ton	0.02	51.4	1.02
Aluminum fluoride transportation (300 miles by rail)	toh-mile	6.0	0.00067	0.00
Fluorspar (CaF_2)	net ton	0.003	1.59	0.00
Electrical energy (including ancillary)	kwh	16,000.0	0.0105	168.00
Subtotal				174.47
Total				243.90

Source: Battelle Columbus Laboratory, *Energy Use Patterns in Metallurgical and non Metallic Mineral Processing*, Phase 4 (Columbus, Ohio: Battelle, June 1975), pp. 9-11.

Table 8-6
Energy Use Distribution
(1972)

| | Btu, percent | | | |
Process	Electricity	Natural Gas	Other Fuels	Totals
Alumina	12.1	80.4	7.5	100.0
Molten Metal	85.2	3.2	11.6	100.0
Hold, Cast, Melt	7.4	80.0	12.6	100.0
Fabrication	38.1	50.8	11.1	100.0
Total for Component:	64.7	24.3	11.0	100.0

Source: Battelle Columbus Laboratory, *Energy Efficiency
Improvement Targets for Primary Metals Industries,
SIC 33*, Vol. 1 (Columbus, Ohio: Battelle, August
1976), p. B-4.

RAW MATERIALS

Almost all bauxite and about one third of the alumina required in
the industry are imported, so that there is a significant savings of
primary energy to the U.S. This amounts to around 7% of the total
primary energy consumption per ton of aluminum product. Because of
the inadequacy of domestic bauxite reserves for meeting projected
aluminum demand, the U.S. will remain a major importer of bauxite.

The threat of the formation of an international bauxite cartel
and the increased cost of these raw materials have interested U.S.
aluminum companies in domestic clays as an alternative raw material.
Kaolin and anorthosite clays are available in abundance and at much
lower costs, based on alumina content, than imported bauxite. The
possibility of producing alumina from the large reservoirs of kaolin
clay in Georgia and South Carolina is being investigated.

ALUMINA PRODUCTION

Of the aluminum ores, bauxite is the richest in aluminum. Untreated bauxite contains 55-65% alumina (Al_2O_3) and requires about four pounds to process each pound of aluminum. If clay or kaolin are used, about 8-9 pounds are required per pound of aluminum. The use of a less rich ore than bauxite requires that more ore be treated, and it follows that more energy would be required to produce alumina. With bauxite becoming less available and more expensive, numerous attempts are being made to develop alternative processes to the Bayer process. A number of them have been proposed, including hydrochloric acid clay leaching, nitric acid clay leaching, Toth chlorination, and an electrothermal process (Pederson process).

Hydrochloric Acid Ion Exchange Process (3)

In the hydrochloric acid ion exchange process, clay is dehydrated, leached with hydrochloric acid, and then settled to separate the residue from the aluminum chloride/iron chloride solution. See Fig. 8-6. The solution is purified in an amine ion exchange system to remove the iron chloride, leaving the aluminum chloride in solution. The aluminum chloride is then crystallized from the solution and decomposed to alpha alumina, and the acid value is recovered.

The Anaconda Company operated a large scale pilot plant using hydrochloric acid leaching at Butte, Montana during the late 1950s and early 1960s. The alumina produced was converted to aluminum in the aluminum smelter at Twin Butte, Montana. The Bureau of Mines is presently testing this process at its laboratory at Boulder, Colorado. The process consumes 134 KWH of power plus 37.8×10^6 BTU of fossil fuels per ton of alumina. The total energy amounts to 39.21×10^6 BTU per ton of alumina, which is considerably higher than the Bayer process.

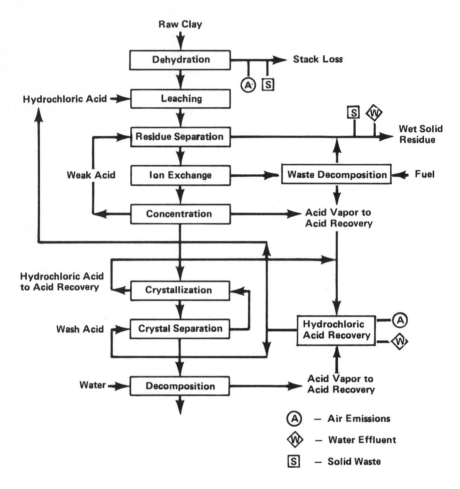

Figure 8-6. Hydrochloric Acid Ion Exchange Process. Source: Environmental Protection Agency, Environmental Considerations, Vol. III, Alumina/Aluminum, p. 24.

Nitric Acid Ion Exchange Process (4)

In the nitric acid ion exchange process, the raw kaolin clay is calcined to make the alumina selectively available for extraction with nitric acid. See Fig. 8-7. The calcined clay is then leached with hot nitric acid at atmospheric pressure to produce a solution of aluminum

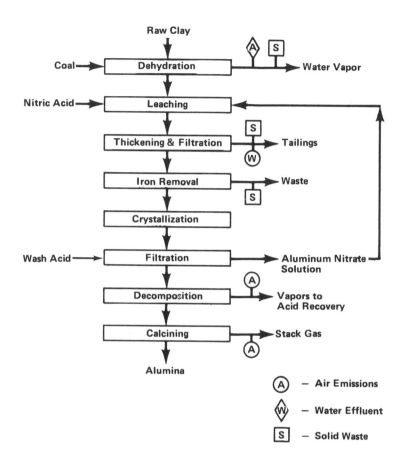

Figure 8-7. Nitric Acid Ion Exchange Process. Source: Environmental Protection Agency, Environmental Considerations, Vol. III, Alumina/Aluminum, p. 24.

nitrate containing a suspension of insolubles, which are separated from the aluminum nitrate liquor in thickeners.

In the next step, the iron and other impurities are removed by use of a liquid ion exchange medium, and then the remaining impurities are removed by vacuum crystallization of aluminum nitrate monohydrate. The alumina is recovered by hydrolysis of the aluminum nitrate under controlled conditions, and the nitric acid is recovered for recycle.

The alumina leaving the final decomposer is sent to a storage silo in which the exothermic phase transformation to alpha alumina occurs.

During the late 1960s, Arthur D. Little, Inc., carried out a non-integrated pilot plant operation using the nitric acid process. Moderate quantities were produced that were considered adequate in quality. This process consumes 139 KWH of power plus 25.3×10^6 BTU of fossil fuels, or a total of 26.76×10^6 BTU per ton of alumina. The energy required is much higher than that required by the Bayer process.

Toth Alumina Process (5)

The Toth process involves the chlorination of aluminum-containing materials in the presence of carbon to produce aluminum chloride vapor and other volatile chlorides. These are purified to eliminate other metal chlorides and then oxidized to produce alumina and chlorine for recycle. See Fig. 8-8. If kaolin clays are used, the steps in the process involve ore drying and calcination, chlorination, separation of the chlorides from the aluminum chloride by fractional condensation and distillation, and separate oxidation of the iron, silicon, and titanium chlorides to their respective oxides for recovery of chlorine for recycle. The aluminum chloride is then oxidized to produce aluminum.

Under development by the Toth Aluminum Company, this process consumes 333 KWH of power and 25.1×10^6 BTU of fossil fuel, for a total consumption of 28.59×10^6 BTU per ton of alumina.

Pederson Process (6)

Several electrothermal processes have been developed for producing alumina, of which the best known is probably the Pederson process. In it, bauxite, coke, limestone, and iron ore are melted in an electric arc furnace to produce pig iron and a calcium aluminate slag. The slag, which contains 30-50% Al_2O_3 and 5-10% SiO_2, is treated with caustic soda to produce sodium aluminate. This is treated to produce

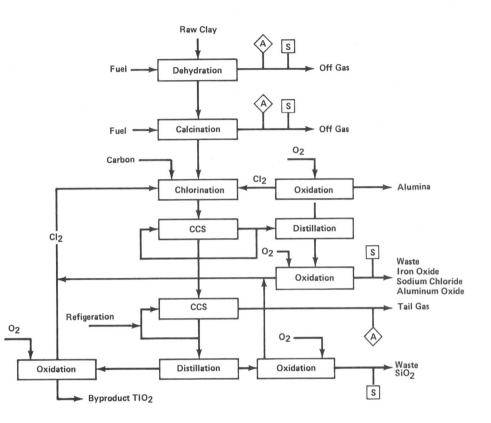

Figure 8-8. Toth Alumina Process. Source: Environmental Protection Agency, Environmental Considerations, Vol. III, Alumina/Aluminum, p. 41.

alumina. The energy required for the Pederson process depends on the credits assigned the pig iron and furnace gases produced in the electric arc furnace. The overall energy requirements for alumina produced by the Pederson process or by the other electrothermal methods are generally greater than the Bayer process.

Implications

A comparison of energy requirements for the present Bayer aluminum process and the hydrochloric acid, nitric acid, and Toth chlorination processes for the production of alumina are presented in Table 8-7. It can be seen that the clay based processes, if implemented, would result in an increase in energy consumption, as compared with the Bayer process, which is the present technology. The major factor that would precipitate a change in clay processing for the production of aluminum is the degree of concern that the U.S. aluminum industry has with respect to its dependence on foreign sources of aluminum-containing raw materials. As foreign bauxite and alumina prices increase, interest in clay as a source of alumina would also increase.

Table 8-7

Energy Consumption Comparison

10^6 Btu per Net Ton Alumina

Process	Bayer Alumina Bauxite	Hydrochloric Acid Leaching Clay	Nitric Acid Leaching Clay	Toth Chlorination Clay
Power — kWh/ton	275	134	139	333
Fuel — 10^6 Btu/ton	11.64	37.8	25.3	25.09
Total — Fossil fuel basis (10^6 Btu/ton)	14.53	39.2	26.8	28.6
Pollution Control (10^6 Btu/ton)	0.05	0.02	0.7	0.3

Source: Environmental Protection Agency, *Environmental Considerations,* Vol. VIII, *Alumina/Aluminum,* p. 85.

FLASH CALCINATION OF ALUMINA

In an intermediate process for refining bauxite, aluminum hydroxide must be calcined to drive off water and form alumina for smelting. An improved process has been developed by ALCOA (7). The ALCOA flash calciner is a fluidized bed that takes advantage of dispersed phase technology to improve heat exchange and reduce radiant losses. This flash calciner reduces energy consumption per pound of alumina from 2000-1400 BTU, or 30% of conventional requirements, and reduces the total energy needed to produce a pound of fabricated aluminum by 1%. The theoretical quantity of heat required to drive water from aluminum hydroxide is 1150 BTU per pound; waste heat during calcination has been reduced from 850-250 BTU per pound, or 70%. Since 1963, ALCOA has installed 24 fluid flash calciners at 7 ALCOA and affiliate company bauxite refining plants throughout the world.

The primary operational improvements over conventional rotary kiln calciners is the improved effectiveness of preheating input materials and the resultant decrease in the exit temperatures of alumina and off gases. The fuel rate requirement of the fluid bed calciner is lower than that of the kiln because of the improved preheating of both the alumina and combustion air. The off gases from this process still provide some of the heat required for the steam generation of the Bayer process.

The energy and availability flows for the rotary kiln calcinating method and the fluidized bed calciner are shown in Fig. 8-9. The energy and availability of both the alumina and the combustion products are lower for the fluidized bed calciner, allowing for reduced fuel input.

SMELTING ALUMINA

Metallic aluminum is produced from alumina by the Hall-Heroult electrolytic reduction process, in which alumina is continuously

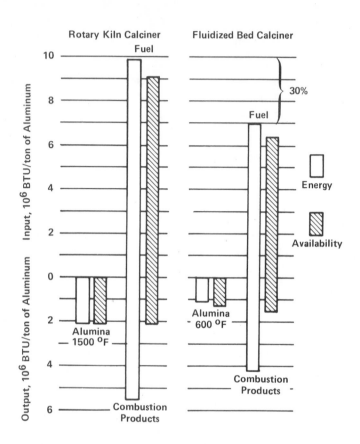

*Figure 8-9. Comparison of Rotary Kiln and Fluidized Bed Calcin-
er. Source: E.H. Hall, "Evaluation of the Potential for Energy Conservation in Industry," Energy Conservation: A National Forum, ed.
T. Veziroghu (Coral Gables, Florida: Clean Energy Research Institute,
December 1975), p. 74.*

dissolved in molten cryolite in the reduction cell. A low voltage
direct current is passed through carbon anodes suspended in the cells,
through the cryolite bath, to the molten aluminum which serves as a
cathode. The cell operates in the temperature range of 1700-1800°F.
In it, the alumina dissociates into aluminum, which is deposited at
the bottom of the cell as molten metal (at the cathode), and oxygen
is released at the anode in the form of carbon monoxide and carbon

dioxide. Molten aluminum is removed periodically to a holding furnace where dross is skimmed off prior to the casting into ingot molds.

Two anode systems are used--prebaked electrodes and the continuous Soderberg electrode. The older system consists of a series of prebaked carbon blocks suspended by iron rods from the anode bus bars above each cell. The blocks are removed from the cell before being fully expended in the process to avoid contamination of the aluminum by the iron bars. The Soderberg system uses an anode which is baked by the reaction heat from the cell itself. Carbon paste is used as anode material and is fed into the top of the anode casing. As the paste moves down, it is baked, forms the anode, and is consumed as carbon dioxide is formed and released. The carbon that is removed is replaced by the paste injected at the top of the anode.

The basic Hall-Heroult process has not really changed since its introduction over 70 years ago. Important design and engineering changes have evolved in which commercial cells have increased in size more than threefold and have diminished 35-40% in power consumption, however. The Hall process is the principal energy consuming unit operation in the production of aluminum ingot. Using about 80% of the total energy requirement for aluminum production it is characterized by an efficiency of 44% on the basis of 3,413 BTU per KWH, or slightly less than 17% on the basis of 10,400 BTU per KWH (8).

Alternative processes for the production of aluminum that are being considered by the aluminum industry are those methods that will significantly reduce the power consumption in the Hall-Heroult electrolytic reduction process. Meanwhile, many improvements are being made in that process. For example, whereas in 1972 the average Hall cell required 8.15 KWH to produce one pound of aluminum (9), the newest cells currently in use require about 6.5 KWH per pound of aluminum (10). It is essential that all future capacity will operate with at least this higher efficiency.

Several promising processes which could not only reduce the fuel requirements of the aluminum industry but also permit the use of low grade bauxite and clay deposits in place of high grade imported bauxite have been proposed. These include the ALCOA chloride process, the

Toth chemical process, and the plasma reduction process. In addition,
two modifications to the Hall-Heroult process would lead to substantial savings: (1) the replacement of the carbon cathode in the conventional cell with refractory hard metal cathodes made of titanium
diboride and (2) the conversion of mercury arc rectifiers to silicon
rectifiers.

ALCOA Chloride Process

One method of producing aluminum with energy savings potential
is the ALCOA chloride process. In the ALCOA process (see Fig. 8-10),
alumina is chlorinated in the presence of carbon to produce a gaseous
mixture of aluminum trichloride ($AlCl_3$), carbon dioxide, and carbon
monoxide. The aluminum trichloride is separated from the gases by
condensation in a fluidized bed of solid $AlCl_3$ particles. The solid
$AlCl_3$ is introduced into a DC electrolytic cell containing a fused
chloride electrolyte at about $1290^\circ F$. Liquid aluminum metal is produced at the cathode and gaseous chlorine, which is recycled to the
chlorination plant, is formed at the anode.

ALCOA uses two electrolytic cell designs, a monopolar cell similar
to the conventional Hall cell, and a bipolar cell containing four bipolar electrodes. Both cell arrangements offer reduced power consumption relative to the standard Hall cells because of greater electrical
conductivity of the chloride electrolyte, compared to that of cryolite,
and because of smaller interpolar separations used in the ALCOA cells
relative to the conventional Hall cell. The advantages of the ALCOA
chloride process appear to be the following (11):

- It is able to use low grade ores.

- The electrical energy requirement is sharply reduced, because the decomposition voltage and the bath resistivity
  are both lower for the chloride melt.

- The chloride process eliminates the need to continually
  fabricate and replace the consumable carbon anodes because
  oxygen is eliminated from the system. Permanent graphite
  electrodes are used, and the expensive energy consuming anode
  baking facilities are eliminated.

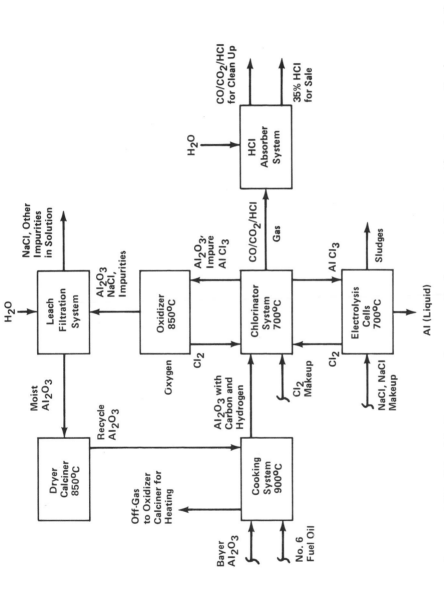

Figure 8-10. ALCOA Chloride Process. Source: Environmental Protection Agency, Environmental Considerations, Vol. III, Alumina/Aluminum, p. 61.

- Because electrodes can now be permanently emplaced, it is possible to design chloride process cells with multiple sheet electrodes stacked one above another so that one cell then becomes equivalent of several single cells, with consequent savings afforded by the much more compact cell design.

- A less expensive electrolyte can be used (NaCl-KCl, rather than cryolite).

- The chloride cell operating temperature is about $1,290^{\circ}$F, rather than the 1800°F temperature of the Hall process.

- The fluoride emissions of the Hall process are completely avoided, since no cryolite or fluoride materials are used in the process.

- There is a less severe pollution problem because the cell operates in a closed system.

- It results in higher purity of output metal.

According to ALCOA, savings of 30% in electricity are achieved with the chloride process (12). Overall fuel savings would be 45×10^6 BTU per ton from the reduction in electricity, and 14×10^6 BTU per ton from the elimination of the oxidation of carbon. This would reduce the total fuel needs for the primary aluminum process by 59×10^6 BTU per ton.

Initial units using the ALCOA process began operation in Anderson County, Texas in 1976. The chloride process was developed over a 15 year period at a cost of $25 million (13).

Toth Aluminum Process

Dr. Charles Toth has proposed a pyrometallurgical process for the production of aluminum as a proposed competitor to the Hall-Heroult electrolytic reduction process. The process is based on the reaction between manganese metal and aluminum trichloride to give manganese dichloride and aluminum metal. This reaction is carried out at 570°F and 220 psig pressure. A simplified flowsheet for the Toth process is shown in Fig. 8-11.

The calcined clay and coke are chlorinated at 1700°F with a mixture of chlorine and silicon tetrachloride. The chlorides are condensed from the carbon monoxide to yield liquid aluminum (Al_2Cl_6),

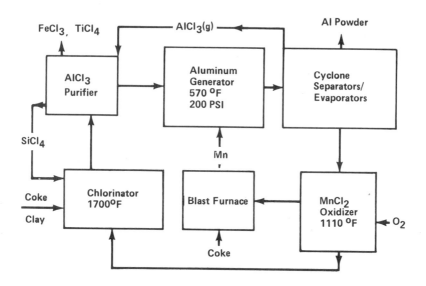

Figure 8-11. Toth Process. Source: Battelle Columbus Laboratory, Energy Use Patterns in Metallurgical and Non-Metallic Mineral Processing, Phase 9 (Columbus, Ohio: Battelle, August 1976), p. 16.

which, in turn, reacts with manganese metal in the aluminum chloride salt mixture. Aluminum chloride is separated from the fused salt by evaporation and subsequent condensation. The remaining solid manganese chloride is oxidized to produce manganese sesquioxide (Mn_2O_3), which is then reduced to manganese metal in a conventional manganese blast furnace (14).

The Toth firm has experienced difficulty in carrying out the reduction of aluminum chloride with manganese to form manganese chloride and aluminum metal. The manganese metal does not completely react; thus, the resulting aluminum product is contaminated with manganese metal. The reaction occurs to the extent that 80% of the manganese metal is consumed, and 20% remains unreacted.

Peacey and Davenport estimate that the Toth process would significantly reduce the electrical energy input for producing aluminum (only about one KWH per pound of aluminum). However, the operation

of a manganese blast furnace and the increased carbon requirement
per ton of aluminum will probably make the energy requirement associated with the Toth process greater than those of the conventional
Bayer-Hall processes. The Toth per unit energy consumption would be
270×10^6 BTU per ton, which is about 80×10^6 BTU per ton greater than
the average Bayer-Hall process.

Plasma Reduction Process

Another interesting process is based upon plasma arc reduction
of Al_2O_3 or of $AlCl_3$ (15). Initial studies show that substantial fuel
reduction relative to the Hall process is theoretically possible.
Three variations in the plasma arc approach are possible:

(1) Reduction using solid carbon:

$$Al_2O_3 + 3C \longrightarrow 2Al + 3CO$$

(2) Reduction using a hydrocarbon:

$$Al_2O_3 + 3CH_4 \longrightarrow 2Al + 3CO + 6H_2$$

(3) Conversion to chloride and reduction of the chloride:

$$Al_2O_3 + 3C + 3Cl \longrightarrow (AlCl_3)_2 + 3CO$$

$$2AlCl_3 + 3H_2 \longrightarrow 2Al + 6HCl$$

Assuming a 75% plasma heat efficiency, calculated electrical consumption is in the range of 5700-9500 KWH per ton of aluminum. This is
well below the best values required for the Hall process. There is
no experimental data available to date.

Electrolysis of Aluminum Sulfide

Electrolysis of aluminum sulfide, rather than alumina, would
provide several advantages, from an energy point of view, over the
Bayer-Hall process. These advantages include electrolysis at a lower
bath temperature (i.e., at $1,560^\circ F$, rather than at $1,740^\circ F$ in a Hall
cell), greater solubility of the aluminum sulfide in the electrolyte
than is possible with alumina in cryolite, and a lower consumption of
anode material because depolarization of oxygen and its subsequent reaction with the anode would be eliminated. However, there are some
technical problems with the preparation of pure aluminum sulfide and

with the purity of the resulting aluminum. There is no estimate available for the potential energy savings with this process (16).

Use of Titanium Diboride Cathodes

Originally, the idea was to eliminate the voltage drop between
the iron cathode support and the carbon and to reduce the voltage drop
between the cathode bus and the aluminum pad by replacing the iron and
carbon with titanium diboride. This would remove the iron and carbon
from the cathode system, and the replacement titanium diboride would
be much more compatible with molten aluminum than the carbon presently
used, because high resistance aluminum carbide forms at this interface.

Interest has now shifted to replacing not only the iron and the
carbon but also the molten aluminum pad, so that the titanium diboride can provide connections between the cathode bus and the electrolyte. One advantage to this arrangement is that aluminum produced at
the cathode can be rapidly removed or drained from the cathode, so
that only a thin film of molten aluminum exists on the titanium diboride. The voltage drop will be significantly reduced through the
molten aluminum. This also permits reducing the distance between the
anode and cathode, which would permit an increase in anode current
density.

It is conservatively believed that a 15-20% savings in power,
along with a simultaneous 30% increase in production, is attainable
with this process. This estimate is based upon some current experience and upon the expectation that satisfactory titanium diboride
cathodes can be made that will survive the atmosphere of the electrolyte in the Hall-Heroult cells for 4-5 years (17). At the same time,
the conversion may substantially increase power input to the cells,
which will result in equivalent increases in production per cell. The
net result of this modification would be a significant reduction of
capital for future expansion.

A comparison of energy requirements for the present Hall-Heroult
process and for the application of titanium diboride cathodes to existing cells is presented in Table 8-8. A savings of 36.5×10^6 BTU per
ton could be achieved by conversion to titanium diboride cathodes.

Table 8-8
Conversion to Titanium Diboride Cathodes

10^6Btu/Net Ton Aluminum

	Existing Hall-Heroult Smelters	Existing Hall-Heroult Smelters TiB$_2$ Cathodes
Power – kWh/ton	15,600	12,480
Fuel – 10^6 Btu/ton	24.02	20.14
Total Fossil Fuel Basis 10^6 Btu	187.82	151.18
Pollution Control 10^6 Btu/ton	1.71	1.31

Source: Environmental Protection Agency, *Environmental Considerations*, Vol. VIII, *Alumina/Aluminum*, p. 88.

Conversion to Silicon Rectifiers

In 1972 about 25% of the primary aluminum smelters were using mercury arc rectifiers to convert AC power to the DC power required for reducing the alumina. The other 75% of the smelters had converted to the more electrically efficient solid state silicon rectifiers. Typically, such a conversion results in an energy savings of 2-3% (18). It is technologically feasible and economically practicable for the remainder of the industry to install the silicon rectifiers, and if they do so, the result will be a savings of 2.6 x 10^6 BTU per ton of metal.

ALUMINUM MELTING

Primary aluminum melting furnaces range from small crucible furnaces for foundry use, called heating pots, to very large reverberatory melters with capacities of 25-80 tons per charge. The various types of furnaces generally used for the melting of aluminum are:

• large and small stationary and tilting reverberatories

• holding furnaces (reverberatory)

• tilting, barrel type furnaces

• stationary and tilting crucible furnaces

• dry hearth melters

Although some continuous melters are in use, the normal practice
is to melt on a cyclic basis. Firing rates on a melting furnace are
usually quite high, in order to obtain the fastest melting rate possible without excessive metal loss or refractory damage. Because of
the major differences between relative inputs for melting and holding
and between the respective combustion requirements, two different furnaces are usually used for the large scale melting installations: one
for melting, called a breakdown melter, and a second for refining and
holding, called a holding furnace.

One of the most important economic factors in aluminum melting
is the metal loss due to oxidation. This loss is critically dependent on melter design, particularly burner type and location, and is
also heavily dependent on the type of charge used, as illustrated
below (19):

Type of Charge	Oxidation Loss, %
Ingots, pigs, sows, and heavy scrap	1 or less
"Normal" recycle scrap	2-2.5
Light scrap and thin sections	20-25
Foil	Up to 50

In a typical plant, substantial quantities of scrap are remelted
with casting, forging, and extrusion operations. If purchased scrap
is used, it is segregated, and the dirty scrap is sold to secondary
metal processors.

Typical Aluminum Melter Efficiencies

Overall melting efficiency for a large aluminum melter is normally 25-30%. However, efficiency might be as low as 15% for some special

alloys. Some of the factors which affect fuel efficiency are type
of alloy, amount of metal loss due to oxidation, type of charge, charging practice, furnace geometry, burner and flue arrangement, and ratio
of melting time to total time between charge and pour. The relation
between fuel consumption per pound of metal heated and fuel efficiency
can be shown as follows (20):

Efficiency, %	Natural Gas, CF/lb
15	3.37
20	2.53
25	2.02
35	1.44
50	1.01

A typical fuel consumption for large melters is two cubic feet of
natural gas per pound of metal, an efficiency of 25%.

Improving Melter Efficiencies

Larson et al., have discussed a number of procedures which can
improve the efficiency of aluminum melters. These include (21):

Stack Charging. In some of the large aluminum melters, the charge
is placed on a sloping hearth at the base of the stack. The high
temperature flue gas transfers heat to the charge of the "dry hearth"
through convection and raises the ingot temperature to the 600-750°F
range before the ingot is pushed into the molten bath. Advantages of
this charging system are an increased melting rate and a higher melting
efficiency. It is estimated that fuel efficiency can be increased by
about 5-10%.

Hot Metal Pumping. Development of hot metal pumps with refractory internals makes possible several improvements in reverberatory
melting, particularly in better utilization of charging wells for high
surface area scrap, such as chips and foil. The increased bath velocity due to the pumping action increases the rate of heat transfer
to the scrap, melting it faster and reducing the time that the surface
area is exposed to oxidation. Metal loss due to oxidation is reduced
by 50%. In addition, either the overall melting rate can be increased

by 15-25% or the fuel efficiency can be increased by 5-6% at the same
melting rate.

Recuperation. Combustion air preheat by recuperation of waste
heat in the flue gases in seldom used in the U.S. Because only 20-40%
of the total melting cycle is required for the actual melting operation, recuperation has been difficult to justify economically. Also,
for melter holders where fluxing is done in the melting chamber, the
stack gases contain hydrochloric acid and other corrosive components.
Wherever feasible, preheating combustion air to the 600-800°F range
will give fuel savings of 20-25%, as the following data demonstrates.

Air Preheat Temperature ^{o}F	% Fuel Saving
400	13.5
600	19.8
800	25.5
1000	30.6
1200	35.0
1400	39.0

Oxygen Fuel Melting. Tests on primary and secondary aluminum
melters have proved the effectiveness of using auxiliary oxy-fuel
burners to reduce meltdown time, metal loss, and fuel consumption.
In remelting, where 97% of the charge is scrap, a reduction in meltdown time from two hours to one hour was achieved. The melting rate
was increased by an average of 55%, and the melt loss reduced by 26%.
Relative combustion efficiency was 48% during the period of oxygen
burning and 29% during air burning. The oxy-fuel, roof-mounted burners
were used as auxiliary burners to back up four air burners.

In another aluminum melting application, an oxy-fuel burner was
located on an end wall of the melter, firing transverse to the direction of flame travel of four normal air burners. Results indicated
that the meltdown rate doubled and fuel consumption during meltdown
decreased by 10-15%.

Infrared Heating-Crucible Melting. Extensive tests run by
Consolidated Natural Gas Service Co. have shown that the efficiency
of crucible melting of lead, zinc, and other metal of low melting

points can be materially increased by using infrared radiant heating
elements. Efficiency with infrared heating has been in the 30-40%
range, compared with the 15-20% normal range obtained with conventional firing methods. The melting temperature of aluminum is somewhat
higher (1,220°F), but a substantial improvement of 20-25% reduction
in fuel requirement should be obtained for aluminum melting.

Improved Fuel/Air Ratio Control. Close control over excess air
is a critical element in maintaining combustion efficiency, particularly for furnaces in which combustion products enter the flue ports
at a temperature above 2000°F. In aluminum melters, flue gas temperatures typically range from 2200-2800°F. With burner designs that give
good control over air gas mixing, only 5-10% excess air is required
to obtain maximum heat release and high flame temperature, but many
melters run with 20% or more excess air. The problem of controlling
excess air is quite difficult because of the large reduction in input
required for firing at maximum rate during melting and for holding at
minimum input after melting is completed.

Fuel inputs on the largest aluminum melters range up to 40 million
BTU per hour, and for many of the large melters, installation of a
control system that maintains precise air/fuel ratio by controlling
stack gas oxygen concentration should be justifiable. Calculations
show that 4.25% of the total input can be saved by reducing excess air
from 20-10%.

Summary of Improvement Methods. The following data summarize the
relative effectiveness of the fuel conservation methods presented
above. These savings are not additive.

| | % Reduction in |
Conservation Method	Fuel Requirement
Stack charging	5-10
Hot-metal recirculation	5-6
Recuperation	20-25
Oxy-fuel melting	10-15
Excess-air reduction (20-10%)	4.25
Infrared heating (Crucible melting)	20-25

Installation of Air Dampers

One of the major primary aluminum producers recently announced
that the installation of air dampers on their holding furnaces was underway, and that the expected energy savings were to be more than
0.25×10^{12} BTU per year (22). It is technologically feasible and
economically practicable for the entire aluminum component to effect
comparable savings of 0.2×10^6 BTU per ton of fabricated products.

CONTINUOUS CASTING

Great potential savings in the manufacture of aluminum is obtainable between the electrolytic reduction step and hot rolling. Often,
the reduction plant casts the aluminum into pigs or sows. These are
remelted, alloyed, and cast into ingots at the finishing plant, where
the ingots are then cooled, scalped, sawed, and heated for hot rolling.
In each of these steps the metal is heated and then cooled with a considerable expenditure of energy.

This energy waste has been reduced to some extent through different methods employed by the aluminum industry. One such method is
to locate the aluminum finishing plant adjacent to the reduction plant.
Hot molten metal can then be transferred, alloyed, and cast into ingots without intermediate solidification and remelting.

Continuous casting and production processes also have capabilities
of reducing energy use and reducing runaround scrap (23). The
Properzi continuous casting process for wire products, such as electrical conductors, provides an excellent example of how energy savings
can accrue. Liquid metal is fed into a continuous casting machine
which feeds hot metal into a special rolling mill where three-eighths
inch diameter redraw rods emerge at $7\frac{1}{2}$ tons per hour. The runaround
scrap is about 8%, compared with 10-15% for older methods. Melting
and preheating for rolling energy are entirely eliminated.

Hunter and Hazelett offer similar potentials for sheet products.
In the Hunter process, liquid metal flows upward through a spreader,

passing between two rolls with a vertical rollparting where metal is
solidified. The end product is a one-fourth inch thick coiled sheet.
The Hazelett method processes metal by passing it into a cavity formed
by two continuous water cooled belts. See Fig. 8-12. Hot slab, ½-1
inch thick, is then reduced while hot to a coiled sheet. The energy
savings from such processes are obvious. To date, these processes have
been successful on only relatively low alloy materials.

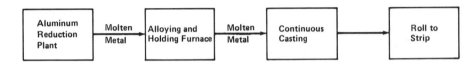

*Figure 8-12. Material Flow in an Integrated Aluminum Plant With
Continuous Casting.*

RECYCLING ALUMINUM

Scrap makes up a large portion of the aluminum charge to a furnace that is used to produce metal for sheet production. Much of this
scrap, called "runaround" scrap, is generated in-plant from ingot re-
jections, ingot scalping, sheet trimming, etc., and its volume is
determined by the efficiency of the fabrication process. As discussed
previously, of 100 tons of metal melted, 45 tons of scrap are generated. This scrap is always recycled within the manufacturing process.

Aluminum scrap returned by customers from their manufacturing
operations is another major source of scrap. Used products with a
very long life, such as aluminum used in buildings and construction,
and those with short life, such as containers and packaging, are usually referred to as old scrap. Old scrap usually requires some processing before remelting, in order to remove impurities such as the lacquer
coating on beverage cans. Since refining of scrap aluminum requires
only 5% of the fuel required for the Hall process, recycling is a highly effective means of reducing fuel requirements.

Figure 8-13 shows how energy demand can drop with increased use
of recycled material.   One major factor is the recycling of aluminum
cans.   Extensive can reclamation programs are presently underway to

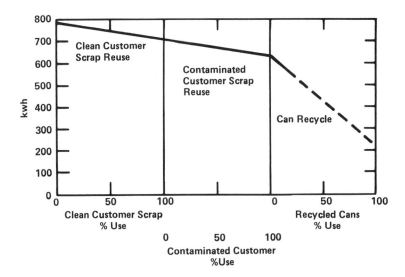

Figure 8-13.   How Energy Demand Drops with Recycled Material.
Source: P.R. Atkins, "Recycling Can Cut Energy Demand Dramatically,"
Engineering and Mining Journal, 174 (May 1973), 69.

insure that future can production will include significant reuse of
previously used cans.   In 1973, approximately 16.6% of all aluminum
cans were returned to collection centers.   One hundred percent recycle
of aluminum cans can result in a reduction of energy consumption in the
aluminum production process of 50% (24).

David P. Reynolds, of Reynolds Aluminum, recently stated at the
annual meeting of the Aluminum Association that there is more than
three billion pounds of scrap per year available in the American solid
waste stream just waiting to be recovered (25).   He suggested that
"our objective should be to increase our annual industry metal supply
through recycling by one billion pounds in 1977."   This goal would:

• Save mining and importing of two million tons of bauxite.

• Save 6.5 billion KWH of electrical energy.

• Save a capital investment of at least one billion dollars at
today's costs.

ENERGY CONSERVATION POTENTIAL

The theoretical minimum energy required to convert bauxite to
aluminum ingot is 28.3 x 10^6 BTU per ton (26). The efficiency associated with U.S. aluminum production is less than 30% on the basis of
3,413 BTU per kilowatt hour, or 14.2% if the electrical input is taken
at 10,400 BTU per KWH (27). The efficiencies of the major energy consuming unit processes for producing aluminum are less than those of
many other major unit operations in the steel, copper, and glass industries. It is, therefore, highly desirable that the industry develop energy saving replacement technologies for the manufacture of
aluminum.

The high and relatively inefficient energy consumption in the
Hall cell has resulted in a continuing conservation effort which has
yielded good results. For example, prior to WWII, 12 KWH were required
to produce one pound of aluminum. In 1972, the average smelter used
8.15 KWH, and the·smelters being constructed today use 6.5 KWH per
pound of aluminum.

But much can still be done. Improved management techniques, as
described in Chapter 2, can realize great savings. In a report prepared for the Federal Energy Administration, Battelle projected a
savings of 8.3 x 10^6 BTU per ton as a result of improved housekeeping (28). In addition, there are many opportunities for conservation
through process modifications and improvements. Table 8-9 indicates
the potential for energy conservation for various alternatives in the
aluminum industry. Use of the techniques discussed which are economical can result in energy savings of 20-30% over conventional practices.

Table 8-9
Summary of Energy Conservation Potential

Conservation Practices or Techniques	Potential Savings 10^6 BTU/ton
Flash calcining of alumina	0.9
ALCOA chloride process	49
Use of titanium diboride cathodes	36.5
Conversion to silicon rectifiers	0.6
Continuous casting	5
Recycling aluminum	up to 95% of energy used
Housekeeping	8.3
Installation of air dampers	0.2
Improved aluminum melting	

NOTES AND REFERENCES

1. C.L. Brooks, "Energy Conservation in the Aluminum Industry," Efficient Use of Fuels in the Metallurgical Industries (Chicago, Illinois: Institute for Gas Technology, 1974), p. 712.

2. Ibid., p. 711.

3. Environmental Protection Agency, Environmental Considerations of Selected Energy Conserving Manufacturing Process Options, Vol. VIII, Alumina/Aluminum Industry Report (Cincinnati, Ohio: Environmental Protection Agency, December 1976), pp. 23-30.

4. Ibid., pp. 30-39.

5. Ibid., pp. 39-49.

6. Battelle Columbus Laboratory, Energy Use Patterns in Metallurgical and Non-Metallic Mineral Processing, Phase 9 (Columbus, Ohio: Battelle, August 1976), p. 11.

7.   Allen C. Sheldon, "Energy Use and Conservation in Aluminum Pro-
      duction," Energy Use and Conservation in the Metals Industry
      (New York: The Metallurgical Society of the American Institute
      of Mining, Metallurgical and Petroleum Engineers, 1975), p. 3.

8.   Battelle, Energy Use Patterns, p. 12.

9.   P.R. Atkins and C.N. Cochran, "Future Energy Needs in the U.S.
      Light Metals Industry," Efficient Use of Fuels in the Metallur-
      gical Industries, p. 722.

10.   Brooks, "Energy Conservation in the Aluminum Industry," Efficient
      Use of Fuels in the Metallurgical Industries, p. 708.

11.   D.E. Kirby, E.L. Singleton, and T.A. Sullivan, Electrowinning
      Aluminum from Aluminum Chlorides, Bureau of Mines report of
      investigation 7343 (Washington, D.C.: Government Printing
      Office, 1970).

12.   "Chementator," Chemical Engineering, 80 (June 11, 1973), p. 61.

13.   Sheldon, "Energy Use and Conservation in Aluminum Production,"
      Energy Use and Conservation in the Metals Industry, p. 3.

14.   J. Peacey and W. Davenport, "Evaluation of Alternative Methods
      of Aluminum Production," Journal of Metals, 26 (July 1974), pp.
      24-28 with permission of American Institute of Mining, Metallur-
      gical and Petroleum Engineers.

15.   R.K. Rains and R.H. Kadlec, "The Reduction of Al_2O_3 to Aluminum
      in a Plasma," Metallurgical Transactions, 1 (June 1970), pp.
      1501-1506. Copyright by American Society for Metals and The
      Metallurgical Society of AIME, 1970.

16.   Battelle, Energy Use Patterns, p. 15.

17.   Environmental Protection Agency, Environmental Considerations,
      Vol. VIII, Alumina/Aluminum, p. 71.

18.   "Kaiser Acts to Effect 2% Energy Savings at Mead," American
      Metal Market, 82 (February 26, 1975), 8. See also "MM Aluminum
      Upgrading Smelters in Northwest," American Metal Market, 83
      (March 30, 1976), p. 6.

19.   Dennis Larson, Mark Fejer, and John Nesbitt, "Improving Energy
      Efficiency in Reheating, Forging, Annealing and Melting," Paper
      presented at Industrial Efficiency Seminar, AGA Marketing Con-
      ference, Atlanta, Georgia, March 5-7, 1975, p. 8.

20. Ibid., p. 9.

21. Ibid., p. 9-13.

22. "Aluminum Faces Power Crisis in Northwest," Metals Week, 47 (May 24, 1976), p. 1.

23. Brooks, "Energy Conservation in the Aluminum Industry," Efficient Use of Fuels in the Metallurgical Industries, p. 717.

24. P.R. Atkins, "Recycling Can Cut Energy Demand Dramatically," Engineering and Mining Journal, 174 (May 1973), p. 69.

25. David P. Reynolds, "Remarks--41st Annual Meeting of the Aluminum Association," October 31, 1974, Release of the Reynolds Metal Company.

26. E.H. Hall, "Evaluation of the Potential for Energy Conservation in Industry," Energy Conservation: A National Forum, ed. T. Veziroghu (Coral Gables, Florida: Clean Energy Research Institute, December 1975), p. 9.

27. Battelle, Energy Use Patterns, p. 10.

28. Battelle Columbus Laboratory, Energy Efficiency Improvement Targets for Primary Metals Industries, SIG 33, Vol. 1 (Columbus, Ohio: Battelle, August 1976), p. III-2.

Chapter 9

THE COPPER INDUSTRY

The primary copper industry in the U.S. is a highly energy intensive industry, ranking twelfth in terms of energy purchased in the
industrial sector of the economy and third in energy consumption per
unit weight among the five major metals (iron, aluminum, copper, lead,
and zinc). In 1972, about 80% of the primary copper production went
to copper and brass mills for fabrication into sheet, plate, rod, extrusions, and mechanical wire. The balance went to foundries and powder plants. More than half of copper mill products is produced by
about ten companies that are affiliated with the major copper producers (1). In addition, there are about 125 large mill fabricators
of copper and brass products that are independent from the major copper mining and refining firms, and several hundred smaller establishments engaged in copper wire drawing (2).

The U.S. is the leading world producer and consumer of refined
copper, accounting for about one third of the free world production
and consumption. See Table 9-1. Refined copper production in the
U.S. is classified into three major groups, depending upon the source
from which the metal is obtained: domestic ores, foreign ores, and
scrap. Figure 9-1 traces the historical trends in total production of
refined copper as well as of copper from each of these individual
sources. Over the past twenty years, world consumption of copper has

Table 9-1
Leading World Producers of Refined Copper
(Short Tons)

Rank	Country	1970	1974
1	United States	2,242,700	2,136,600
2	U.S.S.R.	1,185,000	1,433,000
3	Japan	777,500	1,101,200
4	Zambia	640,100	746,000
5	Canada	543,000	616,300
6	Chile	512,700	578,700

Source: "Non-Ferrous Metal Data 1974," American
Bureau of Metal Statistics, New York, 1975.

been increasing at an average annual rate of 4-4½% per year. The rate of growth is expected to remain fairly stable over the next decade. A recent Bureau of Mines forecast predicts domestic demand for copper in the year 2000 to be two and a half times the current consumption, which is at about six million tons at this time. About four million tons would be supplied by primary copper production, the balance from secondary copper (3).

A number of strikes by mine workers is reflected in Fig. 9-1 by the deep valleys in the graph for the production of domestic ores (e.g., in 1959, 1967, and 1971). This effect also appears in the total production graph because of the large portion of copper derived from domestic ores. In 1972, about 73% of the total refined copper was manufactured from domestic ores; this proportion has steadily increased from about 59% in 1954. The increase has been at the expense of foreign ore, which has declined from about 26% in 1954 to its present level of about 8%. The proportion of the total refined copper derived from scrap has generally remained at about 20% of the total domestic production.

The principal markets for copper in the U.S. are illustrated in Fig. 9-2. Wire products which are used chiefly in electrical applications account for the largest portion of domestic copper consumption,

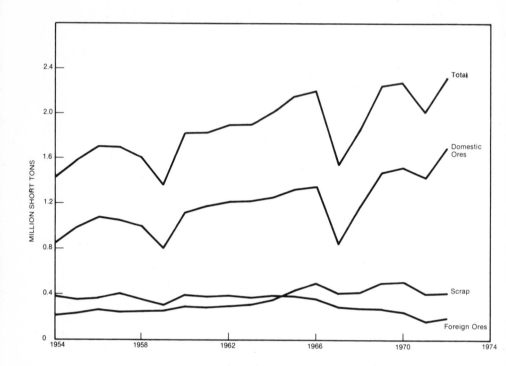

Figure 9-1. Trends in Production of Refined Copper in the U.S.
Source: *Copper Development Association, Inc.*, *Copper Supply and Con-*
sumption (New York: Copper Development Association, 1973), pp. 10-11.

followed by construction and consumer products. Copper is vulnerable
to substitution by aluminum in electrical applications, by steel for
shell casings, and by plastics for plumbing. The most important area
of substitution has been in power transmission and distribution. Aluminum has increased its share of the overhead distribution market from
near zero in the early 1950's to well over 90% in recent years (4).
Lower density and lower cost of aluminum have been the major reasons
for the decline of copper in this use.

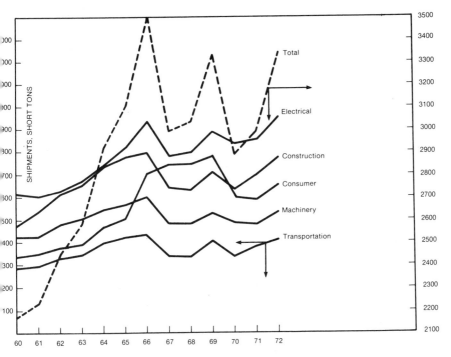

*Figure 9-2. Principal Copper Shipments by End Use Market in the
United States. Source: Gordian Associates, Historical Trends and
Future Projections for Energy Consumption in the Production of Primary
Refined Copper (New York: Gordian Associates, January 1975), p. 33.*

The domestic copper, lead, and zinc industries are interdependent
in that byproducts or residues from one industry provide a part of the
input for the other. The copper, lead, and zinc industries are significant producers of byproducts such as silver, gold, and bismuth.
An important aspect of the entire primary nonferrous industry is that
traditionally, the smelters and refineries have been operated as service operations at a fixed and relatively low profit margin that is
not very sensitive to the price of the finished product.

MANUFACTURING PROCESS DESCRIPTION

Copper production is divided into four segments:

• Mining.   Ore containing 0.5-2% copper is mined.

• Beneficiation.   Copper containing minerals are separated from
  the waste rock to produce a concentrate containing about 25%
  copper.

• Smelting.   Concentrates are melted and reacted to produce pure
  blister copper or anode copper.

• Refining.   Blister copper is refined electrolytically.

Figure 9-3 outlines the production of refined copper by conventional
mining and smelting processes.

Mining and Beneficiation of Ore

Approximately fifteen copper minerals of the 164 known are commonly found in ore deposits.   The ore containing these minerals can
be classified as sulfides and oxides, with most of the copper being
obtained from sulfide ores in the form of chalcopyrite and copper
pyrites.   With the exception of the rich Zambian and Katangan sulfide
ores, which can contain up to 5% copper, most ores have a copper
content of around 1% (5).   In general, the copper content of ores is
declining from about 0.8% in 1954 to 0.55% in 1973, as shown in Fig.
9-4.   Improvements in copper technology, particularly in the area of
ore beneficiation, have compensated for this trend, so that processing
of low grade ores with around 0.4% copper remains economic (6).

Open pit mining accounted for about 89% of the ore produced in
1970 in the U.S.   The balance was mined underground.   In underground
mines, essentially all of the material removed contain more than 0.3%
copper, and it is all sent to concentrators.   In the large open pit
mines, stripping ratios (ratio of waste rock to ore) vary from about
1:1 to as high as 12:1.   Much of the energy required for mining is
consumed in the excavation and hauling of overburden.   In one mine,

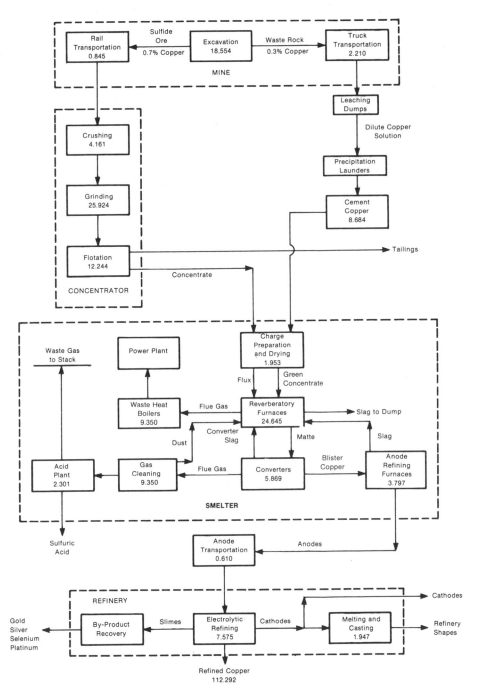

Figure 9-3. *Production of Refined Copper.* Source: *Battelle Columbus Laboratory, Energy Use Patterns in Metallurgical and Non Metallic Mineral Processing, Phase 4 (Columbus, Ohio: Battelle, p. 50.)*

303

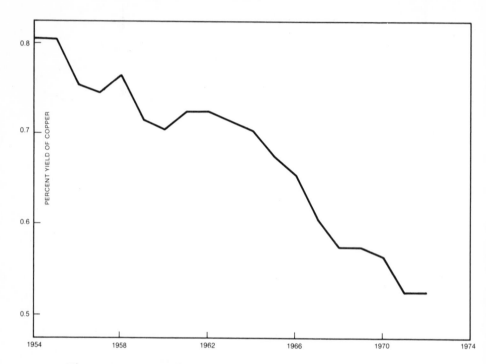

Figure 9-4. Trends in Copper Ore Grade. Source: U.S. Bureau of Mines, Minerals Yearbook, 1954 to 1973 (Washington, D.C.: Government Printing Office, 1954-1973).

during 1974, the total energy consumed was about 40×10^6 BTU per ton of copper. Although the ore averaged 0.7% copper, the stripping ratio was 12:1, which accounted for the high energy consumption. In another mine, the total energy consumed was only 9×10^6 BTU per ton of copper, because although the ore grade averaged only 0.55% copper, the stripping ratio was only 2.5:1 (7).

The ore from the mines is sent to the concentrators, which consist of crushing, grinding, and flotation operations. Crushing is done in gyratory and cone crushers, which consume relatively little energy. Grinding is carried out in rod mills and ball mills and requires most of the energy consumed in ore concentrating.

The major process used for copper ore beneficiation is known as froth flotation. See Fig. 9-5. This involves the creation of a suspension of ore particles in water by means of mechanical or air agitation. The surfaces of desired particles are treated with chemicals called promoters or collectors, which render these particles water repellant. With vigorous agitation and aeration in the presence of a frother, the desired mineral particles float to the top of the flotation machine, where they collect in the froth and are skimmed off. The undesired gangue remains suspended in the bulk of the liquid and is removed as tailings. Typical recovery of copper is over 90% in a concentrate containing between 15-35% copper. The amounts of promoters required is usually very small: a total of six pounds of lime and 0.05 pounds of other chemicals per ton of ore are consumed in the treatment of a 0.9% copper ore yielding 24% copper in the concentrate, at 91% recovery (8).

Production of Cement Copper by Dump Leaching

An alternative method for concentrating copper ores is by dump leaching. The steps in dump leaching are shown in Fig. 9-6. Dilute sulfuric acid is pumped to sprays or ponds on top of the dumps. Acid is formed by the reaction of water with pyrite and with other sulfur bearing minerals. The copper sulfide minerals must be oxidized for dissolution to occur, and the rate of oxidation may be enhanced by the action of bacteria (9).

The only operation that consumes significant quantities of process energy is pumping the dilute acid solution to the tops of the waste dumps. The pumping rates vary widely between mines, e.g., from 1000 gallons per minute to about 45,000 gallons per minute. The only material energy consumption is that used by the shredded and detinned steel scrap for precipitating copper from the pregnant liquor. Steel consumption varies between mines from about $1\frac{1}{2}$-4 net tons per ton of copper. The cement copper which is produced is about 80% copper.

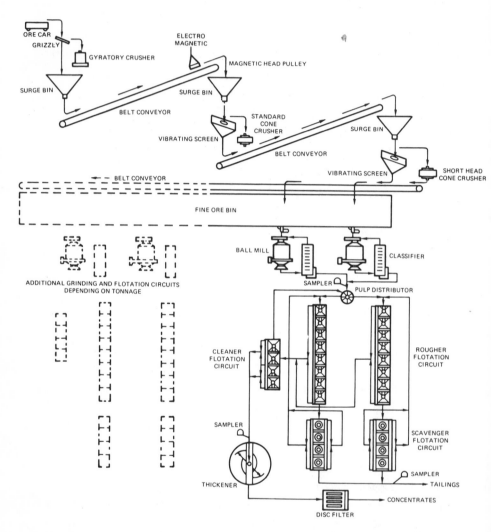

Figure 9-5. Copper Ore Flotation. Source: Environmental Protection Agency, Environmental Considerations of Selected Energy Conserving Manufacturing Process Options, Vol. XIV, Primary Copper Industry Report (Cincinnati, Ohio: Environmental Protection Agency, December 1976), p. 118.

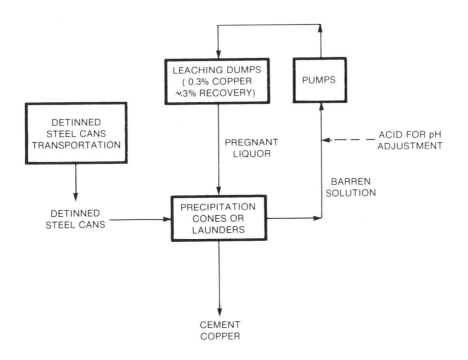

Figure 9-6.   Dump Leaching of Sulfide Copper Ore.

Smelting

Because the concentrates received by the copper smelter can contain 10-15% moisture and because the cement copper can contain as
much as 30% moisture, both concentrate and cement copper are dried
before it is charged to the reverberatory furnace. Drying will reduce
the reverberatory furnace fuel requirement, and the dryer will act as
a blender for homogenizing the charge. Rotary or multiple hearth
dryers are used for drying the feed materials.

In some plants the concentrate is roasted to convert some of the
sulfides to oxides and to eliminate some of the sulfur. The older
smelters use multiple hearth roasters, whereas the new smelters use
fluidized bed roasters. Both types of roasters usually operate

around 1200°F and produce sulfur dioxide in the off gas. These off gases can be used to manufacture sulfuric acid.

In roasting, most of the energy required is obtained from the burning of sulfur in the concentrates. Roasted concentrate, or calcine, is then charged into the reverberatory furnace. In addition to concentrates and cement copper, the reverberatory furnace charge includes silica rock, lime rock, ore as fluxing agents, and copper bearing dust collected from the Cottrell precipitators in the smelter. See Fig. 9-7. The calcines from roasting and the concentrates from beneficiation are fed into a reverberatory furnace typically 110 feet long and 34 feet wide. Heat is supplied by burners located at one end of the furnace. Furnace installations are normally equipped with waste heat boilers, which recover much of the heat of the combustion gases in the form of superheated steam. Between 11-16% of the sulfur in a

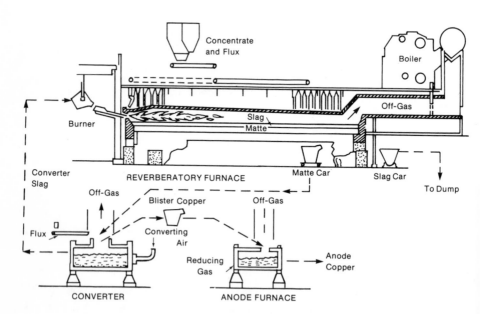

Figure 9-7. Conventional Copper Smelting Process. Source: A. J. Weddick, "The Noranda Continuous Smelting Process for Copper," Efficient Use of Fuels in the Metallurgical Industries (Chicago, Illinois: Institute for Gas Technology, December 1974), p. 654.

roasted charge, or up to 20% in an unroasted charge, is liberated from the reverberatory furnace in the smelting process. Together with the products of combustion, the outlet gas contains between 0.5-1% SO_2 and has a flow rate of 100 million cubic feet per furnace per day (10).

A slag composed of iron, calcium, and aluminum oxide silicates with a copper content between 0.4-1% is tapped off from the furnace. A heavier layer of molten copper matte is formed, typically composed of 30% copper, 39% iron, and 26% sulfur. A copper content of around 30% is desirable because a higher percentage results in higher losses in the slag, while a lower percentage requires more oxidation of iron and sulfur in the subsequent conversion stage. Sulfur elimination in the roasters can be adjusted to give the optimum matte composition (11).

The molten matte is poured into a converter (usually a Pierce-Smith type) consisting of a horizontal cylindrical bricklined vessel about thirteen feet in diameter and thirty feet long. Silica flux is charged along with the matte, and air is blown into the matte through submerged tuyeres, in order to oxidize the iron and sulfur. The silica flux combines with the iron oxide to form a silicate slag, which is skimmed off. The slag is relatively rich in copper and is therefore recycled. Since the reaction of oxygen with iron and sulfur is exothermic, bath temperatures tend to rise above the optimum of $2100^{o}F$-$2200^{o}F$. Because of this, regular addition of cold copper bearing materials are required to limit temperatures. All but about 2% of the iron and much of the sulfur is oxidized prior to skimming off the slag, leaving a "white metal" containing about 70% copper and 24% sulfur. Further air blowing removes most of the sulfur, leaving a final blister copper product containing about 98.5-99.3% copper, 0.3% sulfur, and other minor impurities.

The last operation in the smelter is the fire refining of blister copper to anode copper and the casting of anodes. The blister copper is transferred in the molten state to the anode furnaces, which are fuel fired and either reverberatory or cylindrical furnaces. In this

operation, air is blown through the blister copper to complete the
oxidation of iron and to oxidize minor impurities, e.g., nickel.
After the oxidation step, the slag is removed and returned to the
reverberatory furnace.

The copper, which is now saturated with oxygen, is reduced by
bubbling reformed natural gas through the melt. The deoxidation used
to be carried out by submerging green tree trunks (poles) into the
metal; hence, it is called "poling." Poling is continued until the
oxygen content of the copper is lowered to about 0.17%. The operation
consumes a significant amount of natural gas for "poling" and electricity for blowing. Anodes are cast directly from the anode furnaces
into coated copper molds which are moved into position on large rotating, horizontal casting wheels. The anodes, which contain more than
99% copper, are then shipped to the refinery.

Refining

In the electrolytic refining step, the anodes are suspended alternately with pure copper sheet cathodes (called starter sheets)
in a solution of copper sulfate and sulfuric acid. When an electric
current passes through the system, the anode bars gradually dissolve,
and pure copper is plated out on the cathodes. Most of the remaining
impurities from the anode, such as gold, selenium, silver, and tellur-
ium, fall to the bottom of the tank as a sludge, called "anode mud."
The anode mud is usually further processed to recover the valuable
byproduct metals.

After electrolysis, the copper cathodes are melted in an electric
furnace and cast into bars, cakes, and ingots.

ENERGY CONSUMPTION IN THE INDUSTRY

Aggregate energy usage in the copper industry has been increasing
as production has increased. A review of the historical data of the
copper industry shows a steady increase in the energy used per unit

of output. Figure 9-8 shows the energy consumed in each of the three
major process steps necessary for the production of primary refined
copper. Mining and beneficiation of ore consume about 65% of the
total BTU's required for the complete sequence of operations. Smelting
accounts for about 26%, most of this being in the form of fossil fuels
for furnaces, and the remaining 9% is consumed in fire refining and
electrolytic refining.

The most important factor affecting the energy consumed in copper
production has historically been and will continue to be the grade of
ore processed. Depletion of higher grade ores has generally increased
the unit energy consumption. While it is still economical to process
the currently lean ores (it is feasible to process ore as low as 0.4%

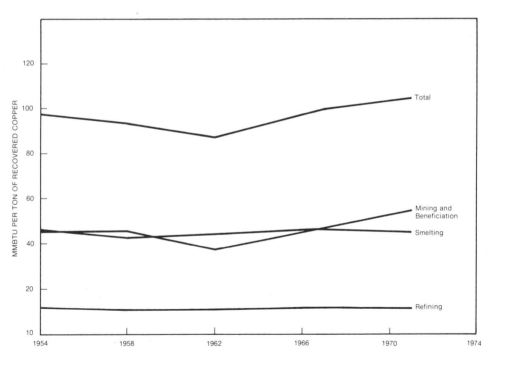

*Figure 9-8. Trends in Energy Consumption in Copper Production.
Source: U.S. Department of Commerce, Census of Mineral Industries
(Washington, D.C.: Government Printing Office, corresponding years).*

Table 9-2
Production of Refined Copper

	Unit	Units Per Net Ton Of Copper	10^6 Btu Per Unit	10^6 Btu Per Net Ton Of Copper
Mine				
Excavation				
Electrical energy	kwh	788.8	0.0105	8.282
Natural gas	ft³	220.5	0.0010	0.221
Petroleum fuels (diesel)	gal	38.62	0.1390	5.368
Coal	net ton	0.0013	25.27	0.033
Coal transportation (500 miles by rail)	net ton-mile	0.65	0.00067	0.000
Explosives	lb	155.0	0.030	4.650
			Subtotal	18.554
Transportation				
Waste rock to dumps (truck-2 miles)	net ton-mile	921.4	0.0024	2.210
Ore to mill (electric rail-8 miles)	net ton-mile	1,261.3	0.00067	0.845
			Subtotal	3.055
Concentrator				
Crushing				
Electrical energy	kwh	396.3	0.0105	4.161
Grinding				
Electrical energy	kwh	2,013.9	0.0105	21.146
Steel balls and rods	lb	273.00	0.0175	4.778
			Subtotal	25.924
Flotation				
Electrical energy	kwh	547.0	0.0105	5.743
Natural gas	ft³	2,465.0	0.0010	2.465
Petroleum fuels	gal	1.62	0.1500	0.243
Steam	lb	366.6	0.0014	0.513
Inorganic reagents	lb	541.0	0.0050	2.710
Organic reagents	lb	28.5	0.0200	0.570
			Subtotal	12.244
Smelter				
Cement copper	net ton	0.10	86.84	8.684
Charge preparation and drying				
Electrical energy	kwh	53.15	0.0105	0.558
Natural gas	ft³	1,395.0	0.0010	1.395
			Subtotal	1.953
Reverberatory furnace				
Reverberatory furnace				

Item	Unit	Quantity	Factor	Energy
Electrical energy	kwh	59.65	0.0105	0.626
Natural gas	ft3	12,799.0	0.0010	12.799
Petroleum fuels	gal	74.8	0.1500	11.220
			Subtotal	24.645
Steam (credit)	lb	6,675.0	0.0014	− 9.350
Converter				
Electrical energy	kwh	38.58	0.0105	0.405
Natural gas	ft3	852.5	0.0010	0.853
Petroleum fuel	gal	0.292	0.1500	0.040
Process steam	lb	3,265.0	0.0014	4.571
			Subtotal	5.869
Anode refining furnace				
Electrical energy	kwh	6.44	0.0105	0.068
Natural gas	ft3	3,139.0	0.0010	3.139
Petroleum fuels	gal	1.50	0.1500	0.225
Process steam	lb	169.22	0.0014	0.237
Coal	net ton	0.005	25.27	0.126
Coal transportation (500 miles by rail)	net ton-mile	2.5	0.00067	0.002
			Subtotal	3.797
Gas cleaning				
Electrical energy	kwh	30.76	0.0105	0.323
Acid plant				
Electrical energy	kwh	257.14	0.0105	2.700
Natural gas	ft3	1,510.00	0.0010	1.510
				4.210
Sulfuric acid produced	net ton	2.3	0.83	− 1.909
			Subtotal	2.301
Refining				
Electrolytic refining				
Electrical energy	kwh	307.49	0.0105	3.229
Natural gas	ft3	1,165.00	0.0010	1.165
Petroleum fuel	gal	19.92	0.1500	2.988
Coal	net ton	0.007	25.27	0.177
Coal transportation (500 miles by rail)	net ton-mile	3.5	0.00067	0.002
Sulfuric acid	net ton	0.0165	0.83	0.014
			Subtotal	7.575
Cathode melting				
Electrical energy	kwh	7.37	0.0105	0.077
Natural gas	ft3	1,869.7	0.0010	1.870
			Subtotal	1.947
Rail transportation anodes (2,000 miles)	net ton-mile	910.0	0.00067	0.610
			Subtotal	0.610
Total				112.292

Source: Battelle Columbus Laboratory, *Energy Use Patterns*, Phase 4, pp. 51-53.

in copper content), the impact of energy consumption in mining and
beneficiation per ton of copper recovered in the future is obvious.

The energy consumed in major process steps varies considerably
from plant to plant in the manufacture of copper. Table 9-2 provides
a detailed estimate of the energy required at each stage of the copper
manufacturing process. It takes an average of 112 x 10^6 BTU to manufacture a ton of copper.

Table 9-3 shows the distribution of types of energy used for the
various processes comprising the copper manufacturing process, using
trade association data. The smelting operations rely heavily on
natural gas (76.2% of total energy). Smelting and refining consume
electric power less extensively than does fabrication, which accounts
for about 23.5% of the total.

Developments in copper smelting during the last two decades have
generally disregarded the need for energy savings. In existing
smelters, roasting operations have been minimized or phased out in
favor of wet charging or reverberatory furnaces, which has lowered
operating and maintenance costs but has increased fuel consumption.
In recent years, concern has focused on the need for the most effective and least expensive air pollution control systems and for overall
cost reductions through process simplifications. Emphasis has been
placed on development of a continuous smelting process to conserve
energy and to minimize air pollution. In other cases, flash smelting
furnaces are being installed, again eliminating roasters but also
replacing the reverberatory furnace with a more efficient unit. Table
9-4 summarizes energy requirements for alternative primary copper
recovery processes.

Because of increasing energy costs, most of the major companies
in the copper industry are well aware of the pressing need for energy
conservation and have implemented energy saving programs. However,
the Subcommittee on Energy Conservation of the American Mining Congress, which submitted its first voluntary report on energy consumption to the Federal Energy Administration on April 5, 1976, concluded
that for 1975, the total energy consumption for the primary copper

Table 9-3
Energy Use Distribution for
Copper Component
(1972)

| | Btu, percent | | | |
Process	Electricity	Natural Gas	Other Fuels	Total
Trade Association Data				
Smelting	2.2	76.2	21.6	100.0
Refining	12.4	40.0	47.6	100.0
Fabrication	23.5	46.1	30.4	100.0
Total for Component:	8.8	62.2	29.0	100.0

Source: American Mining Congress, unpublished data; and
Copper and Brass Fabricators Council, Inc.,
unpublished data.

industry had increased by 21.1% over the base year 1972 (12). The
following factors were listed as contributing to this increased energy
consumption:

• decreasing quality of the ore (lower copper content)

• lower production in 1975

• regulatory changes concerning both the environment and safety

MINING AND BENEFICIATION IMPROVEMENTS

Open pit mines will probably be the major mining method used in
the forseeable future. With the quality of ore grades decreasing,
particular attention must be paid to costs and energy. It is estimated
that approximately 1.55×10^{14} BTU, or 65% of the total energy currently required for U.S. copper production, is consumed in the mineral

Table 9-4

Energy Requirements for Alternative
Copper Recovery Processes
(10^6 BTU/ton)

	Calcine Charged Reverb	Green Charged Reverb	Flash Smelting	Electric Furnace
Roasting	0.52	—	—	—
Drying	—	—	0.52	0.52
Smelting	3.20	5.70	1.80	5.52
Waste Heat, (Credit)	(0.96)	(1.45)	(0.81)	(0.29)
Converting	1.40	1.62	1.33	1.62
Gas Cleaning and SO_2 Recovery	1.86	1.62	2.35	2.32
Leaching	—	—	—	—
Anode Casting	0.78	0.78	0.78	0.78
Electrorefining	2.60	2.60	2.60	2.60
Electrowinning	—	—	—	—
Electrolyte Purge	—	—	—	—
Total	9.40	10.87	8.57	13.07

(a) Electric power converted using 10,500 Btu per kwhr.

Source: J. M. Henderson, "Environmental Overkill—The
Natural Resource Impact," *Mining Congress Journal*,
60 (December 1974), 18-23.

processing steps.   Thus, the potential energy economy in this area is
significant.

Table 9-5 shows the distribution of energy requirements for various unit operations in a conventional copper mining and beneficiation
process.   Excavation, crushing, and grinding together use about 70%
of the total processing energy, while about 30% of the energy is used
in the concentration of the copper sulfide minerals by flotation.
Hence, size reduction and, to a lesser extent, flotation should constitute major targets for effecting energy economies in copper processing.   Reductions in ore grade will undoubtedly call for a finer

Table 9-5

Distribution of Energy Requirements in Mineral Processing

Unit Operation	% of Total Energy Used
Excavation and Transportation	30
Crushing	6
Grinding	35
Flotation and Pumping	29

grind to maintain copper recovery at current levels. A finer grind means a larger energy requirement for size reduction. In addition, reduced ore grade means that a larger amount of waste must be processed in flotation to recover a pound of copper, thereby increasing energy costs per unit of copper concentrate produced by flotation. Fine grinding of a 0.5% ore, for example, requires 15-20% more energy than conventional grinding (13).

Another area in which energy conservation can be achieved is in the drying of the concentrated ore. The drying phase could be phased out in favor of charging the smelter with wet concentrates. Older technology requires a dry charge because of the possibility of flashing and explosion; however, more modern design has resolved this problem by providing safer equipment. The dryers used for eliminating moisture from the concentrates are usually of the conveyer type, requiring about 2000-5000 BTU of fuel (usually natural gas) per pound of water charged and about 60 HP of electricity (14). Moisture content is typically 9-12% of the total slurry weight. On the other hand, flash vaporization of water in the smelter requires only about 1000 BTU per pound of water charged. Thus, the alternative of wet charging appears to be significantly more energy conservative than drying followed by dry charging.

Major technological developments have occurred in the copper mineral process. Energy efficiency in size reduction can now be improved through larger equipment, more efficient classifiers, automatic control, and autogeneous grinding. Energy efficiency in flotation can be improved through larger equipment, reagent changes, and automatic control.

PRODUCTION OF COPPER METAL

Two general methods are used for the production of copper metal: (1) pyrometallurgical or smelting operations, followed by electrorefining and (2) hydrometallurgical operations, in which the ore is treated with a suitable solvent which selectively dissolves the copper and the copper is precipitated by metallic iron or electrowon from the solution. The type of process used depends on the type of ore and its copper content. Smelting is currently by far the more important of the two methods and accounts for about 90% of world production. The hydrometallurgical process is applied primarily to oxide ores and to tailings from previous mining and smelting operations.

Pyrometallurgical Processes

Pyrometallurgical processes employ the same sequence of chemical operations as does the conventional matte smelting process. These are:

- Drying of the concentrate, which may contain about 10% moisture.

- Melting of the concentrate to separate sulfides from gangue.

- Selective oxidation of iron sulfide to form an iron silicate slag and a copper rich matte.

- Recovery of copper from the iron silicate slag formed in the oxidation process.

- Oxidation of the copper matte to blister copper.

The blister copper is then transferred to a fire refining furnace and cast into anodes for electrolytic refining. A number of new technologies, e.g., flash furnaces, electric furnaces, and continuous

process furnaces, have the potential for major energy conservation.
This section will describe and assess these processes with respect to
their energy conservation potential.

Momoda Blast Furnace. The chief competitor to the reverberatory
furnace at this time is the Momoda blast furnace, developed by the
Sumitomo Metal Mining Company and presently in use at two of the
company's smelters in Japan (15). In the Momoda process, plasticized
concentrates are fed directly into the blast furnace. The plasticized
concentrate is typically 50-60% of the feed material and consists of
fine copper concentrates, fine dust, and other fine material in a stiff
plastic mass containing 10-15% water. The kneaded material is dropped
through the furnace as a cohesive mass, along with a mixture of coke,
silica, flux, limestone, and crushed converter slag, which falls away
to the sides. Gas flows up through the coarse material, rather than
through the concentrates, minimizing dust carryover.

The Momoda furnace requires less than two million BTU to smelt
a ton of charge, compared to 6.5×10^6 BTU per ton of wet charge in
a reverberatory furnace (16). At present, this furnace is limited
in size: the standard furnace processes about 500 tons per day,
whereas a typical charge rate in a reverberatory furnace is over
3000 tons per day. However, the outlet gas from a Momoda furnace
contains 7.3% sulfur dioxide, compared to less than 1% sulfur dioxide
in the outlet gas of the reverberatory furnace.

Electric Smelting Furnace (17). Electric smelting furnaces can
be used to replace fuel fired reverberatory furnaces. The electric
smelting furnace has six electrodes in line. The furnace is charged
with dry concentrates to preclude use of expensive electrical energy
for the low temperature drying operation. Because no fossil fuel is
used, the gases from the electric furnace contains a sufficient concentration of SO_2 to send to the acid plant.

Electric furnaces for copper smelting range from 3000-50,000 KVA
ratings, with power supplied to the furnace bath by the Soderburg electrode system. The slag, which floats on top of the molten matte, is
about forty inches in depth for a large furnace. In the reverberatory

furnace, heat must be conducted through the slag, so the slag is maintained at a lower depth to provide high heat transfer rates. The
energy consumed in an electric furnace is about 375-425 KWH of electric
energy per ton of wet calcine charge. This is 4×10^6 BTU per ton of
calcine, compared to 6.5×10^6 BTU for reverberatory furnaces (18).

Flash Smelting. In the flash smelting process, the concentrates
are dried and then fed to the flash smelting furnace, which performs
melting to separate sulfides from gangues in the concentrate and performs some selective oxidation of iron sulfides to form an iron silicate slag and a copper rich matte of approximately 65% copper. The
dried concentrates are burned in preheated, or oxygen enriched cold
air. The process is exothermic, so that it can operate without fuel
and without dilution of off gases with the products of combustion.
Flash furnaces are of two types, as in the prototypes of the Outokumpu
furnace, developed in Finland, and the INCO furnace, developed by
International Nickel Company and in use in Canada.

Outokumpu Flash Smelting. The original flash smelting process
was developed by Outokumpu Oy of Finland for the smelting of sulfide
copper concentrates in 1946-1949. This process was applied successfully to the smelting of nickel concentrates at the Harjavalta Works
in 1959 (19). In the 1960s, a modification of the flash smelting process was implemented, in which elemental sulfur is produced by decomposition of a pyrite concentrate and reduction of sulfur dioxide gases
in the flash smelting furnace (20).

In the Outokumpu process, the conventional methods of separate
roasting, smelting, and converting operations are replaced by a combined process carried out in one unit, the flash smelting furnace.
The heat generated by the exothermic oxidation reactions is used for
smelting the charge, and effective heat recovery systems enable the
minimization of the amount of additional energy used in the whole process.

The flash smelting process at Harjavalta Works of Outokumpu is
shown in Fig. 9-9. Correct proportions of feed materials are fed onto
a belt which transports the charge to a dryer, where the concentrate

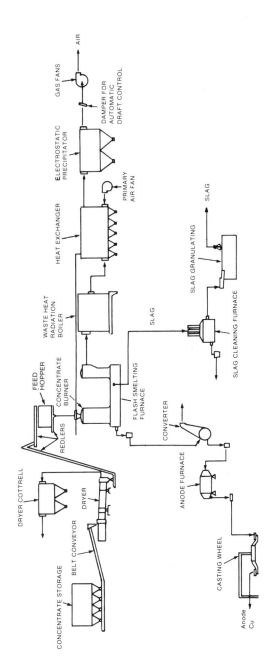

Figure 9-9. Harjavalta Outokumpu Smelter. Source: P. Bryk, J. Ryselin, J. Hankasalo, and R. Malmstrom, "Flash Smelting of Copper Concentrates," Journal of Metals, 10 (1958), p. 395. With permission of American Institute of Mining, Metallurgical and Petroleum Engineers.

is thoroughly dried. The charge is then transported to the feed hopper, from which the dried charge is fed to the concentrate burner and
mixed with preheated air to about $930^{\circ}F$. The charge reacts in the
reaction shaft, and after melting, molten particles are separated from
the gas phase in the settler, forming matte and slag. From the settler
the molten copper matte is tapped and transported in ladles to con-
verters. The blister copper is then processed in an oil fired rotating anode furnace, fire refined, and cast into anodes.

From the uptake, the gases go into a forced circulation type of
water wall radiation waste heat boiler, which cools the gas containing molten dust particles. Leaving this radiation boiler, the gases
at a temperature of $1,290^{\circ}$-$1,470^{\circ}F$ go to the heat exchanger, which
preheats the air needed for smelting to about $930^{\circ}F$. Here, the waste
gases are cooled to $660^{\circ}F$. The waste gases are then led to electro-
static precipitators, where most of the remaining dust is removed.
The flue dust collected from the waste heat boiler and the electro-
static precipitator is returned to the flash smelting furnace through
a separate feed hopper. From the precipitator, the gases go through
the exhaust fans to the sulfuric acid plant.

Because of the high grade of matte, the copper content of the
flash smelter slag is quite high, about 0.8-2.5% copper. It is thus
necessary to recover copper from this slag with either an oil fired
or electric furnace, or using flotation techniques. The flash smelting process is an excellent example of conservation of energy, because
the heat from the exothermic oxidation reaction is used for smelting
the charge and because the heat from the furnace off gases is efficiency recovered as high pressure steam in a waste heat boiler.

The smelting of concentrates at the Harjavalta copper smelter is
performed by using oxygen enriched air preheated to $400^{\circ}F$ with saturated process steam. The process off gases are cooled to $660^{\circ}F$ in a
waste heat boiler, which also generates process steam. This steam is
utilized for preheating process air, producing oxygen and generating
electric power (21).

INCO Flash Smelting. The International Nickel process is very
similar to that of the Outokumpu smelter. This process variation uses

95% oxygen in place of air, producing a more concentrated (i.e., 70-
80%) SO_2 gas. H. Kellogg has indicated that the Outokumpu type flash
smelter requires only about 40% of the energy used in a typical green
charge reverberatory furnace (22). The INCO oxygen flash smelter is
anticipated to provide even greater energy savings, relative to the
conventional reverberatory smelter, i.e., 1.48 x 10^6 BTU per ton versus 5.92 x 10^6 BTU per ton.

Besides improved utilization of energy and diminished gas volume,
the flash smelting practices are considered to yield a higher grade
matte than reverberatory smelting. However, high copper content slags
are produced in flash smelting and must be recovered.

Continuous Smelting. The continuous smelting processes consolidate the conventional reverberatory furnace-converter operation into
a single unit. Through the use of the exothermic reactions from
burning sulfur and oxidizing iron, the process can operate without
using external fuel and without diluting exhaust gases that are rich
in sulfur dioxide with the products of combustion. Continuous smelting should provide an energy saving over the conventional smelter.
A number of continuous smelting processes have been developed to the
pilot plant stage, and a few to the semicommercial stage. Some of
these smelting processes are described below.

The Noranda Process (23). This process, developed by Noranda
Mines of Canada during the 1960s is a continuous smelting process.
The Noranda process uses a single unit which performs roasting, smelting, and converting of copper concentrates. A schematic diagram of
the process is shown in Fig. 9-10. The unit consists of a horizontal
cylindrical furnace with a central depressed area for copper collection and a round hearth at one end for slag removal. An opening in
the roof of the furnace allows off gases to escape into a hood, and
burners are located at each end of the furnace. The unit can be rotated to bring the tuyeres out of the bath if the process needs to
be stopped for maintenance or because of loss of air pressure.

The Noranda furnace uses wet concentrates which are injected with
air and supplementary fuel or oxygen and are burned at one end of the

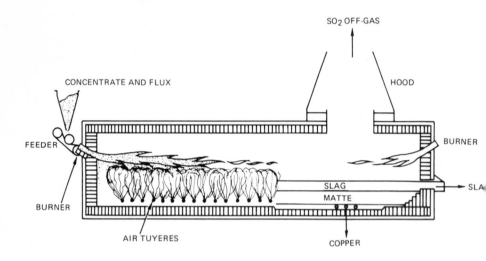

Figure 9-10. The Noranda Process. Source: A. J. Weddick,
"The Noranda Continuous Smelting Process for Copper," Efficient Use
of Fuels in the Metallurgical Industries, p. 655.

cylindrical vessel. This end of the furnace operates like a flash
furnace, making matte and slag. The molten materials move toward the
outer end of the furnace as they are continuously blown with air
through submerged tuyeres. The air is introduced at a sufficient
depth below the surface of the matte to cause 95% or more of the oxygen to react with the matte. The mixing action of the air jets maintains the bath in turmoil and provides a high heat transfer rate
from the copper sulfide matte, where heat is generated by the converting reactions, to the slag and to the concentrate charge on the
surface of the bath.

The blowing action results in 80% copper matte; as the matte
moves near the center of the vessel, it becomes blister copper, which
collects by gravity in a sump which is tapped. The slag, which floats
continues settling out some of the matte particles it contains, but

the slag is still high in copper by the time it reaches the taphole, where it leaves with about 13% copper. The slag is slow cooled, milled, and floated to make a concentrate with about 20% copper and to make tailings with less than 0.5% copper (24).

For some concentrates, the process produces too high a level of harmful impurities in the blister copper. To resolve this problem, a modified operation can be used in which the Noranda furnace produces an 80% matte, which is transferred to a Pierce-Smith Converter for oxidation to blister copper. When the Noranda furnace is operated to make 80% copper matte, control is easier, and the slags contain only 11% copper.

A pilot plant with a rated capacity of 100 tons of concentrate per day has been in operation since May 1968 at the Noranda, Quebec, smelter. An 800 ton per day industrial plant was built in Noranda and became operational in March 1973, using air for converting. Oxygen enrichment of 21-35% more than doubled the instantaneous productivity of the reactor. In addition, Kennecott Copper Corporation is planning the construction of a plant with a rated capacity of 250,000 tons per year of blister copper.

Table 9-6 shows that energy use is decreased significantly for the Noranda process. Savings of 12×10^6 BTU per ton are projected. The Noranda smelter and its associated acid plant recover a higher amount of SO_2 (over 90%), than does conventional smelting (50-70%). The cost of Noranda smelting is lower than a conventional smelter because of its higher energy efficiency.

The WORCRA Process. The WORCRA Process, invented by Dr. J. K. Worner, is a continuous smelting operation with a stationery furnace, similar to the Noranda furnace. It was first implemented in a small pilot plant at Cockle Creek, New South Wales, Australia, and later adapted to a larger continuous pilot unit which operated at Port Kembla (25).

Table 9-6
Projected Energy Balance for a Copper Smelter
Treating Chalcopyrite Concentrates

	Conventional Smelter		Noranda Process (35% Oxygen)	
	Energy kwh/ton Copper	% of Total	Energy kwh/ton Copper	% of Total
INPUT				
Fossil Fuel	5,520	44.9	1,260	17.3
Chemical Reactions	5,240	42.5	3,840	52.6
Electricity at Grid:	670	5.5	1,310	18.0
Generated	(190)		(160)	
Purchased	(480)		(1,150)	
Process Steam	880	7.1	880	12.1
TOTAL INPUT	12,310	100.0	7,290	100.0
OUTPUT				
Anode Copper	200	1.6	200	2.7
Slag	1,300	10.6	1,250	17.2
Off-Gas	3,630	29.5	1,220	16.7
Dust	40	0.3	60	0.8
Heat Losses	4,510	36.6	1,340	18.4
Process Steam	880	7.1	880	12.1
Electricity Generated	190	1.6	160	2.2
Other Losses	1,560	12.7	2,180	29.9
TOTAL OUTPUT	12,310	100.0	7,290	100.0

Source: Weddick, "The Noranda Continuous Smelting Process for Copper," *Efficient Use of Fuels in the
Metallurgical Industries,* p. 653.

In the WORCRA process, molten copper is tapped from one end of
the furnace, slag is removed at the other end, and off gases are collected as a single stream for subsequent recovery. Oxygen is provided
by air lances introduced from the top of the furnace, as opposed to
the air tuyeres in the Noranda furnace. Heat is provided through the
utilization of the exothermic reaction from burning the sulfides, so

that the energy requirement for the WORCRA type unit is expected to
be less than that of conventional processes. Kellogg's analysis of
the operating data for the WORCRA furnace at the Port Kembla pilot
plant, however, did not show a significant energy saving relative to
conventional reverberatory smelting (26). Operation of the WORCRA
type continuous smelter with modifications in its design could lead
to improved energy utilization in the future, which may result in this
process being considered more favorably than the conventional smelting
practices.

Mitsubishi Process. Development of this process began in 1971
by the Mitsubishi Metals Corporation of Japan. It differs from the
Noranda and WORCA continuous processes in that the operation is
carried out in three metallurgical stages. A schematic view is
shown in Fig. 9-11. Copper concentrates, fluxes, and oxidizing air
are introduced into the smelting furnace through lances installed
vertically, reaching just above the bath surface. Mitsubishi
indicates that use of the lances results in high smelting and
oxidation rates and that they simplify furnace design and maintenance. Another advantage is that oxygen or fuel oil can be
introduced through the lances.

In the smelting furnace, the concentrates form a copper matte of
57-63% copper. The molten products, slag, and matte are fed continuously to an electric slag cleaning furnace, where entrapped matte
in the slag is allowed to settle out, and additional iron pyrites and
coke are fed to reduce oxidic copper to matte. The slag is discarded
at a copper content of 0.5%. The matte then flows to a continuous
converting furnace which is also equipped with lances that introduce
air. The remaining iron is oxidized into a slag containing about
10% copper, which is returned to the electric furnace, and the blister
copper formed has a copper content varying from 98-99% and a sulfur
content of 0.4-0.8%. The copper is continuously tapped out of the
furnace (27).

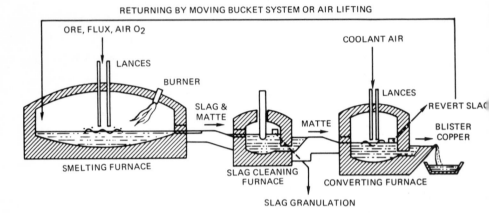

Figure 9-11. The Mitsubishi Process. Source: Environmental Protection Agency, Environmental Considerations, Vol. XIV, Copper, p. 72.

The average gas composition of the continuous smelter is 12% SO_2, which is much higher than that of the conventional smelter. However, the Mitsubishi process generates a smaller gas volume than the conventional smelter, which, from an energy standpoint, means less loss as sensible heat. The continuous stream of SO_2 in the off gases is suitable for acid or sulfur recovery. All the units are interconnected and runners covered, which probably accounts for the good gas quality.

Mitsubishi has built a semicommercial plant which processes 1500 tons blister copper per month and which began operation at the Onohama smelter in November 1971. Steady state operation and control of the Mitsubishi process is achieved by on-line, feed forward computer control of the inputs of air and flux to the converter furnace. A commercial operation producing 50,000 tons per year of blister copper is now being started at Naoshima, and a 130,000 ton per year plant is being designed by Montreal's SNC Consultants for the Kidd Creek smelter of Texas Gulf. The Mitsubishi process potentially has less problems

with fugitive emissions than other new processes, since all the transfer points are covered and hooded.

Energy savings of approximately 12 x 10^6 BTU per ton of copper
can be achieved over the conventional smelter. In addition, the cost
of the Mitsubishi process is lower than for the conventional smelter
because of its increased energy efficiency. Also, because the process
employs a different smelting and converting unit, it should be more
flexible in its ability to treat impure concentrates.

The Queneau/Schuhmann (Q/S) Process. This process is named after
the inventors, Dr. Paul Queneau and Dr. Reinhardt Schuhmann. The Q/S
process is a multistage progressive converter operation that combines
continuous smelting and converting in one furnace. It is like a
stretched out Noranda furnace, but arranged so that metal and slag
are withdrawn at opposite ends, with a provision for treating the slag
with pyrites and coke to lower the copper content to a throwaway level.
An interesting feature is the use of an oxygen-SO_2 gas mixture for
blowing the matte, which results in an essentially pure SO_2 that is
cooled and recycled for temperature control in the furnace. The remainder is sent to the acid plant for recovery. Hence, the system is
pollution free, producing low copper slag blister copper.

Since copper metal is produced, it is likely that similar problems
will arise with impurities as were found in the Noranda furnace. But
even if the Q/S furnace were operated as an 80% copper matte producer,
the slag cleaning feature, if effective, would eliminate the flotation
circuit of Noranda or the electric cleaning furnace of Mitsubishi.
Pilot plant investigations of the Q/S process are currently in the
planning stage. No estimate of the energy requirement is currently
available.

Thermo-Electron Chloride Process. Swinburne and Ashcroft developed a process in which metal sulfides react to form molten chloride
which flows from the reactor and is tapped from the bottom of the
reactor (29). The Bureau of Mines describes a process for chlorination of chalcopyrite, using some of the concepts developed by Swinburne and Ashcroft (30). Initially, chalcopyrite concentrate is

chlorinated at $850^\circ F$ in a shaft furnace. Mixed fused salts of $CuCl_2$
and $FeCl_2$ flow from the bottom of the reactor and are collected. These
salts are pulverized and oxidized with oxygen, which preferentially
oxidizes the $FeCl_2$ and $FeCl_3$, leaving the copper chlorides unreacted.
The copper is then recovered by electrowinning in a diaphragm cell
that regenerates chlorine for the process.

Thermo-Electron has recently proposed a high temperature process
in which first stage chlorination of chalcopyrite is performed at
temperatures of approximately $2,550^\circ F$ (31). Copper concentrate and a
low volatile coke is fed to the reactor. Gas, consisting of a mixture
of chlorine, nitrogen, and carbon dioxide, is introduced near the top
of the reactor, and carbon dioxide is fed into the bottom of the furnace, moving countercurrent to the descending coke. The reaction of
CO_2 with coke is endothermic, moderating the temperature so that the
coke does not react with chlorine, which also moderates the highly
exothermic chlorination reaction. The metal chlorides formed are
swept up in the off gases and are collected in series of condensers.

The first cuprous chloride condenser operates at $800^\circ F$, permitting
the $FeCl_3$ to continue to the second condenser, which operates at $440^\circ F$,
condensing the $FeCl_3$. The cuprous chloride crystals from the first
condenser are oxidized in a second stage at $800^\circ F$ to cupric chloride.
Copper is recovered from the cupric chloride by electrowinning.

The Thermo-Electron chloride process is still in the research
stage. The exact energy savings has not been estimated.

Autogenous Smelting. The Bureau of Mines has been conducting
autogenous smelting work on a laboratory scale (32). Autogenous
smelting is a continuous smelting method in a single unit for the
production of copper directly from concentrate. The furnace combines
flash smelting with converting by means of an oxygen lance immersed
through the slag into the matte. This process appears to offer potential for energy savings, relative to the conventional smelting
processes.

Summary. The new pyrometallurgical processes have two characteristics which make them more energy efficient, compared to conventional reverberatory smelters. They use the heat from oxidation of

sulfur and iron to supply a part of the process energy requirements, and they produce steady concentrated streams of sulfur dioxide that are suitable for the manufacture of sulfuric acid or elemental sulfur.

The processes are quite flexible in their ability to use any form of energy--gas, oil, or coal--and the reduction in energy consumption is significant and can amount to about 30-50%. The reduction in sulfur emissions is also significant. Compared to sulfur capture of 50-70% for conventional smelting, the new processes achieve sulfur capture of more than 90%.

The major shortcoming of the new processes is that their applicability to "impure" concentrates is unproven. Until this issue is resolved, the new processes will be utilized for building large smelters to smelt clean concentrates in regions where acid markets are available.

Hydrometallurgical Processes

In recent years, the general field of hydrometallurgy has received renewed attention. Two major reasons for this interest are (1) consideration for the environment, which has focused strongly on the concerted effort of the minerals and metals industries to find and develop new technology, and (2) scarcity of materials, particularly raw materials sources within the U.S., which has resulted in renewed exploration activity.

Major steps in the hydrometallurgy process are leaching, purification, concentration, and recovery. If these procedures are fully developed, the estimated seventy-five million tons of copper contained in the U.S. ores averaging 0.86% copper would be supplemented by another fifty-eight million tons of copper in ore deposits assaying less than 0.47% (33).

Solution Mining. Many new copper deposits are marginal in grade or poorly situated. Therefore, it would appear that hydrometallurgy should have a considerable impact on the treatment of low grade ores in dumps or in situ. At the present time, copper recovery from western ore deposits by leaching of low grade waste material is estimated to be approximately 195,000 tons of copper per year, or approximately 11.5% of the total production in the U.S.

Recently, in situ leaching of ore deposits has received increased emphasis and appears to be an area in which significant advances will be made in the near future. This type of hydrometallurgical operation is often referred to as solution mining. The direct recovery of metal from an ore deposit will have a significant impact, considering the earth's vast mineral wealth which is not economically extractable by normal techniques. J. B. Fletcher has presented a detailed discussion of in place leaching at Miami Mines, Miami, Arizona (34). Ranchers Exploration and Development Company, using a similar process, used two thousand tons of explosives to fracture the copper ore and, following conventional dump leaching practice, leached the ore by using sprays. After twenty-two months of leaching, 15% of the contained copper was recovered (35).

Lewis and Bhappu have presented economic evaluations of metal values from oxide copper ores (36). Their study considered heap leaching, vat leaching, agitation leaching, and in situ leaching. In situ leaching was the only economically attractive process for oxide ores having a grade of 0.5% copper or less.

Arbiter (37). This process has been developed by Anaconda and has been in operation for several years. In it, copper sulfides are leached with ammonia and oxygen at relatively low temperature, $140^{\circ}F$, and pressure, 5 psig, in tank reactors. The copper and the iron precipitated during the leach are separated by solvent extraction. Part of the ammonia is regenerated and recycled, while stripping of the copper is done with the acid electrolyte recycled from electrowinning. The sulfate is disposed of as gypsum, which permits an additional regeneration of ammonia or ammonia sulfate. The strip extract is then fed to the electrowinner, where the copper is recovered by electrowinning with insoluble anodes. The critical nature of the process appears to be reactor design, specifically, gas dispersion design. The ammonia leach is designed to extract only 80% of the copper values, and the remaining copper is recovered by smelting.

From the energy point of view, the high purity oxygen required,
as compared to the 95% purity normally used in pyrometallurgical processes, represents a high power consumption without the benefit of
energy recovery. In addition, all the oxygen required to oxidize copper, sulfur, and part of the iron must be manufactured, which can represent a considerable consumption of energy. Ammonia is also a high
energy reagent. Although the use of oxygen and ammonia are highly
energy intensive, this process does conserve energy, and the commercial plant which started up recently in Montana will provide data on
the total energy consumption.

Sheritt/Gordon (38). Sheritt/Gordon has been operating a pressure
leach which uses an oxidizing gas, such as air or oxygen enriched air,
and sulfuric acid to dissolve copper. Leaching is accomplished in
autoclaves at 185°F and 110 psig pressure. Under these conditions,
95.6% of the copper is put into solution. The leach liquor is further
processed by oxyhydrolysis to remove unsaturated sulfur compounds.
The soluble copper is then recovered by hydrogen reduction in horizontal autoclaves at 392°F and 500 psi hydrogen pressure. Little dissolution of the iron takes place.

A $10 million pilot plant has been constructed jointly by Sheritt
Gordon Mines, Ltd., and Cominco, Ltd., at Fort Saskatchewan, Alberta.
This plant will produce electrowon copper and elemental sulfur. At
this stage, it is not a proven process, and data is not available to
assess the fuel and energy requirements (39).

Cymet (40). Another process is the Cymet process, which uses a
chloride leach to extract the copper and iron from the concentrate and
produce a copper powder and electrowon iron. The Cyprus Metallurgical
Process Corporation of Los Angeles has indicated that this process will
require minimum amounts of fuel.

Leaching of copper sulfide concentrates is accomplished in an
electrolytic cell at a temperature between 158°F and 195°F, with fer-
ric chloride acting as the oxidant. Some size reduction of the copper

sulfide concentrate is necessary. Leaching is also accomplished electrolytically at the anode and at the cathode, cuprous chloride is reduced to metallic copper powder (99%) at 150 amp/ft^2, and ferrous
chloride is reduced to metallic iron.

A twenty-five ton per day pilot plant is presently under construction near Tucson, Arizona, to evaluate the process and determine the
economics. It should also provide data upon which the fuel and total
energy requirements can be assessed.

Conclusions. Most hydrometallurgical processes end with the production of electrowon copper, as opposed to electrolytically refined
copper from blister. Electrowinning consumes about ten times the energy required for electrolytic refining. Although the fuel consumption
is radically reduced with hydrometallurgy processes, electrowinning
offsets some of this saving. This can be significant when comparing
extraction processes, for the final process selected will depend on
local energy costs.

The most significant advantages of these hydrometallurgical techniques are:

- Costs comparable to pyrometallurgical smelting, with advantages
  anticipated as equipment and technique are improved.

- The feasibility of lower capital investment and smaller plants.

- Reduced energy costs, based on electricity rather than on
  natural gas.

- Flexibility in handling various feedstocks.

OXYGEN USE IN SMELTING

Oxygen enrichment is a practice which can enhance energy efficiency and significantly increase production rates over existing processes. The use of oxygen in smelting decreases fuel requirements and
is energy efficient overall, because the decrease in fuel requirements
is usually larger than the incremental energy required in the oxygen
separation plant.

In the reverberatory process, oxygen enrichment of the air blast shows promise of a reduction in the energy requirements. For example, data from the Copper Cliff smelter in Canada, obtained while smelting copper-nickel concentrates, indicates the following typical figures (41):

Oxygen content of blast	25%
Fuel consumption decrease	13%
Furnace output increase	30%

Tests in the USSR have been performed with oxygen concentrations of up to 40%. Optimal conditions were found around 27%, however, when fuel consumption was decreased 25-30%. Enrichment over 30% led to refractory lining damage. The enrichment not only increases SO_2 concentrations but also increases the scrap melting capacity of the converter.

The specific capacity of furnaces such as the flash smelting furnace can be increased by the use of oxygen. In this case, the increase in energy efficiency is not as large as in the case of converter operations, since waste heat recovery is already practiced. However, the increase in smelting rate significantly decreases the capital costs per unit of output. Oxygen enrichment of Outokumpu furnace gases, for example, has been practiced since about 1972 at Harjavalta. This has enabled almost completely autogenous operations of the furnace and resulted in a large increase in furnace capacity.

COPPER REFINING

The energy consumed in refining of blister copper is currently about 7.5×10^6 BTU per ton of primary refined copper, which is relatively small compared with the consumption in the mining and smelting stages. The method of copper refining has not been significantly altered in the past two decades. This is probably the result of its having a very small portion of the total system energy requirements consumed. Recently, there has been some research directed towards decreasing the energy consumed in refining in Japan and in the USSR.

This work focuses on increasing the current density employed. As much as 250-300 amperes of current per square meter is applied at some refineries in Japan and in the USSR, and in experimental work, current densities of 400-600 amperes per square meter have been demonstrated (42). In general, changes in refining techniques have not been significant.

ENERGY CONSERVATION POTENTIAL

The energy of copper sulfide ore is greater than the energy of copper. Thus, the production of copper theoretically should not require any fuel input. The principal unit operations associated with the production of copper (i.e., reverberatory furnace, converter and electrolytic refining appear relatively energy efficient in comparison with unit operations in other industries. The largest percent of energy usage, about 65%, is for mining, transportation, and concentration of the ore; about 26% is used for smelting the concentrate; and 9% for refining the copper. The reverberatory furnace uses the largest quantity of energy in the smelting operation, requiring about 20% of the total energy used for the production of refined copper.

The availability of alternative energy saving production processes for copper is known, and it is highly desirable that these replacement technologies be installed. Table 9-7 provides a quantitative comparison of base line and of some alternative technologies in the copper manufacturing process. As is readily evident, new technologies can achieve energy efficiency improvements of up to 50%.

Reduced recovery rates from lower grade ores will lead to substantially increased energy consumption per ton of primary copper through the year 2000. Therefore, it is essential that hydrometallurgical techniques be perfected. The potential value of these techniques lies in their ability to efficiently recover copper from low grade ores, including tailings. However, the requirement for more development, in order to perfect the technology, coupled with the heavy

Table 9-7
Quantitative Comparison of Base Line and Alternative Processes in the Copper Industry

	Reverb Smelting (Base Case)	Electro-Refining (Base Case)	Outokumpu Smelting No Oxygen	Outokumpu Smelting Oxygen	Noranda Smelting	Mitsubishi Smelting	Slag Cleaning Electric Furnace	Slag Cleaning Flotation	Arbiter Process
• Environmental Pollution Control Costs ($/ton of Copper)	54–64	—	59	59	46	59	15	1.90	—
• Energy Consumption 10 6 Btu/ton of Copper a	23–26	4.6	15	13.2	12.5	12	1.2 c	0.8 c	61
• Process Economics Investment ($/annual ton)	650–750	450	750	500	750	750	9.5 c	20 c	1000
Pollution Control and Operating Cost ($/ton) b	340–370	230	336	259	336	333	10.47 c,d	12.40 c,d	636

a Includes electricity at 10,500 Btu/kWh; oxygen at 360 kWh/ton.
b Includes pre-tax return on investment — excludes concentrate cost.
c $/ton of slag.
d Equivalent to about $200–$360/ton of recovered copper.

Source: Environmental Protection Agency, *Environmental Considerations*, Vol. XIV, *Copper*, p. vi.

investment in existing processes, may prevent these techniques from having significant impact on the industry through the year 2000.

Table 9-8 indicates the potential for energy conservation through the use of various process alternatives in the copper manufacturing process. Use of the economical techniques described can result in an energy savings of 10-20% over conventional practices.

Table 9-8

Summary of Energy Conservation Potential
(over conventional practices)

Conservation Practice or Technique	Potential Savings (10^6 BTU/ton)
Charge smelter with wet concentrates	1.0
Momoda blast furnace	4.5
Electric melting furnace	2.5
Outukumpu flash smelter	11
INCO flash smelter	12.5
Noranda smelter	12
Mitsubishi smelter	12
Oxygen use in smelting	0.5

NOTES AND REFERENCES

1. Battelle Columbus Laboratory, Study of the Energy and Fuel Use Patterns in the Non Ferrous Metals Industries (Columbus, Ohio: Battelle, December 1974).

2. Department of Commerce, Bureau of Domestic Commerce, 1975 Industrial Outlook (Washington, D.C.: Government Printing Office, 1974), p. 85.

3. R. D. Rosenkranz, Energy Consumption in Domestic Primary Copper Production, IC 8698, Bureau of Mines (Washington, D.C.: Government Printing Office, 1976), p. 14.

4. Council on Environmental Quality, The Economic Impact of Pollution Control (Washington, D.C.: Government Printing Office, March 1972), p. 195.

5. Donald Treilhard, "Copper--State of the Art," Chemical Engineer, 80 (April 16, 1973), p. P.

6. United Nations Industrial Development Organization, Copper Production in Developing Countries (New York: United Nations, October 1970), p. 54.

7.   Battelle Columbus Laboratory, _Energy Use Patterns in Metallurgi-_
     cal and Non Metallic Mineral Processing, Phase 4 (Columbus, Ohio:
     Battelle, June 1975), p. 46.

8.   John H. Perry, _Chemical Engineer's Handbook_, 14th ed. (New York:
     McGraw Hill, 1969), pp. 21-71.

9.   J. T. Woodcock, "Copper Waste Dump Leaching," _Proceedings of_
     Australian Institute of Mining and Metallurgy, No. 224 (Sydney:
     Australian Institute of Mining and Metallurgy, December 1967),
     pp. 47-66.

10. Treilhard, "Copper--State of the Art," _Chemical Engineering_,
     p. P.

11. Ibid., p. Q.

12. "Voluntary Industrial Energy Conservation," _Progress Report 3_
     (Washington, D.C.: Department of Commerce and Federal Energy
     Administration, April 1976).

13. P. F. Chapman, "The Energy Costs of Producing Copper and Aluminum
     from Primary Sources, _Metals and Materials_, 8 (February 1974),
     p. 109, with permission of the Metals Society.

14. Donald Liddell, _Handbook of Non Ferrous Metallurgy: Principles_
     and Processes (New York: McGraw Hill, 1945), p. 16.

15. Treilhard, "Copper--State of the Art," _Chemical Engineering_, p.
     U.

16. "Copper Smelting Today: The State of the Art," Special Edition
     Joint Issue of _Chemical Engineering_ and _Engineering and Mining_
     Journal, Special Section, March 1973, pp. p-z.

17. J. A. Persson and D. G. Treilhard, "Electrothermic Smelting of
     Copper and Nickel Sulfides and Other Metal Bearing Constituents,"
     Journal of Metals, 25 (January 1973), pp. 34-39. With permis-
     sion of the American Institute of Mining, Metallurgical and
     Petroleum Engineers.

18. Treilhard, "Copper--State of the Art," _Chemical Engineering_,
     p. W.

19. T. Niemela and S. U. Harkki, "The Latest Development in Nickel
     Flash Smelting at the Harjavalta Smelter," Paper presented at
     the Joint Meeting MMIJ-AIME, Tokyo, Japan, May 1972.

20. R. O. Argall, "Outokumpu Adds Second Catalyzer to Raise Pyrite
     to Sulfur Conversion to 91%," _World Mining_, 20 (March 1967),
     pp. 42-46.

21.   J. Juusela, S. Harkki, and B. Anderson, "Outokumpu Flash Smelting
      and Its Energy Requirement," Efficient Use of Fuels in the Metal-
      lurgical Industries (Chicago, Illinois: Institute of Gas Tech-
      nology, December 1974), p. 560.

22.   H. Kellogg, "New Copper Extraction Processes, Journal of Metals,
      26 (August 1974), p. 21. With permission of American Institute
      of Mining, Metallurgical and Petroleum Engineers.

23.   N. J. Themelis, et al., "The Noranda Process," Journal of Metals,
      24 (April 1972), pp. 25-32. See also A. J. Weddick, "The Noranda
      Continuous Smelting Process for Copper," Efficient Use of Fuels
      in the Metallurgical Industries, pp. 645-660.

24.   K. N. Subramanian and N. J. Themelis, "Copper Recovery by Flota-
      tion," Journal of Metals, 24 (April 1972), pp. 33-38. With per-
      mission of American Institute of Mining, Metallurgical and Petro-
      leum Engineers.

25.   H. K. Worner and B. S. Andrews, "Integrated Smelting--Converting
      Slag Cleaning in a Single Furnace," Paper presented at TMS-AIME
      Annual Meeting, Dallas, Texas, February 1974.

26.   Kellogg, "New Copper Extraction Processes, " Journal of Metals,
      p. 21. With permission of American Institute of Mining, Met-
      allurgical and Petroleum Engineers.

27.   T. Nagano and T. Suzuki, "Extractive Metallurgy of Copper,"
      TMS-AIME, 1 (1976), pp. 439-57.

28.   R. E. Queneau and R. Schuhmann, "The Q-S Oxygen Process," Journal
      of Metals, 26 (August 1974), pp. 14-16. With permission of
      American Institute of Mining, Metallurgical and Petroleum
      Engineers.

29.   D. M. Lidell, "Chlorine Metallurgy Process," Handbook of Non
      Ferrous Metallurgy, p. 552.

30.   D. H. Yee, et al., "Chlorination Process for the Recovery of
      Copper from Chalcopyrite," SME Transactions, 254 (1973), pp.
      301-303.

31.   Proprietary proposal submitted to Energy Research and Develop-
      ment Administration by Thermo-Electron Corporation, Waltham,
      Massachusetts, 1977.

32.   R. B. Worthington, Autogenous Smelting of Copper Sulfide Concen-
      trate, U.S. Bureau of Mines report of Investigation 7705 (Wash-
      ington, D.C.: Government Printing Office, 1973).

33.   Elias Cyftopoulos, Lazaros Lazaridis, and Thomas Widmer, Poten-
      tial Fuel Effectiveness in Industry (Cambridge, Massachusetts:
      Ballinger Publishing Company, 1974), p. 80.

34. J. B. Fletcher, "In-Place Leaching at Miami Mine, Miami, Ari-
     zona," Transactions of AIME, Vol. 250 (New York: AIME, December
     1971), p. 310.

35. "Rancher's Big Blast Shatters Copper Ore Body for In-Situ Leach-
     ing," Engineering and Mining Journal, 173 (April 1972), p. 98.

36. F. M. Lewis and R. B. Bhappu, "Economic Evaluation of Available
     Processes for Treating Oxide Copper Ores," International Journal
     of Minerals Processing, 3 (1976), pp. 133-50.

37. M. C. Kuhn, "Anaconda's Arbiter Process for Copper," CIM Bulle-
     tin, 67 (1974), p. 62.

38. D. J. Evans, "Treatment of Copper/Zinc Concentrates by Pressure
     Hydrometallurgy," CIM Bulletin, 57 (1964), p. 857.

39. J. C. Taylor, "Recent Trends in Copper Extraction," Efficient
     Use of Fuels in the Metallurgical Industries, p. 642.

40. P. R. Kruesi, "Cymet Copper Reduction Process," Mining Congress
     Journal, 60 (September 1974), p. 22.

41. United Nations Industrial Development Organization, Copper Pro-
     duction in Developing Countries, p. 43.

42. Ibid., p. 49.

Chapter 10

CEMENT INDUSTRY

The hydraulic cement industry is defined as that group of establishments engaged in manufacturing hydraulic cement--portland, masonry
pozzolonic, and natural cements. The U.S. industry consists of fifty-
one companies operating 170 plants in forty-one states and Puerto Rico
Company size ranges from a single plant to as many as fourteen plants,
but no single company accounts for more than 7.5% of the total production capacity. In general, cement companies market their product
locally, where they may compete with as many as 10-20 other companies.
The cement industry, like other mineral commodity industries, is regional in nature (operating within a radius of 100-300 miles from the
plant), due to the relatively short distances which this heavy, low
value product can be transported economically by rail or truck. For
companies with access to water transportation, market areas are expanded considerably beyond this radius.

On the whole, cement plants are fairly old. However, plant age
is difficult to define, since a single plant often has major processin
equipment of different ages. One hundred and sixty-eight U.S. plants
operate 434 kilns, the major piece of equipment, almost half (47%)
of which have been built since 1955, providing 68% of the total cement
producing capacity (1).

Virtually all cement is used for construction. Demand for cement is, therefore, closely tied to fluctuations in the building industry. Because of the relatively high level of fixed costs associated with cement production, the industry's rate of capacity utilization correlates closely with profitability. The 1950s were profitable years for the cement industry. When the rate of utilization peaked at 94% in 1955, the highest rate of return, 18.6%, was achieved. Attracted by the high profits of the 1950s, firms expanded their capacity, but capacity expanded far more rapidly than demand. Between 1950 and 1968, production rose 80% from 220-400 million barrels, while capacity rose 100% to its peak level of 510 million barrels in 1968. See Fig. 10-1.

From 1970-1972, the cement industry operated at nearly 90% of its capacity. By 1974, demand declined, due to depressed housing construction activity, increased inflation, and an uncertain national economy. Faced with the prospect of continued low returns, a growing number of cement firms began to diversify and consolidate. The number of cement companies has been declining steadily, going from 94 in 1923 to 51 in 1974. While the four largest firms account for nearly 24% of the total capacity, they are contributing smaller percentages of the total capacity than they did in the past.

The United States ranks third among world producers of hydraulic cement. In 1973, the U.S. produced 11% of the total world production. U.S. production has grown at 2% per year, whereas world production of hydraulic cement has been growing at 7.9% per year.

MANUFACTURING PROCESS DESCRIPTION

Hydraulic cement is a powder made by heating limestone, sand, clay, and other substances together in a kiln to form a clinker and then grinding the clinker. Table 10-1 outlines the materials used in the cement-making process. Cement reacts with water to form rock, sand, and gravel into concrete. During 1973, 139×10^6 tons of raw materials were used to manufacture 85×10^6 tons of cement. It takes approximately 1.6 tons of raw materials to produce one ton of cement.

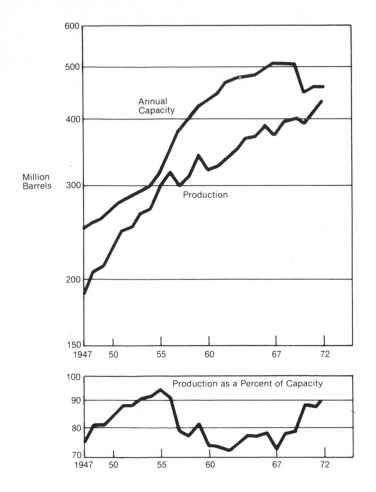

Figure 10-1. Portland Cement Capacity and Production.   Source: U.S. Bureau of Mines, Minerals Yearbook, 1947 to 1972 (Washington, D.C: Government Printing Office, 1947-1972).

The production sequence for the manufacture of cement may be di-vided into the following activities:

- Raw materials acquisition

- Raw materials preparation

- Clinker production

- Grinding and mixing of the cement product

Table 10-1

Types and Quantities of Raw Materials Used in Producing Portland Cement in the United States, 1972-1973 (Thousand Tons)

Raw Materials	1972	1973
Calcareous:		
Limestone (includes aragonite)	84,922	86,699
Cement rock (includes marl)	25,879	26,067
Oyster Shell	5,081	5,144
Argillaceous:		
Clay	8,062	7,931
Shale	4,096	4,099
Other (includes staurolite, bauxite, aluminum dross, pumice, and volcanic material)	110	240
Siliceous:		
Sand	1,993	2,053
Sandstone and quartz	781	748
Ferrous:		
Iron ore, pyrites, millscale, and other iron-bearing material	839	968
Other:		
Gypsum and anhydrite	4,094	4,253
Blast furnace slag	759	682
Fly Ash	271	299
Other	33	5
Total	136,920	139,188

*Includes Puerto Rico

Source: U.S. Bureau of Mines, *Minerals Yearbook, 1973, p. 259.*

There are two common cement processing sequences based on the above production sequences--the wet and the dry processes. In the dry process, grinding and blending of raw materials are carried out with dry materials. In the wet process, grinding and blending are done in a slurry of the materials in water. In other respects, the processes are essentially the same. See Fig. 10-2 for a simplified flow diagram for the manufacture of cement by the wet and dry processes.

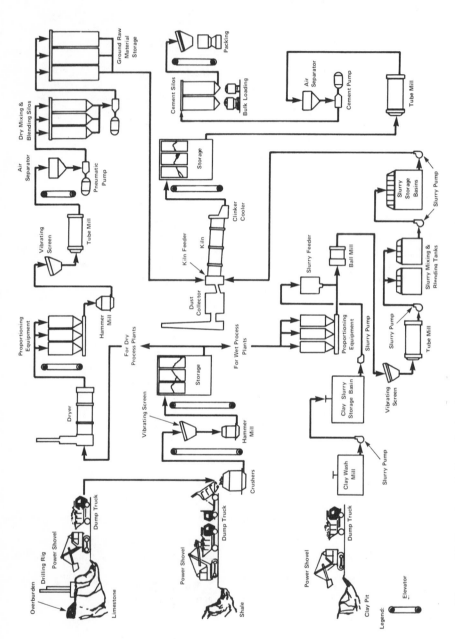

Figure 10-2. Cement Manufacturing Processes. Source: Office of Science and Technology, Patterns of Energy Consumption in the United States (Washington, D.C.: Government Printing Office, January 1972), p. 136.

Raw Materials Acquisition

The first step in the manufacturing process of cement is the mining of raw materials which contain lime, silica, and alumina in suitable proportions. Natural calcareous deposits, such as limestone and shell beds, provide lime. Natural argillaceous deposits, such as clay, shale, and slate, provide both silica and alumina. Natural deposits of argillaceous limestone or marl can provide all three basic ingredients, although not necessarily in the desired proportions. In the Northeast, "cement rock" is found, from which a cement may be produced by merely heating and grinding (2).

Raw Materials Preparation

The next step in the cement making process is simply size reduction. Depending on the raw material and on the design of the grinding system, the crushing system can vary considerably. Two stages are usually employed: the primary crushers produce pieces of rock about six inches in diameter, and the secondary crushers produce pieces under three-fourths inch in diameter (3). The crusher type that is used varies according to the hardness, size, and type of the rock. Primary crushers include gyratory crushers, which consist of a steel cone moving eccentrically inside a housing; jaw crushers, in which reduction takes place between two rolls or between one roll and a breaker plate. Secondary crushers are usually hammer mills which fracture the rock, using heavy hammers swung by centrifugal force from a horizontal shaft. After crushing, the stone is transported for storage according to raw material type.

Milling is the process that prepares the various sizes and mixtures of raw materials. In the wet process, wet raw materials like clays and chalks first require disintegration in wash mills, where a slurry is formed. This wash mill product is then fed into wet mills, along with a proportional amount of limestone and other raw materials. The first stage of grinding occurs in the ball mills, which contain thousands of large steel balls that cascade down onto the materials

being ground. The second stage occurs in tube mills, which are similar to ball mills but are longer and charged with smaller balls. In
many cases, ball and tube mills are combined. The slurry that exits
from the tube mill is stored for feeding into the kiln.

The dry process milling is similar to the wet, except that no
water is added and that the material is ground dry. If necessary,
dryers are used, and energy for them is supplied by their own heat
sources or from hot kiln gases. Air separators classify the milled
product and return the coarse fractions to the milling system. The
finished raw materials are then blended before going to the kiln. Drying and raw grinding can be combined into a single, closed circuit ball
mill system.

In the blending process various proportions of raw materials are
used, according to the desired end product. Table 10-2 provides
descriptions and chemical compositions of the five types of cement
recognized in the U.S. The functions of these chemical compounds are
as follows (4):

- Tricalcium silicate hardens rapidly and is largely responsible
  for initial set and early strength.

- Dicalcium silicate hardens slowly and contributes largely to
  strength increases at ages beyond one week.

- Tricalcium aluminate liberates a large amount of heat during
  the first few days and contributes to early strength.

- Tetracalcium aluminoferrite reduces the clinkering temperature
  in the manufacturing process but contributes little to strength.

A number of cements are known by names which describe their use
or composition (5):

- Masonry cement, which is used in mortars for masonry work.

- Oil well cement, which is designed for use under high temperature and pressure.

- White cement, which is ordinary portland cement with a low proportion of iron oxide, making its color white instead of gray.

- Waterproof cement, which is designed for stucco work and to improve water impermeability.

Table 10-2
Types of Portland Cement

Type of Cement	General Description	Compound Composition (%)			
		Tricalcium Silicate	Dicalcium Silicate	Tricalcium Aluminate	Tetracalcium Aluminoferrite
I	For general purpose use	49	25	12	8
II	For modified general purpose Use	46	29	6	12
III	When high early strength required	56	15	12	8
IV	Where low heat of hydration required	30	46	5	13
V	Where high sulfate resistance required	43	36	4	12

Source: U. S. Bureau of Reclamation, *Concrete Manual*, 7th ed. (Denver, Colorado: Bureau of Reclamation, 1963). p. 642.

• Portland pozzolan cement, which is produced by grinding together cement clinker and pozzolan (a material capable of reacting with lime in the presence of water at ordinary temperature to produce cementitious compounds).

• Portland blast furnace slag cement, which is produced by grinding together portland cement and granulated blast furnace slag. Portland blast furnace slag cement usually contains 35-45% granulated slag by weight. This slag is produced by rapid quenching in water and air of hot slag which is $2500^{\circ}F$ when it comes from the furnace. Because of the rapid cooling, a glass that chemically resembles a low lime clinker is formed.

Table 10-3 shows the relative quantities of the various types of cement produced in the U.S. in 1974.

Table 10-3

Types of Portland Cement Shipped in the United States (1974*)

	Quantity 10^3 ton	Value ($)	Average Value ($/ton)
General use and moderate heat (Types I and II)	73,474	1,927,557	26.23
High-early-strength (Type III)	2,596	71,423	27.51
Sulfate-resisting (Type V)	323	8,653	26.79
Oil-well	989	27,667	27.97
White	474	26,697	56.32
Portland-slag and portland pozzolan	672	16,843	25.06
Expansive	132	4,681	35.46
Miscellaneous**	822	24,385	29.67
Total or average	79,482	2,107,906	26.52

*Includes Puerto Rico
**Includes waterproof cement

Source: U.S. Bureau of Mines, *Minerals Yearbook, 1974,*
          p. 282.

Clinker Production'

The key process in making cement is the "calcining" of the raw materials in a rotary kiln. In the clinkering step, the accurately controlled mixture of raw materials reacts chemically at high temperature in the kiln to produce a clinker, which is then ground into cement.

The kiln is a long cylindrical furnace which rotates slowly on its axis and is mounted at a slight incline to the horizontal. Present commercial kilns vary from about 300-750 feet in length and are about 12-25 feet in diameter (6). See Fig. 10-3. The kiln is lined with refractories to protect the steel shell and to conserve heat. The kiln rotates at about one revolution per minute, and its inclination causes the raw materials to move gradually down the kiln toward the discharge in several hours.

The burner is at the discharge end. As the material travels toward the firing end, it gets progressively hotter. Water evaporates,

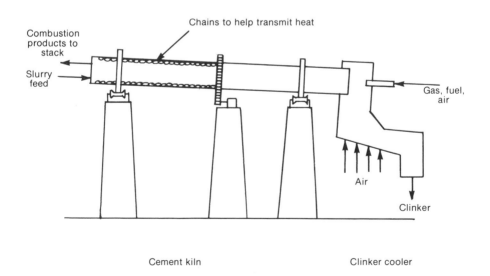

Figure 10-3. Cement Kiln

and the heat causes carbon dioxide to be driven off. The material then enters the hottest zone, where the temperature is about 2800°F and where the main chemical reactions of hot lime with silica, alumina, and iron begin causing the clinker to form. The processed material finally leaves the kiln at the lower end as roughly textured lumps or pellets, ranging in size from one-sixteenth of an inch to three inches in diameter (7).

As noted in Fig. 10-3, many kilns use chains to improve the heat exchange between the hot gases and feed as they move towards the burning zone. Wet process kilns are longer than dry kilns, since part of the kiln is used to evaporate the raw feed slurry water. These kilns are equipped with elaborate arrangements of chains that serve as heat exchangers between the gas stream and slurry. The combustion gases pass through the kiln countercurrent to the material and leave the kiln through its feed end at temperatures between 600°F and 1600°F, depending on the kiln length and the process used. Many modern kilns are also preceded by a preheater system in which hot gases give up heat to the cold incoming raw materials.

After leaving the kiln, the clinker enters coolers that reduce its temperature before storing or grinding and recover its heat for reuse inside the kiln. A number of different clinker coolers are available, some based on moving grate systems and others consisting of tubes mounted along the kiln periphery, close to the clinker outlet. Air is normally introduced to the tubes through induced draft and is preheated before being used in the combustion of the kiln fuel.

Grinding and Mixing of the Cement Product

After cooling, the clinker is transported into storage, where it is segregated, tested, blended, and moved into bins for feeding to the finish grinding mills. The final stage of manufacture consists of grinding the lumps of clinker to a fine powder. The clinker is ground together with gypsum (calcium sulfate), about 3-6% gypsum by weight being added to the clinker to control the time of set of the cement when it is mixed with water in making concrete. Small amounts of other

materials may also be added to facilitate grinding or to impart special
properties to the cement.

The mills used in finish grinding are usually the same as those
used in raw grinding, in which rod, ball, roller, race, and tube mills
are used. Most finish grinding systems are closed circuit systems that
use air separators to provide classification. Fine finished products
are sent to storage while coarser fractions are returned for further
grinding. In some cases, materials such as pozzolans are blended with
portland cement in concentrations ranging from about 10-30% of the
cement by weight.

ENERGY CONSUMPTION IN THE INDUSTRY

Today, roughly 58% of portland cement produced in the U.S. is by
the wet process, with the remainder produced by the dry process. In
1974, the wet process consumed 61% of the total energy consumed in cement manufacture. The dry process consumed 39%. See Fig. 10-4. Since
1950, the wet process has been consuming a greater percentage of the
total energy, peaking at almost 65% in 1969. After 1969, energy use
in the wet process leveled out at around 62%, as the dry process
leveled out at around 38%.

A review of the historical data for the cement industry shows a
steady decrease in the energy used per unit of output in cement manufacture. See Fig. 10-5. Figure 10-6 shows energy use per unit of output under wet process and under dry process portland cement production.
The figure shows that the improvements in energy use per unit of output have been at different rates in each of the processes. The proportion of total output made by dry process should continue to increase.

Total energy use by cement producers has increased more slowly
than output in the past quarter century. See Table 10-4. In contrast,
with the 103% gain in the value of production between 1947 and 1971,
total energy consumed increased by just 53%. Useful energy required

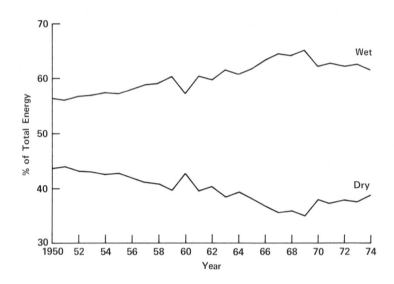

Figure 10-4. Trends in Energy Consumption by Process Step.
Source: U.S. Bureau of Mines, Minerals Yearbook, 1950 to 1974.

per dollar of output has dropped from 458,000 BTU per dollar of output
to 345,000 BTU per dollar of output. This is at an average annual
rate of 1.2% per year.

Gains in energy efficiency have come about mainly through growing reuse of waste heat, economies of scale with the use of increasingly larger plants, advances in mechanization, advances in process control, and a shift from on site power generation to the purchase of
electricity. Energy increases, on the other hand, has been due to
greater per unit energy requirements of increased mechanization, of
finer grinding, and of pollution control equipment.

Form of Energy Use

The cement industry uses all forms of energy: coal, fuel oils
(distillates and residual oils), natural gas, and electricity. Figure 10-7 provides a view of the fuel forms used between 1950 and 1974.
As can be seen, coal usage has been decreasing steadily over the past

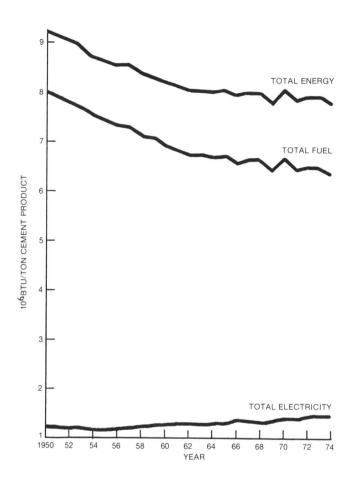

Figure 10-5. Trends in Fuel and Electricity Use. Source: U.S. Bureau of Mines, Minerals Yearbook, 1950 to 1974.

quarter century, and natural gas has been increasing in usage. A substantial reversal in the shift of energy source is expected. Coal's
share, which has been shrinking, is projected to grow appreciably between now and 1985. The types of energy used by the portland cement
industry in 1974 are shown in Fig. 10-8.

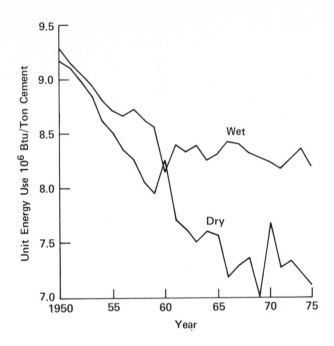

Figure 10-6. Trends in Unit Energy Use for Wet and Dry Process.
Source: U.S. Bureau of Mines, Minerals Yearbook, 1954 to 1974.

Table 10-4
Output, Energy Use, and Energy Use Per Dollar of Output

	1947	1954	1958	1962	1967	1971	1975	1980
Output								
(Million 1967 $)	$684	$930	$1,053	$1,092	$1,246	$1,391	$1,547	$1,767
Energy use (trillion BTUs)								
Useful	313	375	410	428	463	480	496	519
Gross	342	407	449	471	515	536	564	594
Energy use per dollar of output (1,000 BTUs)								
Useful	458	403	390	392	371	345	321	294
Gross	500	438	426	431	413	385	365	336

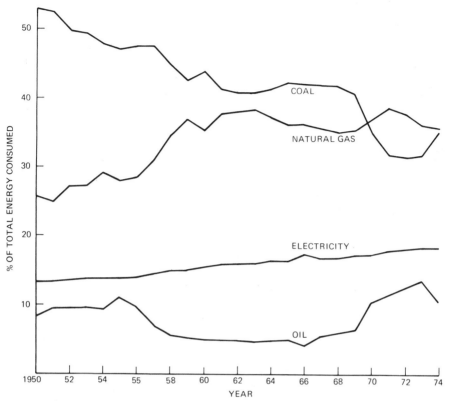

Figure 10-7. Trends in Types of Energy Used. Source: U.S.
Bureau of Mines and Arthur D. Little, Inc.

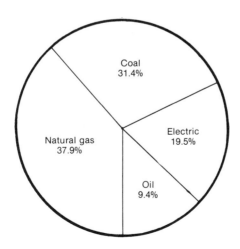

Figure 10-8. Energy Use by the Cement Industry, 1974. Source:
George McCord, Energy Conservation Trends in the Cement Industry,
Society of Mining Engineers of AIME, Preprint No. 75-H-308, 1977, p.20.

Table 10-5 provides a summary of the types of fuels used in the clinkering process. As is evident, in the single fuel plants, coal is predominant, but natural gas provides a substantial amount of the necessary fuel demand. Multiple fuel firing also provides a substantial percentage of the heating requirements.

Table 10-6 provides a breakdown of fuel types used in the wet and dry cement manufacturing processes. As can be seen, coal is used substantially more in the dry process, and natural gas is dominant in the wet process.

Energy Use by Process

Table 10-7 shows the average total energy required to produce a ton of cement at each stage of the manufacturing process. It indicates that the most energy intensive process in cement manufacture is

Table 10-5
Clinker Produced in the U. S. by Kind of Fuel, 1974

Fuel	Number of Plants	Clinker Produced (10^3 ton)	(percent of total)	Fuel Consumed (10^3 ton)	(10^6 cf)	(bbl)
Single Fuel						
Coal	42	19,298	24.8	4,724	—	—
Natural Gas	27	10,980	14.1	—	70,246	—
Oil	10	5,801	7.4	—	—	5,465
Multiple Fuels						
Oil and Natural Gas	31	15,313	19.6	—	74,843	1,902
Coal and Natural Gas	33	12,950	16.6	1,516	47,331	—
Coal and Oil	16	8,465	10.9	1,367	—	2,604
Coal, Oil, Natural Gas	9	5,170	6.6	487	15,962	339
TOTAL		77,977	100.0	8,094	208,382	10,310

Source: U. S. Bureau of Mines, *Minerals Yearbook, 1974*, p. 277.

Table 10-6
Fuel Consumption by Type
Distribution by BTU's of Primary
Fuel Resources
(percentages)

	Wet	Dry
Coal	30.4	42.6
Refined oil products	13.7	8.0
Natural gas	39.9	32.4
Purchased electricity	16.0	17.0
	100.0	100.0

Source: Gordian Associates, *The Potential for Energy Conservation in Nine Selected Industries,* Vol. 3: *Cement* (Washington, D.C.: Government Printing Office, 1975), p. 105.

Table 10-7

Average Energy Use for Cement Manufacture
by Process Step
(10^3 BTU/ton)

Process Step	Wet	Dry	Preheating
Quarrying and crushing	109	109	109
Drying	—	320	320
Raw Grinding	238	222	222
Kiln (Electricity)	287	287	287
Kiln (Fuel)	6491	5449	4562
Finish Grinding	523	523	523
Total	7678	6910	6023

Source: McCord, *Energy Conservation Trends in the Cement Industry,* p. 5.

Table 10-8

Range of Energy Use for Cement Manufacture by
Process Step
(10^3 BTU/ton)

Process Step	Range of Energy Use
Quarrying and crushing	30-250
Drying	0-1200
Raw Grinding	30-700
Kiln (Electricity)	100-626
Kiln (Fuel)	2800-10000
Finish Grinding	265-742

Source: McCord, *Energy Conservation Trends in the Cement Industry,* p. 5.

the clinker operation. This step consumes approximately 80% of the total energy used and is entirely fossil fueled. Electricity, on the other hand, is used mostly for grinding. The table also shows that the wet process uses substantially more energy than the dry process and that the installation of a preheater system would substantially increase energy efficiency.

However, the energy consumption values in Table 10-7 are only industry averages. They do not tell the whole story. Table 10-8 provides a summary of the range of energy usage in each of the cement manufacturing process steps. It indicates that a plant with the worst of all characteristics might use 13,528 x 10^3 BTU per ton of cement, which is double the current average energy use, or that a plant with the most optimal characteristics might consume 3,215 x 10^3 BTU per ton, about half the current average.

CLINKER PRODUCTION

The key process in making cement is the "burning" or "calcining" of the raw materials in a rotary kiln. As noted previously, the kiln, which uses about 80% of the total energy, is the major energy consumer

in the cement manufacturing process. Thus, the energy conservation
potential for the cement industry is largely focused around this system.

   At present, the cement industry predominantly uses the long rotary kiln for clinkering. A number of new systems, e.g., the suspension preheating, flash calcining, and fluidized bed processes, can
significantly reduce the energy consumption in clinkering. This
section will describe each of the major systems available for energy
efficiency improvement.

   The rotary kiln consists of essentially three separate zones:
preheating, calcining, and sintering. The three zones are based on
the nature of the chemical changes or reactions which occur within
them.

   In the preheating zones, the raw material is dried, and its
temperature increases as heat is transferred from the hot combustion
gases to the feed material when the gases exit the cold feed end of
the kiln. The heat transferred is by conduction and convection between the gas and the raw material and between the refractory brick
and the raw material. During its passage through this preheated zone,
the raw material feed is heated to approximately 1400°F.

   The interface between the preheating and the calcining zone is
not a physical one within the rotary kiln. In the calcining zone,
significant thermal composition of the calcium carbonate in the raw
material, which constitutes approximately 75% of the charge, begins.
This thermal decomposition is the first major chemical reaction that
occurs.

   After the calcium carbonate has decomposed to calcium oxide, a
series of reactions occur in the sintering zone between the calcium
oxide and the other components of the raw material. Ultimately, the
result is the formation of tricalcium silicate, dicalcium silicate,
tricalcium aluminate, and tetracalcium aluminoferrite, which are the

four major portland cement compounds. These reactions are exothermic and release a sizeable amount of heat in the sintering zone. This process also generates a sufficient amount of liquid, enabling the materials to form the clinker.

Figure 10-9 compares the three zones of the conventional long rotary kiln with those of a rotary kiln that uses a suspension preheater and a flash calciner. There is no sharp demarcation between adjacent zones in the rotary cement kiln. However, the zones are identified by the temperature profile and by the chemical composition of the raw materials in the kiln.

The best new plans in the U.S. and abroad are those with good preheater systems. Until relatively recently, most of the preheater systems in cement plants existed in Europe and Japan. With rising fuel costs, increasing attention is being given by U.S. cement manufacturers to heat recovery systems and to the use of the dry process wherever possible.

Suspension Preheaters

The suspension preheater is not just a recuperative device added to a kiln but is actually a part of the system where all of the chemical reactions and physical changes occur as the raw materials are processed into cement clinker. Replacing the preheating zone of a rotary kiln, the preheater is an assemblage of refractory lined steel ducts and vessels in which the hot gases leaving the calcining zone of the kiln comes into contact with the incoming raw feed. See Fig. 10-10. This is done by mixing the raw feed into the hot combustion gases flowing at high velocity through the ducts and vessels. Suspension preheating achieves heat transfer from the hot combustion gases that greatly exceed those in the preheating zone of the conventional long rotary kiln.

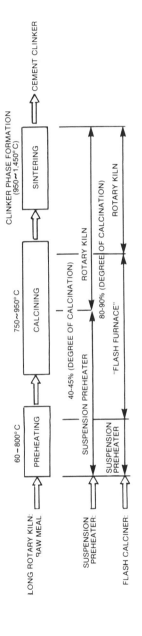

Figure 10-9. Cement Clinker Burning Process. Source: H.M. Garrett and J.A. Murray, "Improving Kiln Thermal Efficiency--Dry Process Kilns," Rock Products, 77 (August 1974), 60.

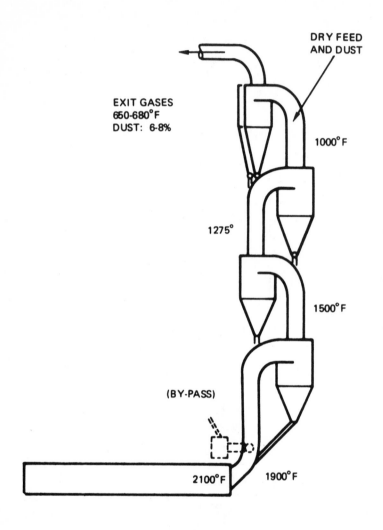

Figure 10-10. Typical Four Stage Suspension Preheater. Source: *Environmental Protection Agency, Environmental Considerations of Selected Energy Conserving Manufacturing Process Options, Vol. X, Cement Industry Report (Cincinnati, Ohio: Environmental Protection Agency, December 1976), p. 20.*

In addition to preheating, reactions in the suspension preheater results in a considerable amount of raw material calcination. Typical suspension preheaters heat the feed to approximately 1400°F, at which point 30-40% of the total calcination is performed. The rotary kiln thus receives hot and partially calcined material.

In the operation of a preheater, the key element is a relatively small cross sectional area duct connected to a cyclone. See Fig. 10-10. In a Humboldt type suspension preheater, the hot combustion gases from the rotary kiln flows up the duct. Raw feed is introduced to the hot gases near the bottom of the duct, resulting in a cloud of fine particles which move upward with the hot combustion gases. After the solid particles have extracted the usable heat from the combustion gas stream, the hot feed particles are recovered from the gas stream through a cyclone. The vertical duct section carrying dust laden gases makes a 90° bend and tangentially enters a cyclone. The heated dust discharges from the bottom of the cyclone, and the gas stream exits through the top of the cyclone.

A four stage suspension preheater consists of four of these duct/cyclone elements assembled in series. This provides four separate heat transfer stages with greater thermal efficiency than the single stage preheater. A number of four stage suspension preheaters, e.g., Humboldt, Polysius, Migg, and Krupp, are on the market. A schematic of these preheaters are shown in Fig. 10-11.

Stage one, located at the top of the unit, consists of two cyclones in parallel. This provides higher velocities, generates higher dust collection efficiency, and minimizes the amount of raw feed carried to the dust collector. The raw feed enters the main vertical duct and is divided into two streams, each of which enters the two first stage cyclones. The gas temperature at this point is low. The partially preheated raw feed collected by the cyclones in the first stage exit through the bottom and drop into the gas stream, leaving the third stage cyclone. This duct is the gas inlet to the second stage cyclone.

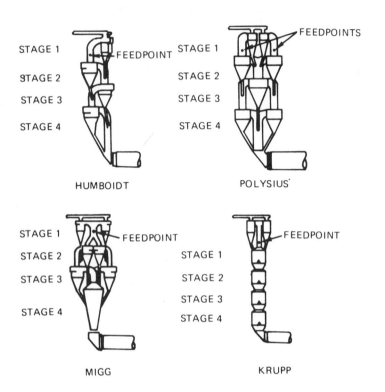

Figure 10-11. Representative Four Stage Suspension Preheaters.
Source: Garrett and Murray, "Improving Kiln Thermal Efficiency,"
Rock Products, p. 58.

The process is repeated four times, after which the final preheated
material, which is partially calcined and is at a temperature of about
1400°F, passes down into the feed end of the rotary kiln.

The suspension preheater was invented in Czechoslovakia in 1933
and was commercialized by Humboldt in 1950 (8). Three other suspension
preheaters were subsequently installed, all in Germany. After introduction, the suspension preheater was quickly adopted by the portland
cement industry in the U.S. and by the other major cement producing
countries of the world, such as Germany and Japan. In the U.S., the
first commercial suspension preheater unit was built in 1953 by the
Fuller Company. But after 1955, the manufacture of Humboldt suspension

preheaters suffered a hiatus because of operating difficulties in the
early units, due to alkalies and to the presence of combustible materials in the raw feed. The last few years have seen a significant
renewal of interest in the suspension preheater in the U.S. The design and operation of this system have evolved along lines that permit
the manufacture of lower alkali cement clinker and a reduction in the
operating problems caused by sticking or clogging of the preheater
system. Also, compared to the wet process and the dry process long
rotary kiln systems, the thermal efficiency of a four stage suspension
preheater with sufficient gas bypass is attractive because it requires
a smaller fixed capital investment and less fuel.

Figure 10-12 shows the relative capital costs of a long kiln versus a suspension preheater kiln. The capital investment for the four
stage suspension preheated kiln is lower than that of a long kiln because the large, heavy, refractory lined rotary kiln is more expensive
than the simpler suspension preheater. Moreover, the most significant
difference in operating costs between the suspension preheater kiln
and the conventional long rotary kiln is in the unit fuel cost.

Table 10-9 compares the energy requirements for a suspension preheater equipped rotary kiln and a long dry process rotary kiln. From
energy and cost standpoints, the installation of a suspension preheater
kiln system in new facilities makes great sense. It also makes great
sense to modify older wet process plants into dry process plants
through the conversion of existing long kilns to suspension preheater
units.

Flash Calciner

A recent modification to the suspension preheating system is the
German and Japanese designed flash calcining systems, which involve
two stages of firing. Fig. 10-13 shows a typical design of a flash
calcining system. The feature which characterizes the flash calciner
kiln is the flash calcining vessel, installed between the rotary kiln
and the suspension preheater.

The combustion gases leaving the rotary kiln pass through the
flash calcining vessel, and the hot raw material leaving the bottom of

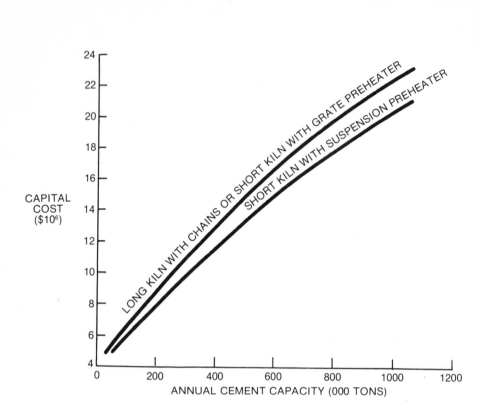

Figure 10-12. Capital Costs of Various Kilns. Source: I.B. Margiloff and R.F. Cascone, Paper presentation at Rock Products Cement Industry Seminar, Chicago, December 8, 1975.

stage 3 of the suspension preheater discharges into the flash calcining vessel. Fuel is burned in the vessel to further calcine and preheat the raw material stream. The combustion gases, combined with kiln gases, carry the raw material from the flash calcining vessel into stage 4 of the suspension preheater, from which the hot and almost completely calcined raw material discharges into the rotary kiln.

   In this system approximately 50% of the total fuel required for the clinker production step is burned at a relatively low temperature and with a low percent of excess combustion air. The combustion gas composition is uniform throughout the combustion chamber. Using this

Table 10-9

Comparison of Typical Energy Requirements for

Suspension Preheater and Long Kiln

(BTU/ton cement)

	Feed Preparation	Clinkering	Finishing	Available Energy Recovery	Net Energy Required After Energy Recovery
	(quarry, crush, dry, mix feed)	(burn, cool)	(grind, pack)	(steam/power generation, dryer fuel savings)	
Preheater, Short Kiln					
Electrical	534,000	374,000	760,000	(346,000)	1,322,000
Fuel	336,000	3,200,000	—	(300,000)	3,236,000
Total	870,000	3,574,000	760,000	(646,000)	4,558,000
Dry, Long Kiln					
Electrical	534,000	315,000	760,000	—	1,609,000
Fuel	336,000	4,600,000	—	(320,000)	4,616,000
Total	870,000	4,915,000	760,000	(320,000)	6,225,000

Source: Margiloff and Cascone, Paper presented at
Rock Products Cement Industry Seminar, Chicago,
December 8, 1975.

system, about 90% of the feed is calcined before it enters the rotating kiln.

Several manufacturers have developed various versions of the flash calciner which differ mainly in gas flow and precalcining vessel locations. These systems are described below.

Suspension Preheater with Flash Furnace (SF)(9).   The Chichibu Cement Company and the Ishikawajima Heavy Industry Company of Japan have developed the suspension preheater flash furnace system illustrated in the schematic flow diagram in Fig. 10-14.   In this system the suspension preheater receives raw feed in the normal manner.   The raw feed progresses down through the first three stages of the preheater, but instead of discharging into the fourth stage, it discharges into the flash furnace, which is fired by oil burners mounted in the furnace roof.   The flash furnace is a cyclonic suspension furnace with

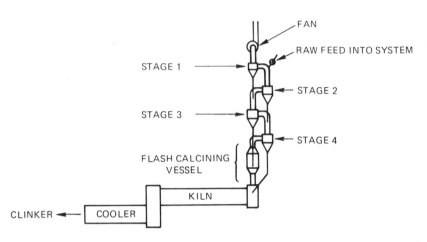

Figure 10-13. Flash Calcining System. Source: Environmental Protection Agency, Environmental Considerations, Vol. X, Cement Industry, p. 36.

gas inlets at the bottom and tangential gas and calcined raw feed outlets at the top. Within the chamber, the burners create a turbulent
calcining atmosphere at approximately 1750°F.

The raw feed is approximately 85% calcined when it leaves the
flash furnace and passes from the fourth stage cyclone into the kiln.
About 60% of the fuel normally required in the clinkering process is
used in the calcining reaction in the flash furnace. The remaining
40% of the normal fuel requirement is provided in the rotary kiln in
the second combustion stage. An important feature of the SF suspension
flash preheater is the addition to the combustion gas of high temperature preheated air taken from the middle of the grate type clinker
cooler. This permits the fuel in the rotary kiln to be burned with
the appropriate minimum quantity of excess air, optimizing the burning
conditions in the kiln.

A disadvantage of providing all of the combustion air to the flash
calcining vessel or furnace by using sufficient excess combustion air
within the rotary kiln is the high volumetric flow rate of gas and its
attendant high spatial velocity within the rotary kiln. Another disadvantage is that the high volume of combustion gas and excess air

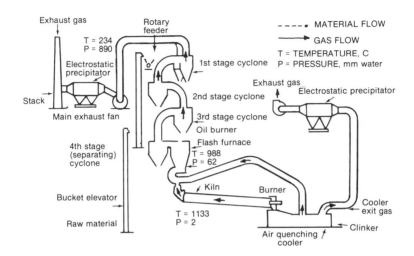

Figure 10-14.   Suspension Preheater with Flash Furnace (SF).
Source: McCord, Energy Conservation Trends in the Cement Industry,
p. 28

passing through the rotary kiln do not concentrate the volatilized
alkalies in that gas stream. The design of the SF suspension flash
preheater avoids these difficulties by taking hot air from the center
of the clinker cooler and conveying it through a separate refractory
lined duct located parallel to the kiln and by mixing this hot air
with the combustion gases leaving the rotary kiln. The proper balance
of air flow to the kiln and to the flash furnace is controlled by
(1) a constricted portion of the kiln extract duct, which functions
as an orifice and achieves a fixed flow resistance on the outlet of
the kiln combustion gases, and (2) an adjustable damper in the secondary air duct, which controls the pressure drop through this duct and
balances the air flow system. The SF preheater system permits operation without requiring a secondary air fan, which generally provides
hot secondary combustion air from the clinker cooler to the rotary
kiln.

The SF design also eliminates the need for a fan to move the hot air from the clinker cooler through the secondary air duct. This results in the hottest available air from the mid-section of the clinker cooler being sent to the flash furnace for the raw feed precalcining, maximizing the heat recuperation from the clinker cooler.

The temperature of the hot combustion gases leaving the rotary kiln in the SF suspension flash preheater system is higher than the temperature of the air in the secondary air duct. The temperature of the hot combustion gases leaving the rotary kiln is about $2,050^{\circ}F$, while the temperature of the hot air in the secondary duct, at the point of mixing, is approximately $1,380^{\circ}F$. The mixture of these two streams results in a cooling of the hot combustion gases from the rotary kiln and causes the solidification of alkali-coated raw material and dust particles. Thus, the design of the SF system prevents the buildup of solid alkali-rich materials.

Reinforced Suspension Preheater (RSP). The reinforced suspension preheater was developed as a joint undertaking by Kawasaki Heavy Industries and Onoda Cement Company. A schematic flow diagram of the RSP system is shown in Fig. 10-15. This system is similar to the SF system in basic concept and end results. The main difference is the manner in which the first stage combustion is accomplished, the swirl burner and swirl calciner in the RSP being equivalent to the flash furnace in the SF system.

The kiln exit gases do not pass through the flash calcining vessel; rather, they mix with the precalcined raw material and with the combustion gases coming from the flash calcining vessel on its way to the stage 4 cyclone. The only gas entering the flash calcining vessel is the hot preheated secondary combustion air coming directly from the clinker cooler, and since it enters without mixing with kiln gases, the higher concentration of oxygen provides a more stable and positive combustion than in the SF system. On the other hand, this higher concentration of oxygen is responsible for a higher concentration of NO_x formed within the flash calcining vessel.

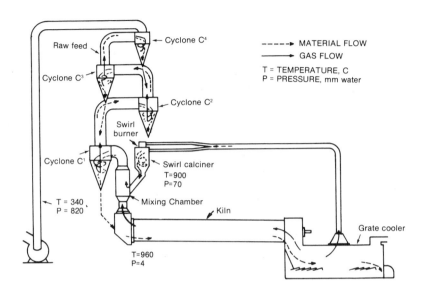

Figure 10-15. Reinforced Suspension Preheater. Source: McCord, Energy Conservation Trends in the Cement Industry, p. 26.

There are two main parts to the flash calcining vessel, a swirl burner and a swirl calciner. The hot air stream from the clinker cooler is divided into two parallel streams, one going to the swirl burner, where an ignition burner is operated, and the other going to the swirl calciner, where the single main firing burner is operated. Following the mixing and the resulting endothermic heat reaction in the swirl calciner, the raw feed moves into the mixing chamber, where exit gases are introduced. After further interchange between raw feed and gases in the mixing chamber, the feed is about 85% calcined. About 55% of the fuel required in the RSP system is used in the swirl burner. Combustion occurring in the rotary kiln uses about 45% of the total fuel requirements.

Another difference between the SF and the RSP systems is that the latter requires an induced draft fan to provide the hot air from the clinker cooler with a high enough pressure for its introduction into the swirl burner and swirl calciner. The operating temperature limitations of the fan in this air stream limits the temperature of the hot air taken from the clinker cooler. This tends to reduce the heat recuperated from the clinker cooler and increases the overall fuel energy required to make cement clinker.

Suspension Preheater with Mitsubishi Fluidized Calcinator (MFC). This system was developed as a joint research effort carried out by Mitsubishi Mining and Cement Company and Mitsubishi Heavy Industries. A schematic flow diagram of the MFC system is shown in Fig. 10-16.

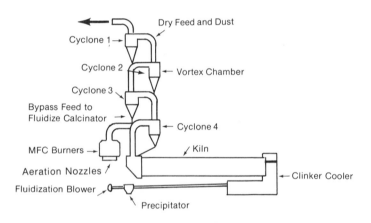

Figure 10-16. Suspension Preheater with Mitsubishi Fluidized Calcinator (MFC). Source: McCord, Energy Conservation Trends in the Cement Industries, p. 27.

The basic characteristics and the process goals of the Mitsubishi fluidized calcinator (MFC) are equivalent to both the SF and the RSP flash calciner processes. However, this system has some significant differences which put it in a different category. Like the SF and RSP systems, MFC uses the two stage combustion concept, but unlike the other two systems, the MFC calcines the preheated raw material in a fluidized bed instead of in a vortex type suspension vessel. Moreover, only a fraction of the preheated raw feed enters the precalcining or

flash calcining bed vessel. Whereas the SF and RSP processes are designed to operate with 100% of the preheated raw feed entering the
flash calcining vessel, in the MFC process only 20% of the total raw
feed is diverted to the fluidized bed calcining vessel. The precalcined mixture which is fed to the rotary kiln has been 55% calcined.

About one third of the air required for complete combustion of
the fuel oil in the fluidized bed is introduced, along with the oil,
through burners that are submerged below the fluidized bed surface.
The remainder of the required combustion air is introduced above the
bed, enabling a significant amount of combustion to take place above
the fluidized bed. Fluidization of the preheated raw material in the
fluidized calcinator is done with hot air from the clinker cooler,
which first passes through a cyclone type dust collector to remove
the fine clinker dust.

An advantage of the MFC system is that coal can be used as the
sole fuel in the fluidized calcining vessel. Other materials containing fuel value but not normally used for fuel can also be burned in
the fluidized calcinator.

Actual commercial operation of this system began in December 1971,
increasing kiln capacity and system availability significantly.

Polysius System. The Polysius Corporation, designer of the "Le-
pol" grate preheater and of the "Dopol" suspension preheater kilns,
has developed the "Prepol" precalciner kiln design shown schematically
in Fig. 10-17. In the Polysius system all of the air required for combustion in the precalciner is contained in the rotary kiln exit gases
as excess air. The kiln exit gases enter from the bottom of the calcining shaft, where a number of burners supply up to 50% of the total
process heat.

The Polysius system was developed jointly with Portland Zement-
werk Dotternhausen expressly for the purpose of utilizing an oil shale
raw material feed to effectively reduce the amount of purchased fuel
used in the rotary kiln. About 12% of the raw feed to this kiln is
oil shale burned in the precalcinator.

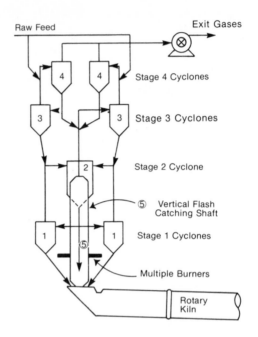

Figure 10-17. Polysius Flash Calcining Process. Source: McCord, Energy Conservation Trends in the Cement Industry, p. 27.

The Polysius system boasts a number of salient features (10):

• Coal can be utilized in the flash calciner.

• No kiln bypass duct is used.

• Planetary coolers can be used, since all of the combustion gas is conducted through the rotary kiln.

• Raw material containing fuel values, such as oil shale, can be successfully used.

• Six years of actual plant experience have been obtained.

Pyroclon. Humboldt-Wedag has developed a precalcining system called "Pyroclon," which is similar to the Polysius system. In the Pyroclon design, a burner is placed at the bottom of a vertical shaft incorporated within the preheater system. This burns about two-thirds of the fuel used in the plant, the other third being burned in the conventional manner in the kiln itself. The feed is about 90% calcined before entering the rotating kiln.

Clinker cooling air can be directed to the precalciner through the kiln (Pyroclon S) or through an auxiliary duct outside the kiln (Pyroclon R). Precalcining systems which bypass the clinker cooler air and feed directly to the precalcining chamber provide an opportunity to reduce the kiln size for equivalent production and minimize the energy cost of kiln gases that must be wasted for alkali control.

F. L. Smidth System. In Denmark, F. L. Smidth & Company has also been developing a flash calcining system. Most of the combustion of fuel and calcination of the raw material occur in a toroidal recirculation zone in the precalciner, where the raw feed particles, after being preheated, are calcined in suspension.

The kiln exit gases move through a conventional four stage suspension preheater. The preheated air from the clinker cooler goes through a secondary air duct which is parallel to the kiln to enter a second and separate suspension preheater equipped with a flash calciner. The preheated raw material from the discharge of the fourth stage of the suspension preheater system exits into the flash calciner and is precalcined along with the preheated raw material entering from the third stage of the flash calciner. In this manner, the two high temperature gas streams are separated until they have passed the induced draft fans. The distribution of the combustion air from the kiln and the secondary air from the clinker cooler can be regulated through the two separate fans.

Advantages of New Precalcining Systems. The flash calciner represents a substantive change in the de. ign of the rotary kiln, or cement reactor, more so than the suspens.... preheater rotary kiln. Almost 50% of the total fuel is burned in the flash calciner, and up to 90% of the total calcination of the raw material is accomplished before the raw material enters the kiln. The advantages of precalcining, or two stage combustion system, compared with conventional systems, appear to be as follows:

- Significantly increased rotary kiln capacity. When the kiln
feed is about 90% calcined, the production capacity of a given
kiln size will be two to two and a half times that obtained
from a standard four stage suspension preheater kiln without
additional firing of the kiln.

- Increased kiln size. Kiln systems of 9000 tons/day and possibly
higher outputs can be built with preheaters containing a first
stage combustion area for burning part of the fuel requirements.
The largest reported flash calciner system is at Chichibu,
Japan; it has a production capacity of 8500 tons per day of
cement clinker, and a heat consumption of less than 2.6×10^6
BTU/ton. This is a record, not only for the largest daily
productive capacity from a single kiln facility, but also for
the lowest reported heat consumption.

- Reduced fuel consumption. The heat losses through the rotary
kiln shell are less than those of a conventional rotary kiln.
The cement produced per square foot of kiln shell area is very
high. Fuel requirements are reduced by about 5% over the conventional suspension preheater system.

- Improved kiln operation. A more stable kiln operation will
result from uniformly precalcined raw mix and from higher kiln
speeds that permit more efficient heat transfer.

- Reduced nitrogen oxide emissions. Since half of the total fuel
burned in this system is burned in the flash calciner at a
temperature of 1500°F, the nitrogen oxide formed in the calciner will be considerably less than what is formed in the high
temperature (2700°F), free standing flame of the rotary kiln.
Also, the combustion gases stand in the burning zone for a
shorter period of time than in the conventional kiln.

- Reduced capital and operating costs. The fixed capital investment of a precalcining system should be slightly less than for
a suspension preheater. This is due to the fact that a stationery and smaller precalcining vessel can replace a section of
the rather expensive rotary kiln. Operating costs should be
slightly lower in a precalcining system, to the extent of a
smaller kiln requirement, better refractory life, and, possibly,
more stream days per year.

- Improved kiln availability. Flash calcining systems use conventional sized kilns which exhibit a refractory life considerably
in excess of the large rotary kilns required for equivalent
production capacities.

• More efficient alkali removal. Since only half of the total
fuel is burned in the rotary kiln and since the sintering zone
of the rotary kiln is the only place where the alkali compounds
are volatilized, the alkali compounds are more highly concentrated in a smaller quantity of gas, and bypassing of a smaller
quantity of gas is needed to eliminate more alkali.

Disadvantages of the Precalcining Systems. All of these systems
are attractive for very large kiln systems which can achieve maximum
fuel savings. In Europe and Japan, where small market areas consume
large quantities of cement, large kiln capacities are a tremendous advantage. In the U.S., cement plants with low density markets may
not be able to economically utilize the advantages of precalciner
systems. It has been reported that the precalcining systems with air
supplied through the kiln are not economically justifiable until production capacity is at least 2500 tons per day. The average kiln
capacity in the United States in 1974 was less than 1000 tons per
day (11).

FLUIDIZED BED PROCESS

An alternative to the rotary kiln is the fluidized bed process,
which was developed in this country and has been successfully demonstrated in a semi-commercial scale plant of 100 ton/year capacity.
This process utilizes a fluidized bed reactor, rather than a rotary
kiln, for the production of portland cement clinker. It differs from
other conventional processes only in the clinkering step.

In a fluidized bed reactor, the raw material is introduced through
the bottom of the bed of fluidized cement clinker particles. See Fig.
10-18. This bed of clinker particles is fluidized by hot combustion
gases produced by the introduction of preheated combustion air through
an air distribution grid. Any hydrocarbon fuel, such as natural gas,
oil, or coal, can be used. The bed operates at a temperature of
2400°F, and its extremely large heat transfer coefficients can quickly
heat the incoming raw material particles to clinkering temperature.

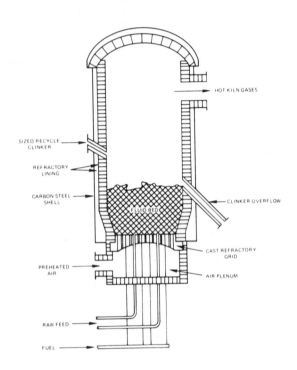

Figure 10-18. Kiln for Fluid Bed Cement Process. Source: Margiloff and Cascone, Paper presented at Rock Products Cement Industry Seminar, Chicago, December 8, 1975.

The distribution of the optimum particle size of the bed material is maintained by the continuous removal of bed material and by the reintroduction of crushed and screened clinker product which act as nuclei for continuing growth of new particles (12). Figure 10-19 illustrates the recycle of both the fine fraction of extracted bed material and the finer product obtained by crushing coarse particles removed from the bed to act as new "seed" particles. The values shown in this figure illustrate a simple material balance around the reactor.

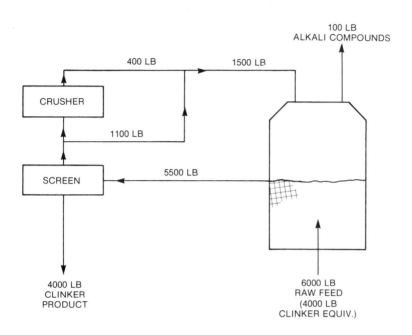

Figure 10-19. Material Balance Around the Reactor. Source:
Environmental Protection Agency, Environmental Considerations, Vol.
X, Cement Industry, p. 43.

The 4000 pounds of clinker generated from the raw feed stream are
deposited on the surface of the clinker which constitutes the fluidized
bed. The bed increases in total volume, raising the upper surface of
the bed. As the surfaces rises, the clinker particles spill through
the overflow outlets of the reactor. Only the largest diameter particles are considered finished product. They are separated from the
overflow stream of reactor material by a screening step. A total overflow of the reactor contents equal to 5500 pounds is screeded to remove
the 4000 pounds of clinker product. The 1500 pounds of finest material
are then returned to the reactor for further "growth."

Figure 10-20 shows the Scientific Design Company fluid bed cement
process. As is evident in the figure, the hot combustion gases leaving
the reactor at 2400°F pass through a heat exchanger which heats the

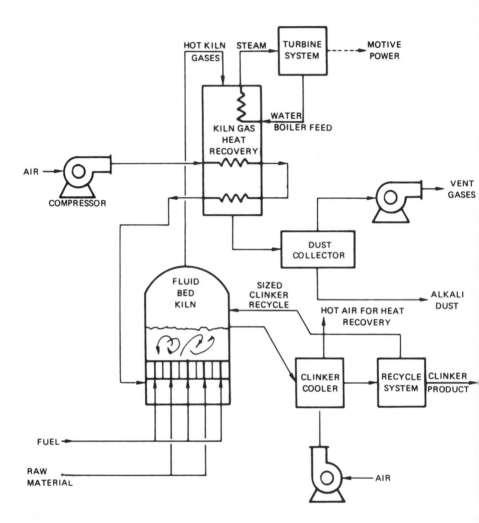

Figure 10-20. Scientific Design Fluid Bed Cement Process.
Source: Margiloff and Cascone, Paper presented at Rock Products
Cement Industry Seminar, Chicago, December 8, 1975.

incoming cold combustion air as well as water for the generation of
steam. The heat exchanger is divided into two sections. The first
section receives the hottest gas and is designed to exchange radiant
heat energy from the hot gas to the water contained in alloy tubes for
the generation of steam. After the combustion gases leaving the reactor have been cooled to a temperature sufficiently low, they pass

through a conventional preheating unit which initially heats the combustion air. The steam generated could either directly drive the air blowers required to supply the pressurized fluidizing air to the reactor or generate electrical energy. In fact, sufficient steam is available to provide electrical energy, not only for grinding all of the raw material required by the fluid bed reactor but also for driving the finish cement grinding mill.

Table 10-10 shows the comparative energy use of the conventional rotary kiln versus the fluidized bed. The fourth column of Table 10-10 indicates the excess heat from clinkering, which is recoverable for power generation. The last column of the table shows the net energy required, including energy recovery.

On a total energy basis, the process design studies conducted by Scientific Design Company indicate that the fluidized bed cement process with proper heat recovery requires significantly less total energy than either the conventional wet or dry long rotary kiln. Its energy requirements are close to the preheater equipped short rotary kiln, which is the most energy efficient cement clinkering process available to the cement industry today.

Advantages of the Fluid Bed Process

- Reduced capital requirements. Scientific Design Company envisions a significantly lower fixed capital investment for the fluidized bed process, as compared to capital investment for the long kiln. See Fig. 10-21.

- Higher yield. In a rotary kiln, the typical loss of potential clinker through the kiln dust is about 8% of the raw material. In the fluidized bed reactor, only about 3% of the raw material is lost. Therefore, less raw material is required to produce a ton of cement by the fluidized bed process than by the conventional rotary kiln process.

- Higher efficiency. This process is reported to be equivalent in overall thermal efficiency to the suspension preheater equipped rotary kiln, which uses about 4.2×10^6 BTU/ton cement as compared to about 6×10^6 BTU/ton used by the conventional long rotary kiln.

Table 10-10
Comparison of Typical Energy Requirements
for Fluidized Bed Process and Long Kiln
(BTU/ton Cement)

	Feed Preparation (quarry, crush, dry, mix feed)	Clinkering (burn, cool)	Finishing (grind, pack)	Available Energy Recovery (steam/power generation, dryer fuel savings)	Net Energy Required After Energy Recovery
S.D. Fluidized-Bed Kiln					
Electrical	490,000	—	760,000	(1,400,000)	(150,000)
Fuel	310,000	5,000,000	—	(180,000)	5,130,000
Total	800,000	5,000,000	760,000	(1,580,000)	4,980,000
Dry, Long Kiln					
Electrical	534,000	315,000	760,000	—	1,609,000
Fuel	336,000	4,600,000	—	(320,000)	4,616,000
Total	870,000	4,915,000	760,000	(320,000)	6,225,000

Source: Margiloff and Cascone, Paper presented at
Rock Products Cement Industry Seminar, Chicago,
December 8, 1975.

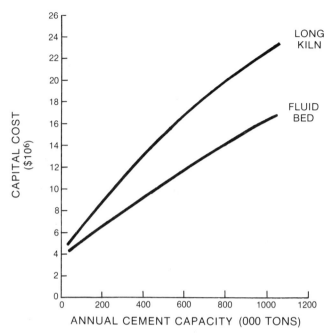

Figure 10-21. Capital Requirements. Source: Margiloff and Cascone, Paper presented at Rock Products Cement Industry Seminar, Chicago, December 8, 1975.

• Lower alkali concentrations. The fluidized bed will produce cement clinker of significantly lower alkali concentration than will any of the rotary kiln type clinkering processes, all other things, such as the chemical and physical characteristics of the raw material, being equal.

• Low emissions. Actual data obtained from the operation of a pilot scale fluidized bed cement reactor show a significantly lower NO_x concentration in its combustion gases when what is found in the gases from an equivalent rotary kiln process. The data also indicate that the combustion gases leaving the fluidized bed reactor are as low in SO_2 as those of a rotary kiln and also significantly lower in particulates.

• Fuel substitution. The fluidized bed process can use coal and waste products as fuel.

Implications

It appears that the U.S. cement industry is seriously considering the fluidized bed cement process as an alternative, at least for converting waste kiln dust from rotary kilns into marketable cement. A major barrier to the wholesale adoption of the fluidized bed process is the fact that the technology which constitutes this alternative is foreign to the cement industry. However, the recent engineering design studies and Scientific Design's commercial offer of the process are factors which might induce the cement industry to accept this new technology.

CRUSHING AND GRINDING

Most of the electricity used in cement plants is for grinding the clinker. The energy used for size reduction of raw material in the wet process is slightly less than that used in the dry system. Despite the fact that about 1.6 tons of raw material are processed for every ton of clinker, about 60% of the electrical energy used in the cement-making process goes for clinker grinding. This is due to the hardness of the clinker and to the need for grinding it to a very fine particle size in the final product. The grinding process is very inefficient, as over 90% of the energy is converted into heat, while under 10% is actually used in the production of additional surface area (13). Ther should be an incentive for the development of more efficient grinding systems and for the incorporation of air classifiers and drying systems, using kiln waste gases.

Most of the cement mills in the U.S. today are closed circuit bal mills for grinding the raw materials. The air circulated through thos ball mills and then through the air separator or classifier is usually heated by an oil fired furnace. This is done to dry the materials for proper grinding and classifying. However, a ball mill is not as effective as a roller mill in gas handling capability and drying ability A roller mill can team up with a flash calciner especially well. Afte the kiln off gases have heated raw materials entering the kiln, the waste gases can be further utilized in a roller mill for removing moisture from other raw materials. Energy savings with a roller mill are between 25-34% over today's ball mill (14).

OXYGEN ENRICHMENT

A way to increase cement capacity in an existing rotary kiln is by oxygen enrichment of the combustion air. If heat losses through the exterior surface of the kiln remain constant and the clinker output is increased, oxygen enrichment should result in decreased unit fuel consumption.

Tests on a number of kilns, with oxygen rates varying from about 150-1,060 SCF per ton of clinker, gave the results shown in Table 10-11. The corresponding energy for producing the oxygen range from about 40-240 BTU per ton of clinker, which indicates that a substantial net energy savings may be achievable.

Oxygen enrichment will permit the use of special fuels, such as high sulfur petroleum coke, which may have an impact on the quantity of sulfur in the kiln gases and in the cement product. In addition, oxygen enrichment may permit the maintenance of proper burning characteristics of the kiln flame when using large quantities of recovered dust (15).

FUEL SUBSTITUTION

In the past, the greater part of cement manufacture was accomplished with coal as the major fuel source. In the past quarter century, the use of coal within cement plants has been decreasing steadily, and the use of fossil fuels, such as oil and natural gas, has been increasing. With the growing problems of oil and gas availability, a major occurrence in cement plants is the rapid reconversion from gas and oil firing of rotary kilns to direct coal firing.

Table 10-11

Energy Efficiency Gains Through Oxygen Enrichment

Wet or Dry	Fuel- Coal, Oil or Gas	Fuel used in kiln, MMBTU/ton of clinker		Fuel Use Reduction %
		Without oxygen	With oxygen	
W	C	6380	5320	17
W	O	10100	9570	5
W	C	7980	6920	13
W	C	7980	7450	7
W	O	5850	4790	18
W	C	6380	5320	17
W	C	6380	5320	17
W	G	6380	5590	12
D	C	5850	5000	15
D	G	4260	4260	0

Source : R. D. Stirling, J. C. Blessing, and S. L. Fredericks,
"Oxy-fuel Burners Streamline Cement Production,"
Rock Products, 46 (November 1973), 44.

Since the kiln is by far the largest consumer of energy in the manufacturing process, a reconversion to coal in the cement industry can have a major impact on scarce oil and gas savings. All rotary kilns are suitable for coal firing, and only changes in the ancillary facilities, such as coal storage handling and grinding, must be added to a plant which is switching from gas or oil to coal firing. This, however, may require a major capital investment.

A large fraction of the coal ash from a coal fired cement kiln is combined with the clinkering raw materials. Therefore, an adjustment must be made in the proportions of the various raw material components going to the kiln raw feed mixture to account for the additional iron, silica, and alumina values coming from the coal ash. Most of the sulfur values contained in the coal are absorbed by the lime in the rotar kiln and become part of the cement produced.

The two main sources of sulfur in cement are the clinker and the gypsum that are added to the clinker when the finished cement is ground. The gypsum is added to control the setting time of the cement. As the sulfur content of the clinker increases, the quantity of gypsum that can be added to control the physical characteristics of the cement must decrease. Thus, excessive sulfur contained in the clinker has a detrimental effect upon the cement, because it limits the amount of gypsum that can be added to produce the desired characteristics of the finished cement.

COLD PROCESSING

Cold processing represents a radical departure from today's cement manufacture in that no high temperature processing is involved. Its inventors are C. J. Schifferele and J. J. Coney (16). In their methodology, quicklime and an argillaceous component, such as shale or clay, are chemically combined by grinding them in a conventional ball mill.

Cold processing has been demonstrated in a pilot plant and is reported to have produced a hydraulic cement with characteristics and properties which compare favorably with portland cement. The obvious advantage of the technique is the significantly reduced fuel energy requirement, since the only thermal processing necessary is for calcining limestone by conventional means. Electrical energy required is used for grinding the quicklime with the other components. This process, if commercialized, represents an entirely new direction in conventional cement-making technology.

ENERGY CONSERVATION POTENTIAL

The cement industry has been taking steps to become more energy efficient, in order to attain higher productivity and to become more cost competitive in the marketplace. Over the past 25 years, the industry has reduced the average per unit energy consumption from about 8.8 x 10^6 BTU per ton to about 7.3 x 10^6 BTU per ton. However, there is still much that can be done. From an availability analysis point of view, the theoretical minimum fuel consumption is 0.8 x 10^6 BTU per ton. The current most efficient plant, the Chichibu plant, is operating at a kiln consumption of 2.6 x 10^6 BTU per ton, which is remarkable.

The following presents a summary of specific actions that can be taken to achieve energy conservation gains.

Replacing Obsolete Plants

Older plants generally are significantly less energy efficient than newer plants. Figure 10-22 shows the average total BTUs consumed per ton of cement produced for plants built before 1935 (40 or more years old) compared with plants built since 1965 (ten years old or less). In 1974, older plants consumed more than 8.4 x 10^6 BTUs per ton, whereas the newer plants consumed about 7.1 x 10^6 BTUs per ton using the wet process--a difference of almost 21%.

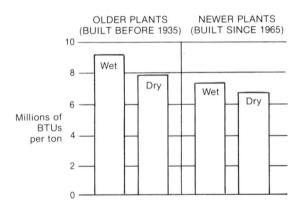

Figure 10-22. Energy Efficiency by Plant Age. Source: McCord, Energy Conservation Trends in the Cement Industry, p. 27.

When replacing older plants, consideration should be given to
the installation of a suspension preheater, flash calciner system over
the conventional long kiln. Table 10-12 provides a comparative analysis of the energy savings that can be accrued with the various options.
A savings of 25% over the long kiln can be achieved.

Switching to Other Hydraulic Cements

There are several types of hydraulic cements that require less
energy in their manufacture than does portland. These cements are used

Table 10-12
Summary of Results of Process Options in the
Portland Cement Industry
(Basis: 1350 Tons Cement Per Day)
Base Line: Long Kiln (Oil) – Energy Consumption from
3.4 to 6 × 10^6 BTU/ton

	Long Kiln (Oil)	Suspension Preheater & Flash Calciner	Fluidized Bed	Long Kiln (Coal)
PRODUCTION FACILITY				
Fixed Capital Investment ($$10^6$)	42.	40.	38.	45.
Production Cost ($/ton)	47.81	43.71	44.30	45.56
Energy Requirements (10^6 Btu/ton)	5.6	4.2	5.0	5.6
ENVIRONMENTAL CONTROL FACILITIES				
Fixed Capital Investment ($$10^6$)	1.6	1.2	1.9	2.0
Operating Cost ($/ton)	1.97	1.40	2.10	2.27
Energy Requirements (10^6 Btu/ton)	.069	.047	0.102	.069
PRODUCTION PLUS ENVIRONMENTAL CONTROL FACILITIES				
Fixed Capital Investment ($$10^6$)	43.6	41.2	39.9	47.0
Operating Cost ($/ton)	49.78	45.11	46.40	47.83
Energy Requirements (10^6 Btu/ton)	5.7	4.2	5.1	5.7

Source: Environmental Protection Agency, *Environmental
Considerations*, Vol. X, *Cement Industry*, p. 8.

extensively in both South America and Europe and have physical properties which are desirable and quite competitive with portland cement.
These cements include:

- Pozzolanic cements. Pozzolans react with lime in an aqueous
  solution to produce a material with hydraulic cementitious
  properties. It can be used as an additive to portland cement.

- Slag cement. Blast furnace slag from the iron and steel
  industry can be used as an additive to portland cement,
  producing the same result as pozzolanic cement.

The primary energy savings in the manufacture and use of these
cements occur only when blast furnace slag or other pozzolanic
materials enter the cement manufacturing process at the final grinding step. Doing so, the large energy consuming processing steps of
clinkering in the rotary kiln and raw material grinding are avoided.
A shift toward the increased use of such hydraulic cements would have
a significant energy conserving impact.

Converting from Wet to Dry Process

On the average, wet process plants are less energy efficient than
dry process plants. Figure 10-6 provides a comparison of the energy
consumption of the two processes. Of the 1975 U.S. production capacity, about 58% is by the wet process, and 42% is by the dry process.
In 1974, wet process plants consumed an average of 8.0×10^6 BTU per
ton of finished cement, compared with an average of 6.8×10^6 BTU
per ton consumed by dry process plants--a difference of 15%.

With rising fuel prices, the industry should use the dry process wherever possible, bearing in mind raw material limitations
(17). Conversion of wet process plants to the dry process is possible for specific plants, although the economics of such a change,
with due allowance for the kiln down time required, may sometimes
be questionable. A number of these conversions have been performed
in Europe, with significant fuel savings being achieved. The following example shows this (18):

	Production Ton/Day	Kiln Fuel Consumption MBTU/ton
Before (wet)	360	6300
After (dry)	750	3530

Retrofitting of Existing Wet Process Plants

Most of the plants in the cement industry are old and fairly inefficient, but it would be unreasonable to expect that new plants be built where existing facilities are providing a reasonable rate of return. There have recently been some modifications of older wet process cement plants to dry process through the conversion of existing long kiln to a suspension preheater. A. D. Little presents a logical plan for converting an old wet process plant with several relatively small kilns:

- Convert the old wet-process rotary kiln to a four-stage suspension preheater. This would be done by cutting the kiln approximately in half and removing the feed end. The suspension preheater tower would be constructed adjacent to the longitudinal axis of the rotary kiln, allowing for the construction of a second preheater tower adjacent to the first, and on the other side of the kiln axis, thereby providing symmetry in plan view. The purpose of this offset preheater tower is for the addition of a second preheater at a future stage of capacity expansion. The major impact of this first step conversion to a four-stage suspension preheater kiln is the significant reduction in fuel energy. Where the older and relatively small wet-process kiln may have been operating with a fuel consumption on the order of 6×10^6 BTU/ton, the new four-stage preheater kiln should have a fuel requirement of approximately 3×10^6 BTU/ton; this reduction of fuel consumption by 50% provides high motivation for such a conversion. In addition, there will be a modest increase, 20-30%, in the production capacity of the kiln.

- Add a flash-calcining vessel. In this second step, the flash calciner is added between the rotary kiln and the four-stage suspension preheater. The design of the suspension preheater should allow room for this flash-calcining vessel. The flash calciner would increase the production capacity of the total facility by about 25%, and should again slightly decrease the quantity of fuel required for clinker production. This increase in capacity results from the combustion of part of the required fuel in the flash calcining vessel, and not within the rotary kiln; this provides almost totally calcined feed to the rotary kiln. The rotary kiln would probably be operated at a higher speed to maintain proper bed depth and residence time.

- Add a second calciner to the kiln. This third step employs the the construction of the second suspension preheater and flash calciner in the space initially provided for this tower. This flash calciner and suspension preheater would be identical with the existing one, and would serve the same rotary kiln. The combustion gas leaving the rotary kiln would be divided into two streams and would be fed to the two flash calciner units operating in parallel. Half of the total raw feed to the rotary kiln will go to each flash calciner. The major effect of this third step would be essentially a doubling in kiln capacity, or a 100% increase in the capacity of the kiln with only a suspension preheater installed (19).

Increasing Kiln Speed

A technique for improving fuel efficiency may be increasing the rotating speed of the kiln to, perhaps, 2.5 RPM from around 1 RPM. To maintain the correct residence time for the raw materials, the kiln inclination must be reduced. Modern suspension preheater kilns are designed for 2.5-3% inclination, compared with the 4-5% of older kilns (20).

Improved Insulation

The heat loss from the kiln through the walls is not insignificant, representing about 10% of the kiln heat consumption (21). Improved insulation materials and standards can obviously produce considerable fuel savings.

System Implications

Table 10-13 indicates the potential for energy conservation in the cement industry. The technology for efficient fuel management is readily available, in most cases, from overseas equipment manufacturers. Our reluctance to use it comes from the relatively high capital investment required and from obstacles in terms of cement manufacture and construction technology. One such obstacle is the requirement for low alkali contents in portland cements.

Table 10-13

Summary of Energy Conservation Potential

Conservation Practice or Technique	Potential Savings* 10^6 BTU/ton
Substitute coal for oil and natural gas	Fuel Substitution
Install suspension pre-heaters	2.0
Install precalciner system	2.0
Conversion from wet to dry process	1.4
Use fluidized bed process	0.8
Oxygen enrichment	0.4
Use waste heat for drying incoming wet materials	0.3
Use of a roller mill instead of a ball mill	0.2

*These savings are not additive and are based on efficiency improvements over 1974 conventional practices

If U.S. industry were to adopt the most modern foreign practice, average input to the kiln would be reduced from 6.6 x 10^6 BTU per ton to 3.6 x 10^6 BTU per ton of cement (22). Use of these techniques could result in the reduction of fuel needs by at least 30% and will increase productivity.

NOTES AND REFERENCES

1. Environmental Protection Agency, Environmental Considerations of Selected Energy Conserving Manufacturing Process Options, Vol. X, Cement Industry Report (Cincinnati, Ohio: Environmental Protection Agency, December 1976), p. 13.

2. George Troxell, Harmer Davis, and Joe Kelly, Composition and perties of Concrete, 2nd ed. (New York: McGraw Hill, 1968), p. 17.

3. Concrete Information (Skokie, Illinois: Portland Cement Association, 1971), p. 5.

4. Ibid., p. 5.

5. Environmental Protection Agency, Environmental Considerations, Vol. X, Cement, p. 88.

6. Troxell, et al., Composition and Properties of Concrete, p. 19.

7. Ibid., p. 19.

8. Environmental Protection Agency, Environmental Considerations, Vol. X, Cement, p. 25.

9. George McCord, Energy Consumption Trends in the Cement Industry, Society of Mining Engineers of AIME, Preprint No. 75-H-308, 1977. See also Environmental Protection Agency, Environmental Considerations, Vol. X, Cement.

10. Environmental Protection Agency, Environmental Considerations, Vol. X, Cement, p. 114.

11. McCord, Energy Conservation Trends, p. 20.

12. A.M. Sadler, Paper presented at AIChE meeting, New York, November 30, 1967.

13. The Making of Portland Cement (Skokie, Illinois: Portland Cement Association, 1964). See also V.A. Kaiser, "Computer Control in the Cement Industry," Proceedings of the IEEE, 58 (January 1970), p. 76.

14. Environmental Protection Agency, Environmental Considerations, Vol. X, Cement, p. 83.

15. R.D. Stirling, J.C. Blessing, and S.L. Fredericks, "Oxy-fuel Burners Streamline Cement Production," Rock Products, 76 (November 1973), p. 44.

16. Ibid., p. 85.

17. Bureau of Mines, Minerals Yearbook, 1971 (Washington, D.C.: Government Printing Office, 1973), pp. 278-79.

18. Gordian Associates, The Data Base: The Potential for Energy Conservation in Nine Selected Industries, Vol. 3, Cement, (Washington, D.C.: Government Printing Office, 1975), p. 108.

19. Environmental Protection Agency, Environmental Considerations, Vol. X, Cement, p. 22-25.

20. P.K. Mehta, "Trends in Technology of Cement Manufacture," Rock Products, 73 (March 1970), p. 85.

21. M. Kunnecke and B. Piscaer, "Choosing Insulation for Rotary Kilns," Rock Products, 76 (May 1973), p. 138.

22. E.P. Gyftopoulos, L.J. Lazaridis, T.F. Widmer, Potential Fuel Effectiveness in Industry (Cambridge, Massachusetts: Ballinger Publishing Company, 1974), p. 86.

Chapter 11

GLASS INDUSTRY

The manufacture of glass and glass products is a widely diversified industry in the United States. The establishments classified
in SIC 3211 are those that are primarily engaged in manufacturing flat
glass, including sheet, plate and float, and laminated and tempered
automobile glass. SIC 3221 comprises establishments primarily engaged
in manufacturing glass containers for commercial packing and bottling
and for home canning. The major types of glass containers are food,
beverage (including beer, wine, and liquor), and pharmaceutical glass.
Pressed and blown glass, SIC 3229, include tableware, TV sets, lamp
enclosures, tubing, etc.

Table 11-1 shows the relative size of the basic glass industries
as a percent of total shipments. As is evident, the glass container
industry is now about 60% larger, in terms of output, than the pressed
and blown glass industry, and more than twice as large as the flat
glass industry.

Since the end of World War II, growth of the basic glass industries has been less rapid than that of the economy as a whole. Output
by the three basic glass industries combined increased 116% between
1947 and 1971, an average of 3.3% a year. During the same period,
real gross national product expanded by 141%, or at an average annual
rate of 3.7% (1).

397

Table 11-1
Relative Sizes of the Basic Glass Industries
(As a per cent of total shipments)

Year	Flat Glass (SIC 3211)	Glass Containers (SIC 3221)	Pressed and Blown Glass n.e.c. (SIC 3229)	Total Basic Glass Industries
1947 ...	22.4%	50.8%	26.8%	100.0%
1958 ...	22.0	48.5	29.5	100.0
1971 ...	21.7	48.0	30.2	100.0

Source: From ENERGY CONSUMPTION IN MANUFACTURING. Copyright 1974. The Ford Foundation. Reprinted with permission of Ballinger Publishing Company.

Table 11-2 gives a summary of the product mix in the glass industry. The companies in the glass industry generally participate in one or, sometimes, two categories of glass production, but rarely does a single company have broad participation in more than two categories. In the flat glass category, the industry is concentrated in such a manner that seven companies dominate this market. The concentration is lower in the glass container industry, with some 130 plants being operated by approximately 30 companies. The 8 largest of the 30 firms produce 75-80% of the product. There is a large range of plant capacity, and present plants are tending toward larger sizes.

Soda-lime silica glass accounts for about 90% of the glass melted in the U.S. All flat glass, container glass, incandescent lamps, and fluorescent lamp envelopes and tubing and a large fraction of the machine pressed and blown ware are basically soda-lime glass, with only a small difference in composition among the different product areas (2). Borosilicate glass makes up approximately 2-3% of the glass melted and is used principally in heat resistant laboratory ware and consumer ovenware. Other glass compositions are lead glass for electrical applications, optical glasses, and handformed artware, which altogether represent 0.5% of total glass production. Opal glass, a

Table 11-2

Product Mix of Individual Basic Glass Industries

	Per Cent of Total
Flat Glass (SIC 3211) 1967	
Sheet (window) glass	22%
Plate & float glass	30
Laminated glass	13
Other & n.s.k.	20
Secondary products	15
Glass containers (SIC 3221) 1971	
Food	35
Medicinal, toiletries, cosmetics	15
Household & industrial	2
Beverages (including beer)	33
Liquor & wine	13
Secondary products	2
Pressed and blown glass, n.e.c. (SIC 3229) 1967	
Table, kitchen, art, & novelty	30
Lighting and electronic	37
Textile fiber	14
Other & n.s.k.	15
Secondary products	4

	Total (millions of dollars)
Flat glass, 1967	$ 611.3
Glass containers, 1971	1,352.4
Pressed and blown glass, n.e.c., 1967	886.2

n.s.k. — not specified by kind.
n.e.c. — not elsewhere classified.

Source: Bureau of the Census, *Census of Manufacturers,
         1967;* "Glass Containers," Summary for 1971,
         Current Industrial Reports.

modified soda-lime silica glass, accounts for an additional 1% of the
glass melted.

In 1972, the U.S. industry shipments were valued at more than
$4 billion and were from approximately 400 plants employing an esti-
mated 15,000 workers. Table 11-3 provides typical plant characteristics for the glass industry. As can be seen, the bulk of the glass
production in the United States is concentrated in the East North
Central, Middle Atlantic, and Pacific regions. The major glass producing states are Illinois, Ohio, Pennsylvania, New York, West Virginia, New Jersey, and California.

The markets for the industry are very diverse. Flat glass shipments are heavily dependent on the automotive and construction industries, and pressed and blown ware are closely tied to consumer spending
for television, lighting, and household goods. The container glass
market is highly influenced by the beverage market, where glass com-
petes with aluminum and steel.

Table 11-3
Typical Plant Characteristics
(1975)

Industry Segment	Location	Age (yrs)	Capacity	Glass	Sales Revenue/Plant ($10^6)
Flat Glass					
Float	Pa. - Ohio	10	400,000 tpy	Soda-lime	72.
Sheet	Ohio	30 - 35	165,000 tpy	Soda-lime	29.
Laminated	Mich.	25	$18.0 \times 10^6 \text{ ft}^2/\text{yr}$	Soda-lime	26.
Tempered	Mich. - Ohio	30	$37.5 \times 10^6 \text{ ft}^2/\text{yr}$	Soda-lime	18.7
Glass Container	East North Central	15 - 20	75,000 tpy	Soda-lime	14.
Pressed & Blown					
Tubing	Mid-Atlantic	30	30,000 tpy	Borosilicate	12.
T.V.	Ind. - Pa. - NY	10	85,000 tpy		35.
Incandescent lamp	Ohio - Pa.	25	65,000 tpy	Soda-lime	30.
Machine Ware	Ohio - Pa.	35	55,000 tpy	Soda-lime Borosilicate Lead	
Fiber Glass					
Textile	S.C. - N.C. - Tenn.	15	$45 \times 10^6 \text{ lb/yr}$	Low alkali Borosilicate	15.
Wool	Ohio - Kan. - N.J.	20	$50 \times 10^6 \text{ lb/yr}$	Borosilicate	15.

Source: Arthur D. Little, Inc., estimates.

MANUFACTURING PROCESS DESCRIPTION

Glass is a mixture whose basic ingredients are silica (from sand), calcium monoxide (from limestone), and sodium monoxide (from soda ash). In addition to the major raw materials, small quantities of feldspar and salt cake are used to produce soda-lime glass. Feldspar provides a source of metallic oxide and retards devitrification of the glass, while salt cake acts as a "sponge" in furnaces by absorbing "scum" (3).

Some types of glass use other ingredients, such as oxides of aluminum, lead, and boron. Table 11-4 provides the composition of commercial glasses. A material always included in the manufacture of glass is cullet, composed of crushed glass from imperfect production, trim, and other in-house waste glass. Cullet assists in the melting process because it liquifies at a lower temperature than do the virgin materials, providing a molten bath which helps transfer heat to the other materials. To assure themselves of its composition and purity, manufacturers generally prefer using internally generated cullet over purchased cullet.

The production sequence for glass making involves the following principal operations:

- Mining of sand

- Mining of limestone

- Production of soda ash

- Raw materials handling and batch preparation

- Melting and fining the ingredients in the furnace

- Forming the product by such means as drawing, blowing, molding, pressing, floating, and casting

- Annealing, or reheating and slowly cooling the products to relieve stresses caused by the unavoidable cooling that takes place during the forming process

- Inspection and packing

Table 11-4

Composition of Commercial Glasses by Weight Per Cent

| Component | Containers | Soda-Lime Glass | | Borosilicate Glass | Lead Glass |
		Plate and Window Glass	Tableware	Specialty Glassware	Specialty Glassware
SiO_2	70-74	71-74	71-74	70-82	35-70
Al_2O_3	1.5-2.5	1-2	0.5-2	2-7.5	0.5-2.0
B_2O_3	0	0	0	9-14	0
Na_2O	}13-16	}12-15	}13-15	}3-8	4-8
K_2O					5-10
CaO	}10-14	}8-12	5.5-7.5	}0.1-1.2	0
MgO			4.0-6.5		
BaO	0	0	0	0-2.5	0
PbO	0	0	0	0	12-60

Source: Source Assessment Document No. 3, Glass Manufacturing Plants, Monsanto Research Corporation,
for EPA, Contract No. 68-02-1320, November 1974.

Figure 11-1 provides a schematic diagram of the glass manufacture process. Figure 11-2 is a flow diagram of glass container manufacture.

Mining of Sand

Most of the sand mined in the U.S. is surface mined and is used
in the construction industry (4). This sand is very coarse and is used
in paving and in buildings. The glass industry requires a finer
material known as "glass sand," some of which is imported from Australia. Since silica is found naturally pure, no beneficiation is required.

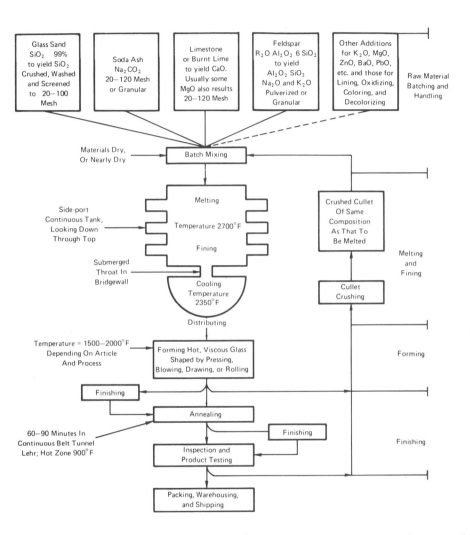

Figure 11-1.   Glass Manufacturing Process.   Source: Environmental Protection Agency, Environmental Considerations, Vol. XI, Glass, p. 83.

Mining of Limestone

Most crushed and broken limestone is mined in surface quarries, with underground mining accounting for only 4% of the total crushed stone (5). Following blasting from the rock face, the broken stone is crushed and screened.

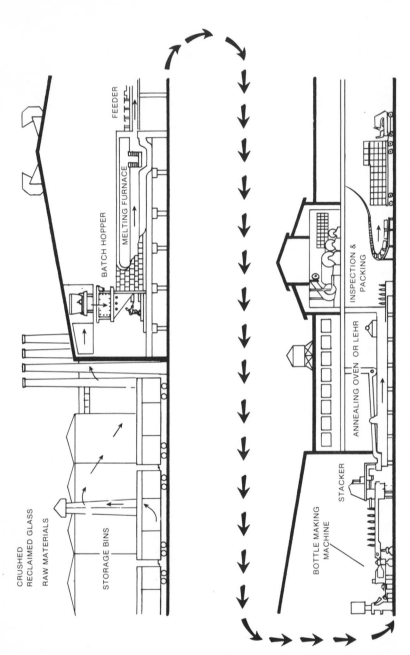

Figure 11-2. Flow Diagram of the Manufacture of Glass Containers. Source: Glass Container Manufacturers Institute.

Production of Soda Ash

Soda ash is obtained from both synthetic and natural sources.
Figure 11-3 shows the historical trends in producing soda ash, half of
which has historically been consumed by the glass industry (6). The
figure indicates that most soda ash has been synthetically produced,
but in recent years there has been a sharp increase in natural soda
ash production. The decline of the synthetic soda ash industry is primarily due to poor economics and to pollution problems (7).

Soda ash is synthesized in the Solvay process, which uses limestone, rock salt, and coke. A purified salt solution is contacted

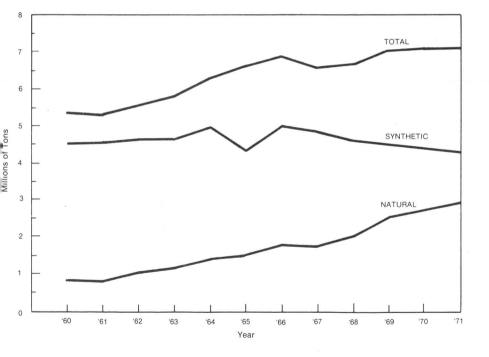

*Figure 11-3. Trends in Production of Soda Ash. Source: Gordian
Associates, The Data Base, Vol. 7, Glass (Washington, D.C.: Government Printing Office, 1975), p. 120.*

with ammonia in an internally cooled absorption tower. The brine is
then fed into carbonating towers, where it is contacted with gaseous
carbon dioxide to form a sodium bicarbonate slurry. The slurry is
vacuum filtered, and the recovered solid is heated in a rotary calciner
to expel carbon dioxide and water vapor. The overall process is the
combination of calcium carbonate (soda ash) and waste product sodium
chloride. Ammonia acts as a catalyst (8).

The natural soda ash used for glass manufacture occurs as trona
ore. Trona consists of sodium sesquicarbonate, a combination of water
with sodium carbonate and bicarbonate. Trona is mined using the room
and pillar method. At the surface, the ore is dissolved in water, and
insoluble impurities are removed by clarification and filtration. The
sesquicarbonate is recovered by vacuum crystallization for subsequent
calcining in the production of sodium carbonate (soda ash). A small
portion of natural soda ash is produced from complex brines by processes involving carbonation, evaporation, and cooling (9).

Raw Materials Handling and Batch Preparation

In the batch preparation process the mixture of sand, soda ash,
limestone, feldspar, saltcake, and cullet is charged to the glass
melting furnace. The batch preparation process is not a significant
energy consuming step in the glass making process.

Melting and Fining

Figure 11-4 shows a melting operation in a continuous glass tank.
Usually, these tanks are rectangular, about 60 feet long and 40 feet
wide, and are divided into two compartments, a large melting compartment and a smaller cooling or refining compartment. A crown above
the tank walls provide space for combustion. The molten batch is
heated to about $2800^{\circ}F$ by directing gas flames across its surface,
melting the cullet first and subsequently dissolving the other materials. High flame temperatures are achieved by preheating the combustion air.

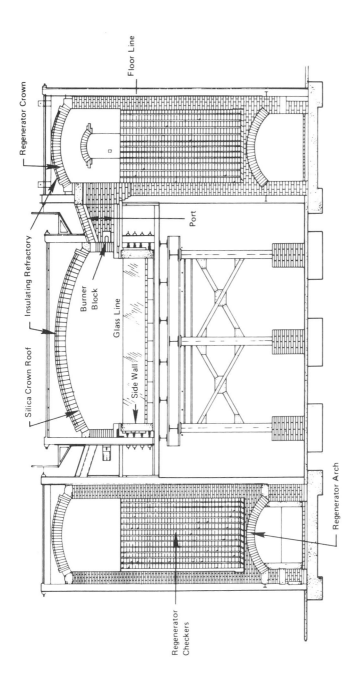

Figure 11-4. Typical Side Port Regenerative Furnace. Source: Environmental Protection Agency, Environmental Considerations, Vol. XI, Glass, p. 20.

Considerable energy savings are effected by using "regenerative furnaces" which operate in two cycles. At opposite sides of the furnace are chambers known as "checkers." The combustion gases, after passing over the molten batch in the furnace, continue out through one set of checkers, heating them up. At regular intervals, the flow is reversed, and the cold fuel air mixture passes through the heated checkers, then out the checkers at the other side. The cycle is repeated regularly.

After the charge melts, time is allowed for bubbles to rise and leave, or dissolve in the glass, a phase known as "fining." Melting of the raw materials to form a viscous glass melt consumes approximately 70-75% of the total energy used in glass production. The refining step, which homogenizes the melt accounts for about 5-10%.

Efficiencies of furnaces vary widely, depending on their basic design and age, the type of glass being melted, and the end use of the product. Flat glass, for example, cannot have any noticeable amounts of bubbles, unreacted raw materials, or other substances. The addition of color affects the absorptivity of the mixture, affecting the rate at which it melts.

Furnaces are also subject to considerable stress. Because molten glass is highly corrosive, the insides of furnaces and tanks must be lined with materials that retard erosion. These refractories are made of composite clays, which may contain alumina, silica, or zircon. Gradual erosion does occur, however, and small fragments of tank lining occasionally break off into the molten glass. Eventually, the furnace lining becomes deeply scored and has to be renewed.

Forming the Product and Annealing

As the molten glass continuously flows from the furnace, it drops into a feeder which forms the product. In the case of glass container manufacture, as molten glass flows from the furnace, it drops through a "gob feeder," a trough with an orifice at the lower end. The glass drops through the orifice and is cut into pieces (gobs) of exactly the desired size by mechanical shears. The gob passes into a mold in which the bottle is shaped with the aid of air injection.

After shaping, the bottle is annealed (10). This occurs in two steps: (1) holding the glass above its critical temperature long enough to reduce internal strains, and (2) cooling to room temperature slowly enough to hold the strains below a predetermined maximum. Annealing takes place in a specially designed heater chamber (lehr) where the rate of cooling can be carefully controlled. Following the annealing, the bottles undergo finishing operations, such as cleaning, polishing, and coating.

ENERGY CONSUMPTION IN THE INDUSTRY

Aggregate energy usage in the glass industry has been increasing as production has increased. See Fig. 11-5. However, a review of the historical data for the glass industry show a steady decrease of the energy used per unit of output in glass manufacture. See Fig. 11-6. Based on the historical perspectives, it appears reasonable that further declines in energy use per dollar of output will continue, but at different rates. Flat glass seems to have the highest potential reduction in energy use, with the glass container industry next. Figure 11-6 indicates the regression lines for projecting energy use per unit of output.

The projections indicate a decrease in per unit consumption in flat glass production from 84,900 BTU per dollar of output in 1971 to 75,000 BTU in 1980. Energy consumption by glass container manufacturers is projected to decline from 92,600 BTU per dollar of output to 83,500 BTU, and pressed and blown glass energy use is projected to drop from 73,200-69,000 BTU. As in most industries that are heavy energy users, a continuation of the recent increase in real prices of energy and the threat of supply difficulties will provide cost incentives to lower energy use per unit of output.

A number of technological developments have resulted in reduced energy savings, including economies from the use of increasingly larger furnaces, growing reuse of waste heat, and the introduction of an auxiliary heating unit inside the body of the molten glass in the furnace.

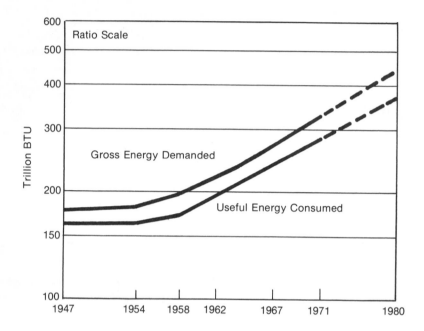

Figure 11-5. Energy Used by the Glass Industry. Source: From ENERGY CONSUMPTION IN MANUFACTURING, Copyright 1974, The Ford Foundation. Reprinted with permission of Ballinger Publishing Company.

Increases in energy consumption have been caused by pollution control equipment installations and increased mechanization.

There has been a sharp drop in the use of coal and oil by the industry over the past quarter century, which was offset by a large increase in the use of natural gas and a smaller increase in electricity use. Energy use by fuel type is provided in Table 11-5. As can be seen, natural gas has been used as the primary source of energy for many years because it was a low cost, easily available, clean, and convenient fuel. In 1973, it accounted for about 70% of the total energy used in a glass plant. Electricity accounted for about 20%, and the remainder was largely fuel oil. The principal use of natural gas is in the melting function, which consumes 70-75% of the total energy used in a typical glass plant. In addition, natural gas usage for annealing constitutes up to 10% of the total plant energy consumption. Natural gas is also used for space conditioning.

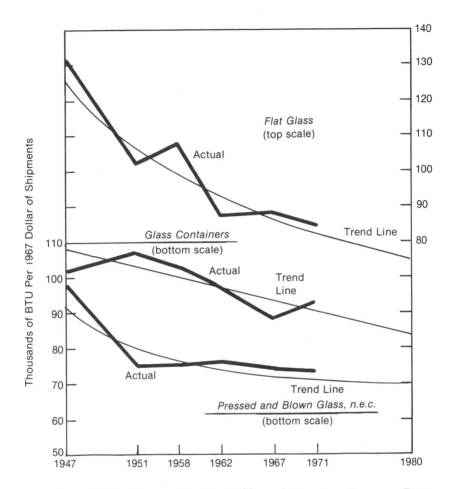

Figure 11-6. Energy Use Per Dollar of Output. Source: From ENERGY CONSUMPTION IN MANUFACTURING, Copyright 1974, The Ford Foundation. Reprinted with permission of Ballinger Publishing Company.

Over recent years, some shifts in the patterns of use of the various forms of energy have occurred. The relative fuel oil consumption has increased, and the percent of natural gas usage has declined, reflecting the supply problems facing the industry. Since natural gas has been available, in some cases, only on an intermittent basis, standby oil and propane have been necessary. Many plants now have procedures for using oil, particularly during the winter months in the

Table 11-5

Projections of Aggregate Energy Use
Basic Glass Industries Combined

	Coal	Oil	Gas	Electricity	Total*
		Trillion BTU			
Useful energy consumed					
1967...................	10.7	10.9	184.8	15.4	238.4
1971...................	3.8	10.8	229.3	20.1	274.4
1980...................	7.6	66.7	259.5	38.0	381.4
Gross energy demanded					
1967...................	10.7	10.9	184.8	47.1	270.1
1971...................	3.8	10.8	229.3	61.9	316.1
1980...................	7.6	66.7	259.5	117.0	460.7
Per cent distribution of energy consumption (useful) by source					
1967...................	4.5%	4.6%	77.7%	6.5%	100.0%
1971...................	1.4	3.9	83.6	7.3	100.0
1980...................	2.0	17.5	68.0	10.0	100.0

Source: From ENERGY CONSUMPTION IN MANUFACTURING. Copyright
1974. The Ford Foundation. Reprinted with permission of Ballinger
Publishing Company.

New England and Atlantic states. Table 11-6 provides a summary of the
relative use of the different energy forms in the various glass industry segments. As indicated, the pressed and blown glass segment has
the largest energy usage per ton of product.

The energy consumed in major process steps varies considerably
from plant to plant in the making of identical products. Table 11-7
provides a summary of the average energy usage in these process steps.
Because the melting of glass is the single greatest consumer of energy,
the most significant changes in energy efficiency must come in this
area. Many factors affect the energy use in melting glass, e.g.,
type of glass, type and size of furnace, method of operation, type of
fuel, pull rate, furnace age, and specific furnace construction. Furnaces are designed and constructed to burn a single fuel--usually
natural gas--and the use of an alternate fuel usually results in a
loss in efficiency.

Table 11-6
Relative Use of Different Energy Forms
(1973)

Segment	Energy Used % (10^9 Btu)	Type of Fuel (%)	10^6 Btu/$ Shipment	10^6 Btu/ton
Flat Glass Sheet Float	20 (65,660)	79. Gas 14.9 Electric 1.1 Coal 6.4 Fuel Oil	63	17.2
Glass Container	48 (155,230)	70. Gas 17. Electric 4. Fuel Oil 7.8 Mid. Dist.	74	13.7
Pressed and Blown	26 (79,920)	69. Gas 25. Electric 3. Fuel Oil 2.6 Mid. Dist.	64.7	43.8
Industry	100	70. Gas 20.6 Electric 3.9 Fuel Oil 4.6 Mid. Dist. 0.3 Propane 0.4 Coal	56,200	

Source: Battelle Columbus Laboratories, *Final Report on Industrial Energy Study of the Glass Industry* (Columbus, Ohio: Battelle, December 1974).

GLASS MELTING

In the manufacturing of glass products, the melting unit process is by far the most energy intensive. This operation accounts for 70-75% of the energy consumption in a glass plant; therefore, it is the unit process which offers the industry an opportunity to make a significant change in the energy consumption patterns.

Table 11-7
Per Cent of Total Plant Energy Consumption by
Industry Segment and Process Step

Segment	Batch Handling	Melting & Fining	Forming	Post Forming	Product Handling	Space Conditioning
Flat Glass						
Sheet	1.2	75	2.0	2.0	1.3	18.5
Float	1.2	76	5.0	7.0	1.3	9.5
Glass Containers	2.0	70.0	1.5	9.5	2.5	4.5
Pressed and Blown						
Machine Ware	2.0	55.0	15.5	20	7.0	6.0
Lamp Envel., T.V. & Tubing	1.0	65.0	14.0	15	2.0	3.0
Glass Fibers						
Textile	2.0	45.0	28.0	10.0	3.0	12.0
Wool	2.0	40.0	38.0	12.0	2.0	6.0

Source: Environmental Protection Agency, *Environmental
Considerations*, Vol. XI, *Glass*, p. 97.

In the glass industry about 90% of all glass is melted in regenerative furnaces. Two types of regenerative furnaces are used:
side port and end port. The side port regenerative furnace has higher
pull rates because of its ability to incorporate a greater checker
volume than the end port furnace. Initially, the end port furnace
consumes less energy, but as the checker volume decreases through plug-
ging, melting efficiency over the life of the furnace is somewhat less
than that of the side port furnace. Because of this more glass is
melted in the side port furnace.

In a side port regenerative glass melting furnace the regenerative
checkers and the burner ports are on opposite ends. See Fig. 11-4.
In the end port regenerative glass melting furnace, the burner ports
and checker structure are at the same end. The checker structure of
both furnaces is a lattice of brickwork which allows the passage of
air. Incoming air for combustion is heated as it passes through the
hot checker system to meet the fuel in the burner port. The flame
then burns over the surface of the glass within the space under the
melter crown. The combustion gases leave the melting area through the

opposite burner ports and through the other checker system, thereby heating these checkers. The furnace is operated in this mode for about 20 minutes before the flow is reversed by an air reversing valve through the other checker and port system. The checkers use waste heat to heat the incoming air for combustion, providing higher flame temperatures and greater melting efficiency. Figure 11-7 shows the fuel savings which can accrue due to preheating combustion air through a regenerator.

Glass flow is perpendicular to the checkers and burners. The incoming raw material, which is a carefully premixed formulation of silica, feldspar, dolomite, limestone, and soda ash, is fed across the melt at one end of the furnace. The glass moves through the melter from the feed end to the finer. The highest temperature used in the melter is about $2900^{\circ}F$, since at approximately $2950^{\circ}F$, the silica crown roof begins to soften and drip, causing inhomogeneities which lead to rejects in the final glass product. Typically, the furnace has about five feet of melting area per ton of glass pulled per day.

About one third of the energy consumed is required to melt the raw materials entering the furnace, one third is lost through the refractories and one third is lost up the stack. Previously, energy savings were made possible through better use of insulation and through better burner and furnace design. Further improvements through increased insulation is minimal, because the outside walls of the tank have to be water cooled to extend the life of the furnace. Thus, increased utilization of waste heat appears to be a reasonable approach to improving melting efficiency.

In the analysis of the glass manufacturing process, two factors indicate the likelihood of process changes that might be instituted by the industry because of energy considerations. First, the melting process is by far the most significant energy consuming unit process in glass manufacturing. Second, the availability of natural gas has been enough of a major problem for glass companies for the past few years so that alternative fuels, principally oil, have been required

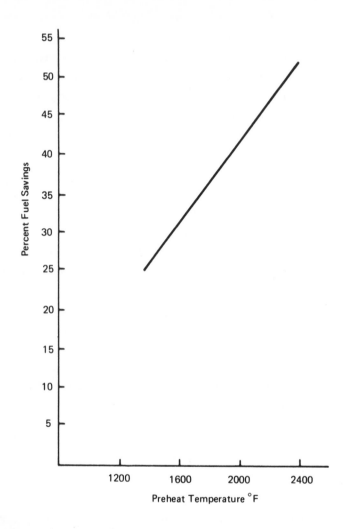

Figure 11-7. Fuel Savings Due to Regeneration. Source: Environmental Protection Agency, Environmental Considerations, Vol. XI, Glass, p. 21.

to maintain production. Alternatives to natural gas fired melting processes are of great importance to the future viability of the industry. The use of fuel oil will continue to be a substitute to natural gas. However, it is not a panacea, since the availability, cost, and supply of oil is certainly not assured. Thus, the only fuel option which reduces the long term uncertainties associated with the present fuels is coal. A number of process options to using coal will be discussed. The remainder of this section will discuss energy efficiency improvement measures as well as fuel substitution alternatives.

Proper Burner Position and Minimizing Excess Air

Proper burner position and control of excess air is an important variable in increasing production and improving fuel utilization. In an experiment performed on a Leone furnace, it was found that both increased production and better fuel utilization occur simultaneously when techniques for proper burner positioning are applied (11).

Minimizing the amount of excess air used for combustion, while holding the gas input constant, increases the amount of available heat for melting. For the production melter, proper burner positioning allows a reduction in the excess air from 18% (3.5% oxygen in the flue) to about 8% (1.5% oxygen in the flue). The amount of natural gas that is required to maintain proper temperature decreases substantially with even a small decrease in the amounts of excess air used.

In their study, Larson and Fejer found that the 3.5% oxygen in the flue was not necessary (12). Excess air could not be completely eliminated, but 1.5% oxygen in the flue (about 8% excess air) provided equally good melting and ensured a low hydrocarbon flue gas and low carbon monoxide content over a wide range of operating conditions. Some excess air will always be necessary in a conventional glass furnace because of the type of combustion being used. The investigation also showed that the potential savings in gas would justify the cost involved in minimizing the excess air. In the furnace used for the test work, the minimum required level of excess air after burner positioning was maintained by using a hydrocarbon and oxygen analyzer.

When the burners are properly positioned (proper angle between the burner axis and glass surface), significant changes can occur in

the amount of natural gas required to melt a ton of glass. Figure
11-8 shows the typical effect of small adjustments of the burner on
the gas consumption, with the gas input held constant and the excess
air maintained at 8%. At the start of the Leone test, 6.8 million BTU
were required to melt each ton of glass. As the burners were adjusted,
gas consumption per unit of output changed significantly. Precise
and proper burner positioning achieved a 13% increase in the production
rate and increased fuel efficiency.

During the course of these tests, the following results were obtained by using a normal 85 ton/day container glass melting furnace:

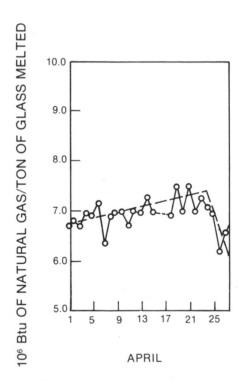

*Figure 11-8. Natural Gas Required for Ton of Glass Melted for
Typical Change in Burner Position. Source: D.H. Larson and M. Fejar,
Improving Glass Melting Furnace Operations Using Improved Combustion
Control Methods (Chicago; Institute of Gas Technology, May 1973),
p. 7.*

• Production rate was increased 13% by precisely adjusting
  existing burner equipment to tailor the combustion to the
  melting process.

• An additional 12% increase in production was achieved using
  a production boosting method which involves increasing or
  enriching the oxygen content of the combustion air from 21-23%.

• A 14% decrease in the amount of fuel necessary to melt a ton
  of glass was achieved. The tests showed that only very small
  burner angle changes of two or three degrees are required to
  achieve the improvements in melting.

• There was no increase in the amount of air pollution emissions
  from either the burner positioning or oxygen enrichment methods.

• Both glass quality and refractory life were unchanged by the
  positioning or oxygen enrichment methods.

Oxygen Enrichment

Oxygen enrichment has been used by some glass manufacturers as a
means of increasing the production of a glass furnace. Oxygen enrichment increases the flame temperature so that a higher melting rate is
achieved and more glass is produced from a given furnace. It has been
reported that by increasing the oxygen content of the combustion air
from 21-23%, an increase in glass production of about 12% can be
achieved (13). The purpose in shifting to oxygen enrichment, then,
is to increase capacity, with a corresponding savings in fuel consumption per ton of glass, for short periods of time. The use of oxygen
also appears to be a viable technique for reducing natural gas
consumption per ton of glass. The principal deterrents to the increased use of oxygen enrichment of fossil fuel firing are the cost
of oxygen and the possible effect of the higher flame temperatures on
the furnace life, particularly on the silica crown roof.

From an air pollution control point of view, sulfur emissions will
not be altered, but NO_x could increase because of the increased flame
temperature, and particulate emissions would be increased. Oxygen
enrichment probably would increase baghouse costs by 10-15% (14).

CULLET

The consumption of cullet is a significant factor in the production of glass. An average of 20% of the glass produced from the furnace originates as cullet: the range is from 10-30%. Changes in this feedstock to the furnace can have a major impact on energy consumption in the melting process.

Today, "house cullet" accounts for 10-15% of the furnace charge, and most glass container manufacturers purchase another 5%. It is estimated that an increase in the percentage of outside cullet from the average (20%) to 50% would reduce glass melting furnace energy requirements about 6.8% (15). The potential savings are 0.51 million BTU/ton of glass, or about 3% of the total energy currently used.

Present day concern with resource recovery has led to renewed interest in the use of outside cullet. Recycling postconsumer cullet by separate collection has continued to increase each year. In 1974, 2-3% of the glass containers were collected and recycled. About 30-35 major resource recovery facilities are currently in operation, under construction, or in the planning stage in this country. Nevertheless, in 1980 no more than about 4% of the glass container production will be recycled by separate collection (16). The future availability of post consumer cullet from mechanical resource recovery systems remains an unknown quantity for the following reasons:

- The high capital investment necessary to install a glass recovery system.

- The questionable economics of a mechanical glass recovery system.

- The high costs of transportation.

- The need for stringent cullet specifications and the absence of a commercially operating facility that produces acceptable cullet.

Because of the energy savings potential and the potential for increased production with increased cullet use, it is expected that the average glass container manufacturer might use about 20-25% cullet by 1985, instead of 15-20%.

ELECTRIC BOOSTING OF MELTER

The mix of fossil fuel and electric energy may see changes in the near future, since the use of electric energy for melting is growing in the industry. Many plants have installed electric boosters to aid in increasing glass output in their existing furnaces without extensive capital expenditures. These boosters consist of a pair of electrodes specially sited in the wall or bottom of the furnace to generate strong convection currents of hot glass, which can increase furnace throughput from 10-40% (17). This is said to increase furnace output by much more than the relative amount of additional energy; in addition, it helps to mix the molten glass and consequently, to improve its homogeneity.

The energy introduced by electrical boosting is small, relative to the total energy input; however, it is utilized at something approaching 100% efficiency. Experience varies with specific installations, but the boosting energy required to increase throughput by one additional ton of glass is 600 KWH on average (18). It is interesting to note that the marginal throughput resulting from this electrical energy boosting system is greater than the average throughput for the same quantity of electric energy expended in an electric furnace, which consumes around 900-1100 KWH per ton of glass output (19). Increased use of electric boosting would shift the energy consumption to a less critical type of fuel--coal.

ELECTRIC MELTING

The electric melting process appears to be a technically and economically competitive alternative to fossil fuel melting. Complete
electric melting has been successfully used in the U.S. The melting
process using electricity is significantly different from that using
fossil fuels. The former directly heats the glass melt by passing
high current through the conductive glass melt via electrodes inserted
in the wall or bottom of the furnace, while the latter relies on radiation above the melt as the principal heat transfer mechanism. Because
the top of the glass melt is covered by the raw batch, high heat
losses from the hot melt are reduced.

The high cost of electric energy and the present limitations on
the size of all electric furnaces have resulted in relatively few installations. All electric furnaces involve complete new furnace construction rather than retrofitting, as can be done with electric
boosters. However, once installed, the efficiency is high, with
claims of 80% (800 KWH/ton) being made.

By using the alternative of electric melting, a more available
form of energy replaces a less available one, which is a form of energy conservation. No overall reduction in the environmental problem
and control is anticipated, since this approach essentially shifts
the environmental problems from the glass furnace to the electric power
station.

BATCH PREHEATING

An important process improvement currently being studied by the
glass industry is the preheating of the batch by using exhaust gases
from the furnace. Depending upon the specific operating conditions,
from one quarter to one half of the energy necessary to melt glass
could be obtained from waste heat. By heating the glass batch outside

the furnace, better utilization can be made of the energy. But such
a preheat operation is difficult to accomplish without modification
in the batch handling methods.

Figure 11-9 shows a general schematic of the batch preheating
system. A number of storage bins hold the raw material. The raw
materials are weighed individually, fed to a collecting belt, and
conveyed to a mixer. A pan mixer is used to blend the dry materials.
From the mixer, the blended batch is transferred to a surge hopper and
feeder. The material is then fed to a pelletizer, where water is
added to about 4% by weight as a binder for the pellets. The pel-
letized material is then conveyed through a high temperature continuous dryer which is heated by the waste gases of the glass melting
furnace. Before being fed into the glass tank, the pellets are heated
to about $1300^{\circ}F$ (20).

The thermodynamic analysis of the melt furnace and its regenerator indicates that there is enough energy availability in the furnace
exhaust gases to preheat the incoming combustion air to present levels
and to preheat the batch to $1400^{\circ}F$. Figure 11-10 shows an energy
and availability summary for a combined melting furnace and regenerator and for such a combined melting furnace with air and batch preheaters. The reduction in fuel consumption for such an arrangement
amounts to 14.8% (fuel input of 6.36 x 10^6 BTU versus 7.465 x 10^6 BTU).
The overall efficiency predicted with batch preheating, 38.6%, approaches the maximum achievable by melting furnaces of conventional
design (21). If raw materials are heated by the waste gases from the
furnace to about the $1,290^{\circ}F$ temperature range, this would correspond
to an energy savings of more than 1 x 10^6 BTU per ton or about 20%.

DIRECT COAL FIRING

The only fuel alternative that reduces the long term uncertainties associated with the present fuels is coal. Pulverized coal has
been used in cement processing and in some metallurgical processes but

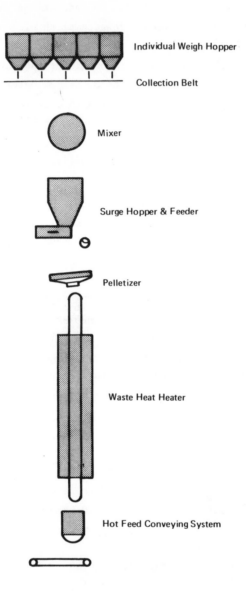

Figure 11-9. Batch Agglomeration: Preheating. Source: Environmental Protection Agency, Environmental Considerations, Vol. XI, Glass, p. 59.

never in glass making. One exception is the Coors plant in Colorado, which replaced several of their gas burners in one furnace with coal burners.

The technical feasibility of direct firing of the glass melting furnace with pulverized coal has not been determined. Two technical problems can be anticipated: (1) the effect of the fly ash on the life of the refractories, and (2) reactions between the fly ash and the refractories which may result in excessive seeds in the finished

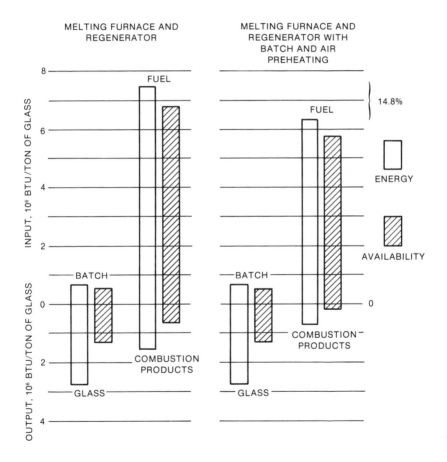

Figure 11-10. Effect of Batch Preheating on Energy Requirements. Source: E. Hall, "Evaluation of the Potential for Energy Conservation in Industry," Energy Conservation: A National Forum, T. Veziroghu, ed. (Coral Gables, Florida: Clean Energy Research Institute, December 1975), p. 21.

product. Whether the entire furnace can be converted to direct coal
firing or whether only a portion of the melting can be done with coal
is questionable, because the adverse effect on glass quality is still
not known.

The characteristics of coal in terms of heating value, ash content, and sulfur are major considerations because variable coal feed
would make furnace operation difficult. Direct firing of glass furnaces, if proved feasible with purverized coal, would be a form of
energy conservation because it utilizes a less critical form of fuel
than do present natural gas processes.

COAL FIRED HOT GAS GENERATION

Wormser Engineering, Inc. is conducting a pilot plant feasibility
study on a new direct combustion system called COHOGG (coal fired
hot gas generator). See Fig. 11-11. COHOGG is a conversion burner

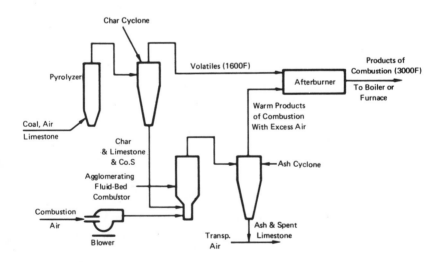

*Figure 11-11. Flow Chart for Industrial COHOGG. Source: En-
vironmental Protection Agency, Environmental Considerations, Vol. XI,
Glass, p. 50.*

that may be substituted directly for oil or gas burners in furnaces, boilers, and other direct heating processes (22).

The system generates a hot gas by burning char, mixing its products of combustion with the volatiles, and then burning them together. A pneumatic conveying system feeds powdered coal to a pyrolyzer, along with limestone or a mixture of limestone and sodium chloride. The coal entering the system is heated along with the limestone and air, thereby forming a gaseous product and a solid char. These products are then separated, the char going to the char burner, and the volatiles to an afterburner.

Excess air is mixed with the combustion air in the fluidized bed reactor to maintain the temperature of the fluidized bed at about $1600^{o}F$. The gases from burning the char leave the fluidized bed reactor, are combined with the gaseous products, or volatiles, from the pyrolyzer, and are burned along with excess air in the afterburner. The outlet temperature of the gases leaving the afterburner is $3000^{o}F$. These hot gases are transferred from the COHOGG system to the glass melting furnace.

Theoretically, this approach to glass melting appears feasible. The most important questions concern the reliability of the system of the size required for glass melting and the adaptability of this technique to the melting of glass. A potential drawback of the process, which may hinder its acceptability in the glass industry, is the absence of a luminous flame. Because the system is still in the pilot plant stage, it is unlikely that any glass company would seriously consider adopting the process within the next 10-15 years.

SUBMERGED COMBUSTION

Submerged combustion experiments have been attempted by placing the burners in the bottom of the glass furnace. This results in greatly improved heat transfer and vigorous convective stirring of the melt. The potential impact on the glass industry is a 50% reduction in the projected energy consumption, or about 160 trillion BTU reduction annually, assuming 100% acceptance. Since submerged combustion consumes only natural gas, the reduction in energy consumption could also result in a change in fuel mix, perhaps eliminating electric melting and fuel oil combustion altogether.

At the present time, the quality of the glass produced by the submerged combustion techniques has been poor because of the large number of bubbles produced. This requires extensive refining to produce an acceptable glass.

ORGANIC RANKINE BOTTOMING CYCLE

About one third of the energy used in the melter escapes through the exhaust flue after passing through the regenerator. This gas is often at a temperature of 800°-1000°F, so that there is much work potential remaining in the waste exhaust gases. The installation of an organic Rankine bottoming cycle engine is a viable option for generating electricity which can be used in the plant. Small systems have been developed by companies such as Sundstrand. A potential energy utilization improvement of the glass melter of 15% or about one million BTU per ton could be achieved.

ENERGY CONSERVATION POTENTIAL

The theoretical minimum energy required to produce glass for containers is 2.15×10^{6} BTU per ton of glass melted. The glass container industry currently uses 13.7×10^{6} BTU per ton (23). Gains in energy efficiency in the past have come about mainly through economies of scale from the use of increasingly larger furnaces, through advances in process control, through the introduction of an auxiliary heating unit inside the body of molten glass in the furnace, and through the growing reuse of waste heat.

Much additional savings can be accomplished in the glass industry through improved management techniques as described in Chapter 2. In addition, there are many opportunities for conservation through process modifications and improvements. Table 11-8 indicates the potential for energy conservation for various alternatives in the glass industry. Use of techniques which are economic can result in energy savings of 20-30% over conventional practices as well as in substantial productivity gains.

Table 11-8
Summary of Energy Conservation Potential
(over conventional practices)

Conservation Practice or Technique	Savings* 10^6 BTU/ton
Proper burner position & minimizing excess air	0.9
Oxygen enrichment	0.7
Increase cullet use	0.5
Batch Preheating	1.3
Submerged Combustion	3.5
Electricity generation through Organic Rankine Bottoming Cycle	1.0
Improved management and waste heat recovery	1.0

*These savings are not additive and are based
on efficiency improvements over 1974 conventional
practices.

In addition, Table 11-9 provides a summary of energy consumption in the melting process for the various fuel options discussed. In many of the systems, particularly in the coal based options, energy conservation in the form of fuel substitution is gained, rather than in the form of energy efficiency per se.

Table 11-9

Summary of Energy Consumption in Glass Melting Processes
(basic 200 TPD of glass)

Energy Source	Energy Used 10^6 BTU/ton	Difference with Natural Gas Firing (BTU/ton)
Natural gas firing	7.0	0
Direct Coal Firing	7.0	0
Coal gasification	8.6	+1.6
Coal Fired Hot Gas Generator	8.6	+1.6
Electric	8.2	+1.2
Batch Preheating	5.7	−1.3

NOTES AND REFERENCES

1. John Myers et al., Energy Consumption in Manufacturing (Cambridge: Ballinger Publishing Company, 1974), p. 327.

2. Environmental Protection Agency, Environmental Considerations of Selected Energy Conserving Manufacturing Process Options, Vol. XI, Glass Industry Report (Cincinnati, Ohio: Environmental Protection Agency, December 1976), p. 79.

3. R.N. Shreve, Chemical Process Industries, 3rd ed. (New York: McGraw Hill, 1967), p. 195.

4. U.S. Department of the Interior, Minerals Yearbook, 1971 (Washington, D.C.: Government Printing Office, 1973), p. 1045.

5. Ibid., p. 68.

6. Gordian Associates, The Potential for Energy Conservation in Nine Selected Industries, Vol. 7, Glass (Washington, D.C.: Government Printing Office, 1975), p. 100.

7. "Are Solvay Plants on the Way Out?" Chemical Week, 112 (March 7, 1973), pp. 40-41.

8. Shreve, Chemical Process Industries, p. 226.

9.   W.L. Faith, D.B. Keyes, and R.L. Clark, Industrial Chemicals, 3rd ed. (New York: John Wiley & Sons, 1965), pp. 666-67.

10. J.F. Hanlan, Handbook of Package Engineering (New York: McGraw Hill, 1971), pp. 6-10.

11. D.H. Larson and M. Fejer, Improving Glass Melting Furnace Operations Using Improved Combustion Control Methods (Chicago, Illinois: Institute of Gas Technology, May 1973), p. 7.

12. Ibid., p. 8.

13. Ibid., p. 8.

14. Environmental Protection Agency, Environmental Considerations, Vol. XI, Glass, p. 102.

15. Gordian Associates, The Data Base, Vol. 7, Glass (Washington, D.C.: Government Printing Office, 1975), p. 106.

16. Environmental Protection Agency, Environmental Considerations, Vol. XI, Glass, p. 105.

17. Larry Penberthy, "In Rebuttal: Factors which Justify the Increased Cost of Electricity Over Natural Gas for Glass Making," The Glass Industry, 47 (June 1966), p. 319.

18. M. Fort, "Some Practical Aspects of Electric Boosting," Glass Technology, 5 (October 1964), p. 200.

19. J.P. Heu, "Heat Balance and Calculation of Fuel Consumption in Glassmaking--Conclusion," The Glass Industry, 52 (February 1971), p. 61.

20. Environmental Protection Agency, Environmental Considerations, Vol. XI, Glass, p. 58.

21. E. Hall, "Evaluation of the Potential for Energy Conservation in Industry," Energy Conservation: A National Forum, ed. T. Veziroghu (Coral Gables, Florida: Clean Energy Research Institute, December 1975), p. 10.

22. Ibid., p. 48.

23. Ibid., p. 9.

Chapter 12

THE PULP AND PAPER INDUSTRY

The pulp and paper industry is extremely diversified, employing numerous manufacturing techniques for the production of over two thousand primary products. It ranks fourth among all industry groups with regard to energy consumption. In 1973, the industry consumed approximately 2.2×10^{15} BTU of energy, with pulp mills, paper mills, and paperboard mills accounting for approximately 90% of this total.

The pulp and paper industry in the U.S. is made up of 407 companies and 744 operating mills. In 1972, there were 212 mills with capacities of 20,000 tons or less per year, and these mills accounted for 3.4% of the industry's total capacity. At the other end of the spectrum, there were eleven mills whose capacities were over 500,000 tons per year, and these mills accounted for 11.4% of the industry's capacity (1).

Pulp and paper are manufactured in both integrated and non-integrated mills. Over two thirds of the U.S. production capacity consists of integrated mills which tend to be relatively large, with average capacities on the order of 300,000 tons per year. Integrated mills have both pulp and papermaking facilities; non-integrated mills (forming facilities only) tend to be smaller, with capacities ranging from 3,000-30,000 tons per year. Since non-integrated mills rely heavily on waste paper as their source of fiber, they are located close to the more industrialized areas of the country.

432

The pulp and paper industry is mature in that the demand for its products have grown at about the same rate as the GNP in real terms (see Fig. 12-1). Total paper and paperboard mill capacity has increased significantly in recent years to satisfy growth demand. Pulp and paper production figures for 1972 are shown in Table 12-1. Total wood pulp production in 1972 was 46.8×10^6 tons. With imports of 3.7×10^6 tons and exports of 2.2×10^6 tons, the total wood pulp supply was 48.3×10^6 tons. Waste paper consumption in 1972 was 12.4×10^6 tons. However, the projection is that the industry may grow more slowly than the GNP in the future because of substitution by competing products and saturation of per capita consumption potential. Growth in demand for paper and paperboard products is expected to drop to 3%, in contrast with 4.5% between 1960 and 1970, because of a slowdown in GNP growth, tight supply/demand balance, and rapid price escalation (2).

The most significant of the long run trends in pulp production is the tremendous increase in kraft pulp. This increase is shown in Table 12-2. In 1920, the production of kraft pulp was relatively unimportant; by 1950, it was half of the total output, and by 1970, 70% of the total. It is probable that in the future the production of kraft pulp will continue upward at a steady rate.

PAPER MANUFACTURING PROCESS

Papermaking is an ancient art. As far back as the fourth century, paper was made employing a technique still in use today for handmade paper. The first advance in this technique came in the eighteenth century--the continuous forming device. In principle, the process is unchanged today: the stages of papermaking are essentially still pulpmaking, forming, pressing, and drying. What has changed are the scale and speed of operation and the capital investment requirements.

Figure 12-2 is a diagram of the papermaking operation. The manufacture of paper is a two stage process: first, the cellulosic raw

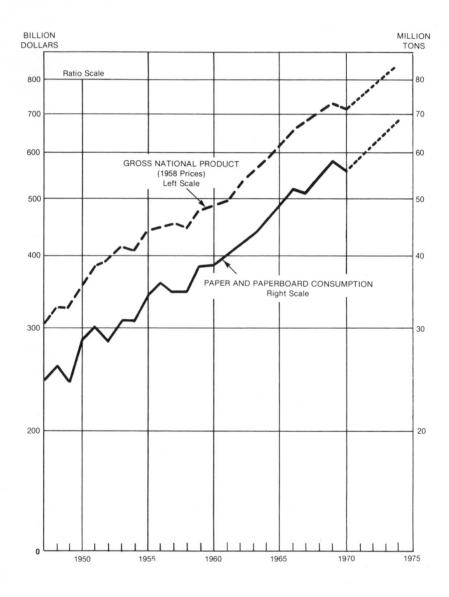

Figure 12-1. Paper and Paperboard Consumption Gross National Product (1958 prices). Source: American Paper Institute for Paper and Paperboard Consumption; Department of Commerce for Gross National Product.

Table 12-1

Paper and Paperboard Production—1972

Category	Integrated Mills (tons)	Non-integrated Mills (tons)	Total (tons)
Printing and writing papers	12.1×10^6	3.6×10^6	15.7×10^6
Packaging and industrial converting paper	4.5×10^6	1.2×10^6	5.7×10^6
Tissue and other creped paper	1.9×10^6	2.1×10^6	4.0×10^6
Paperboard	21.8×10^6	6.7×10^6	28.5×10^6
Construction paper and board	3.5×10^6	1.8×10^6	5.3×10^6
Market	5.0×10^6	—	5.0×10^6
	48.8×10^6	15.4×10^6	64.2×10^6

Source: Elias Gyftopoulos, John Dunlay, and Sander Nydick, **A Study of Improved Fuel Effectiveness in the Iron and Steel and the Pulp and Paper Industries** (Waltham, Massachusetts: Thermo-Electron Corporation, March 1976), p. 3-5.

material must be processed to yield a uniform product that possesses the necessary color, texture, strength, and matforming properties. This step, which forms pulp, may be mechanical, chemical, or a combination of both. Pulp has no direct use except in the manufacture of paper or paperboard. The second step consists of forming slurried pulp in a thin, porous mat and then pressing, drying, and coating or impregnating it to form the desired product.

Table 12-2

Production of Pulp by Major Grades, 1920-1970

	Kraft		Sulphite[a]		Groundwood[b]		Semichemical		Other[c]		Total Pulp	
	Thous. of Tons	%	Thous. of Tons	%	Thous. of Tons	%	Thous. of Tons	%	Thous. of Tons	%	Thous. of Tons	%
1920	189	4.9	1586	41.5	1584	41.4			463	12.1	3822	100
1925	410	10.3	1403	35.4	1612	40.7			537	13.6	3962	100
1930	950	20.5	1567	33.8	1560	33.7	30	.6	523	11.3	4630	100
1935	1468	29.8	1580	32.1	1356	27.5	67	1.4	455	9.2	4926	100
1940	3748	41.8	2608	29.1	1633	18.2	165	1.8	806	9.0	8960	100
1945	4472	44.0	2360	23.2	1826	18.0	295	2.9	1214	11.9	10,167	100
1950	7506	50.5	2844	19.2	2216	14.9	686	4.6	1597	10.8	14,849	100
1955	11,577	55.8	3251	15.7	2729	13.2	1408	6.8	1775	8.6	20,740	100
1960	15,034	59.4	3272	12.9	3292	13.0	1991	7.9	1727	6.8	25,316	100
1965	21,146	63.5	3643	10.9	3920	11.8	2885	8.7	1702	5.1	33,296	100
1970	29,408	69.7	3287	7.8	4393	10.4	3339	7.9	1789	4.2	42,216	100

[a]Includes dissolving and special alpha pulp.
[b]Includes exploded wood pulp 1930-40.
[c]Includes soda, screenings, off-quality, and miscellaneous.

Source: American Paper Institute, **Wood Pulp Statistics,** 35th ed. (New York: American Paper Institute, 1971).

Pulping

Papermaking starts with pulpwood. The log is de-barked, the bark being collected and used as a fuel. Pulp can then be made either by a mechanical or chemical process. Wood consists of small cellulose fibers which are bound together by a gluelike substance called lignin. When the sap, resin, lignin, and other matter in the tree have been separated from the cellulose fibers by chemical means, the remaining fibers are called chemical pulp.

In chemical pulping, the log is reduced to chips by a mechanical chipper which uses an electric drive. The chips are then mixed with chemicals and cooked in large vats called digesters under controlled pressure, temperature, liquor composition, and time. The cooking dissolves the lignins and frees the fibers, suspending them in water. In the second stage of the process, the pulp is blown under pressure from the digesters to separate the fibers, washed to remove the chemicals and other materials from the fibers, and then sent to the beaters.

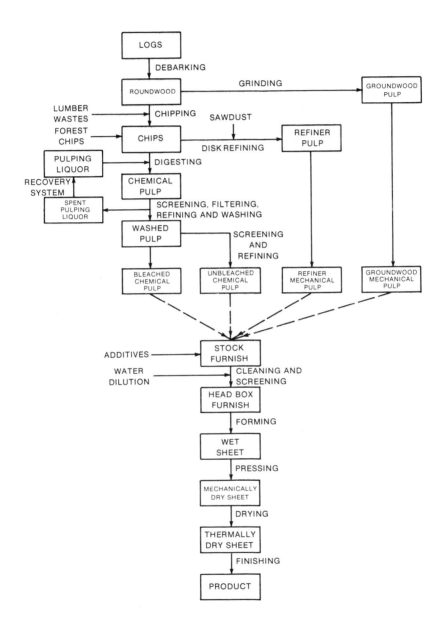

Figure 12-2. Basic Operations in the Making of Paper. Source:
A. Kouvalis, Industrial Practices and Technology Review: The Pulp
and Paper Industry (Argonne, Illinois: Argonne National Laboratory,
April 1977), p. 25.

The waste fluid, or "black liquor," which is the chemical solution
containing the lignin compounds, is removed for use as a significant
source of energy in the pulp and paper industry. Pulp to be used for
white papers is bleached before entering the beaters.

The most dominant type of chemical pulps are kraft pulp, sulfite
pulp, pretreated and untreated refiner pulps, and neutral sulfite
semichemical (NSSC) pulp. Figure 12-3 shows a diagram of the kraft
digester and recovery cycle. Figure 12-4 shows the sulfite digester
and recovery cycle.

In the mechanical process, wood is reduced to small particles by
rubbing against huge grindstones revolving at high speeds or by ex-
plosive decompression. Mechanical pulping wastes less raw materials
and potentially produces fewer effluent solids than the chemical process. The pulp produced lacks the strength of most chemical pulps
but has qualities found useful in high speed printing. Most of the
newsprint throughout the world is primarily from groundwood pulp.

Papermaking

The papermaking process consists of four major stages: stock
preparation, sheet formation, water removal, sheet finishing. See
Fig. 12-5. Before the pulp is fed to the papermaking machine, it
must be converted to an aqueous suspension. To assure a product of
uniform density and strength, the physical characteristics of the
fibers must be kept within close tolerances. Consequently, the pulp
is usually subjected to further mechanical operations prior to sheet
formation.

If dry market pulp or recycled materials are used as a fiber
source, the pulp material must be resuspended in water in a pulper,
which disintegrates the material into its component fibers, resuspends
it in a water slurry, and drives the suspension through a perforated
outlet plate. If the feedstock is recycled fiber, the pulpers may be
equipped with mechanical means for removing debris. Where recycled
printed matter is used to make a product requiring a clean, bright
surface, it must be deinked by cooking for 1-2 hours in an alkaline

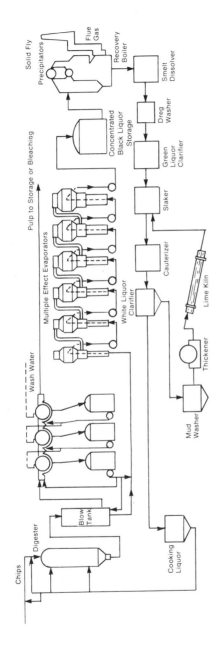

Figure 12-3. Kraft Digester and Recovery Cycle. Source: Kowalis, Industrial Practices and Technology Review, p. 28.

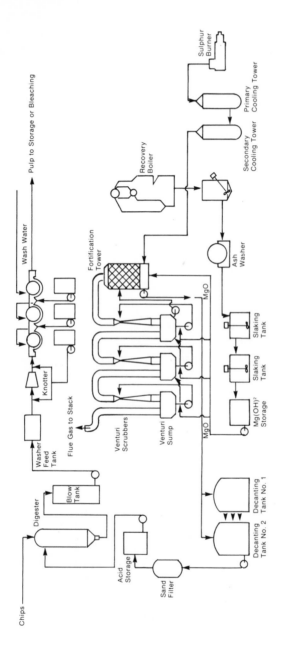

Figure 12-4. Sulfite Digester and Recovery Cycle. Source: Kouvalis, Industrial Practices and Technology Review, p. 28.

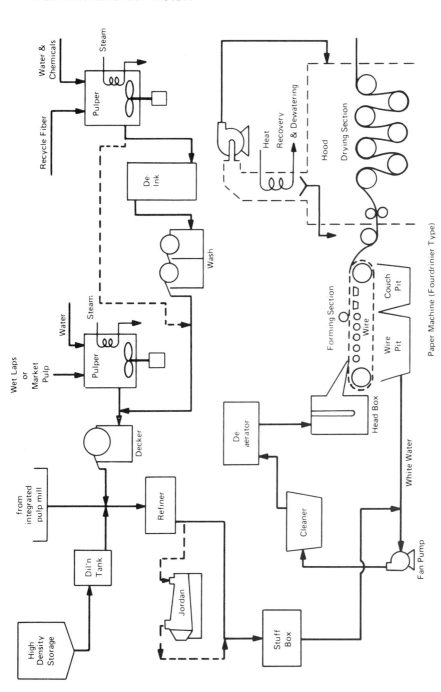

Figure 12-5. The Papermaking Process. Source: S. I. Kaplan, Energy Use and Distribution in the Pulp, Paper and Boardmaking Industries, (Oak Ridge, Tennessee: Oak Ridge National Laboratory, August 1977), p. 43.

solution, followed by washing (3). Deinking is not required for
other products such as utility packaging materials.

The ability of the pulp fibers to form strong interlocking bonds
is very much influenced by the degree of fraying of the fiber surfaces. Preparation of the fiber surfaces to the desired condition
is done by beating, which is a mechanical abrasion process that splits
the outer covering of the fibers, thereby increasing the surface area.
The beating action is comparable to, but less vigorous than, the refining action employed in mechanical pulp preparation and is performed
in a Jordan, which is a special type of refiner in which a conical
plug rotates within a conical shell. The Jordan can also be used to
chop the fibers to a desired size by adjustment of the plug clearance.

After the pulp is beaten to the desired level, its consistency
is adjusted, and it is diluted with returning white water from the
web-forming section. It then proceeds to the papermaking machine.
All papermaking machines have three basic functional units: the web-
forming unit, the press section, and the dryer. The press section
and the dryer are basically the same on all machines. There are,
however, two principal types of web-forming-units--the Fourdrinier
machine (see Fig. 12-6) and the cylinder machine (see Fig. 12-7). If
the machine is a Fourdrinier type, the feed is usually pumped through
a bank of liquid cyclone cleaners, proceeding then to the headbox.
In the case of cylinder type machines, centrifugal screening alone
is ordinarily considered adequate. Liquid cyclone cleaners require
considerable energy to operate.

In the Fourdrinier machine, a continuous, dilute pulp suspension
or "stock" consisting of 99.5% water is sprayed from the headbox onto
a wire, a horizontal endless moving belt of bronze wire screen or
plastic thread. The fibers are retained on the belt to form a mat
or web, while the bulk of the white water drains by gravity through
the wire into "pits" underneath. The milkiness of the water is caused
by small fibers and other materials which are washed through the
wire. As the web is carried forward on the moving wire about two-
thirds of the way, it passes over slotted suction boxes with large

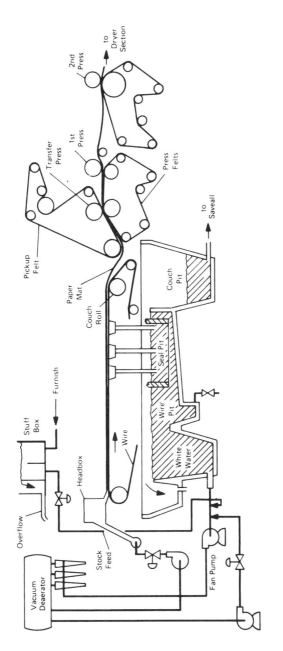

Figure 12-6. Fourdrinier Machine. Source: Kaplan, Energy Use and Distribution, p. 47.

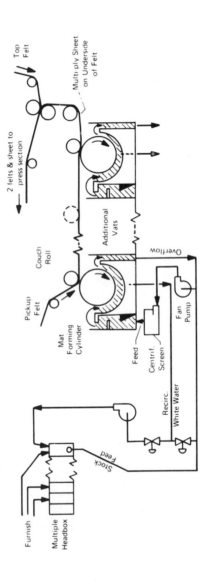

Figure 12-7. Cylinder Machine. Source: Kaplan, Energy Use and Distribution, p. 47.

vacuum pumps which drain additional moisture from the fibers. At the
end of the wire section, the wire bends around the couch roll, which
is also connected to a vacuum pump, and then the web is brought into
contact with a moving belt or felt. The fiber mat clings to the felt
and is lifted from the couch roll. The felt and web together pass
through a set of press rolls to squeeze water out of the paper and
into the felt. The web is transferred to a second felt and pressed,
and the sheet is then led to a third press, at which time the paper
web is for the first time self supporting. In the third press, the
felt is on the other side of the paper to even out the effect of
pressing on the smoothness of the finished paper. After the pressing
steps, the sheet is looped in an over and under pattern on an array
of heated rolls for final drying. In most applications, the sheet
is held against the rolls by endless belts of heavy felt.

The cylinder machine, which is primarily used for making paperboard, differs from the Fourdrinier in one essential particular. In
place of the endless wire belt, the wet end consists of from 1-9
vats filled with flowing pulp stock suspension, each containing a
partially submerged, revolving wire mesh covered cylinder. The cylinders are drained from the inside as the cylinder rises with its fiber
layer, so that water flows through the submerged parts of the screen
while the stock is strained out to form a web covering the cylinder.
As the rotating cylinder surface rises from the stock, it is contacted
by an overhead moving belt which continuously picks the web from the
cylinder. Starting with the rearmost web-forming cylinder, the belt
and its attached sheet pass over the remaining cylinders, picking up
an additional ply of sheet emerging from each. The composite then
passes through press rolls and is dewatered in the same manner as in
the Fourdrinier machine.

The next stage in the process is the dryer section, which usually
uses steam heated iron cylinders for drying. These dryer drums are
3-6 feet in diameter, and there may be over 100 of them in a large
machine. In modern machines, the entire dryer section is enclosed
in a hood. This improves comfort in the machine room, conserves

energy, and helps even out the drying across the width of the machine.

Paper shrinks when it dries, so the speeds of the dryer drums have to be controlled. The properties of the finished paper are very sensitive to the tension during drying. If the tension is kept high, then the finished paper does not have as much stretch, but it has a lot more strength. Drying is an expensive proposition which uses about 2 pounds of steam to dry a pound of paper.

ENERGY CONSUMPTION IN THE INDUSTRY

Energy, wood, and water are the three commodities used in quantity in making paper; thus, energy cost and supply are of great concern. The pulp and paper industry ranks among the top four U.S. manufacturing industries in terms of total energy consumption, but more than 45% of its energy requirements are met by self-generated fuels. See Table 12-3. Fossil fuel and purchased energy use per unit of output has declined by approximately 11% between 1972 and 1976. The table shows that the percent of energy derived from self-generated and waste fuels have been increasing from 42-45% of the total energy consumed. Per unit of output, it has increased by about 4%.

Energy consumption also varies significantly with geographical location. Table 12-4 shows energy usage in the 6 paper producing regions of the U.S. In terms of purchased energy, wide variations are seen. For example, in New England, residual fuel oil is dominant. In the Middle Atlantic and South Atlantic regions, fuel oil is important, but coal is also important. In the North Central and Mountain and Pacific region, fuel oil and natural gas are predominant. In terms of waste fuels used, the South Atlantic, South Central, and Mountain and Pacific regions derive more than half of their energy through these sources. On the other hand, the Middle Atlantic and North Central regions derive less than 20% of their energy from this source. New England derives about one-third of their energy from waste fuels.

Table 12-3

U.S. Pulp, Paper and Paperboard Industry

Estimated Fuel and Energy Use

SOURCES	UNITS	1976*			1975**			REVISED 1972***		
		ESTIMATED USE	BILLION BTU'S	% OF TOTAL+	ESTIMATED USE	BILLION BTU'S	% OF TOTAL+	ESTIMATED USE	BILLION BTU'S	% OF TOTAL+
Purchased Electricity	MMKWH	32,462.4	110,371.4	5.2	26,948.3	91,623.2	4.8	27,145.8	92,428.6	4.2
Purchased Steam	MM lbs.	15,215.4	17,974.1	0.9	13,608.0	16,135.9	0.8	18,667.3	21,763.6	1.0
Coal	M tons	8,660.0	212,306.1	10.0	7,190.1	174,049.1	9.1	8,950.7	221,991.2	10.1
Residual Fuel Oil	M 42 gal. BBL	76,944.3	482,957.6	22.9	69,702.9	437,852.2	22.9	73,975.4	466,529.8	21.3
Distillate Fuel Oil	M 42 gal. BBL	2,984.3	17,672.7	0.8	2,280.8	13,527.1	0.7	3,403.3	20,274.4	0.9
Liquid Propane Gas	M gal.	18,289.6	1,676.7	0.1	14,205.2	1,292.6	0.1	24,323.3	2,281.5	0.1
Natural Gas	MMCF	320,648.1	327,603.4	15.5	340,897.9	347,634.3	18.1	453,445.1	461,920.9	21.0
Other Purchased Energy			2,532.7	0.1		1,019.0	0.1		207.3	
Energy Sold			(−18,692.2)			(−17,151.8)			(−14,143.0)	
Total Purchased Fossil Fuel & Energy			1,154,402.5	55.5		1,065,981.6	56.6		1,273,254.3	58.6
Hogged Fuel (50% Moisture Content)	M tons	9,179.5	75,578.9	3.6	6,971.0	56,083.9	3.0	5,507.8	44,794.3	2.0
Bark (50% Moisture Content)	M tons	10,547.0	94,992.0	4.5	9,138.2	82,818.7	4.3	11,139.8	102,642.2	4.7
Spent Liquor (Solids)	M tons	60,994.9	754,337.3	35.7	54,415.6	675,788.4	35.3	59,181.1	748,354.3	34.1
Self Generated Hydroelectric Power	MMKWH	2,830.5	9,555.4	0.4	2,686.3	9,132.4	0.5	2,573.1	8,750.9	0.4
Other Self-Generated Energy			6,245.9	0.3		6,055.4	0.3		3,448.9	0.2
Total Self-Generated & Waste Fuels			940,709.5	44.5		829,878.8	43.4		907,990.6	41.4
Total Energy			2,095,112.0	100.0		1,895,860.4	100.0		2,181,244.9	100.0

* Based on a sample of 85% of total dried pulp, paper and paperboard production.
** Based on a sample of 80% of total dried pulp, paper and paperboard production.
*** Based on a sample of 81% of total dried pulp, paper and paperboard production.
+ Determined by using "Total Energy" + "Energy Sold" as demoninator.

SOURCE: American Paper Institute, New York, August 13, 1977.

Table 12-4

Energy Use by Region

% OF TOTAL ENERGY INPUT* BY SOURCES

	NEW ENGLAND			MIDDLE ATLANTIC		
	First Six Months			First Six Months		
SOURCES	1976	1975	1972	1976	1975	1972
Purchased Electricity	3.1	3.7	3.3	7.2	8.0	7.0
Purchased Steam	0.4	0.3	0.4	0.2	0.2	0.3
Coal	0.1	0.1		20.2	20.9	19.0
Residual Fuel Oil	61.6	63.6	66.7	44.7	44.1	49.3
Distillate Fuel Oil	0.1	0.3	0.1	1.3	1.6	0.6
Liquid Propane Gas	0.2	0.3	0.3	0.1	–	0.2
Natural Gas	0.2	0.3	0.1	7.0	7.2	10.0
Fossil Fuel & Purchased Energy	65.7	68.6	70.9	80.7	82.0	86.4
Hogged Fuel	2.3	1.8	1.4		0.1	
Bark	3.9	3.0	1.5	3.0	2.4	1.4
Spent Liquor	23.2	22.0	22.0	15.3	14.7	11.5
Self-Generated Hydro-Electric Power	4.9	4.5	4.2	1.0	0.8	0.7
Other Self-Generated Energy	–	0.1		–	–	
Total Self-Generated & Waste Fuels	34.3	31.4	29.1	19.3	18.0	13.6
Total Energy	100.0	100.0	100.0	100.0	100.0	100.0
Paper and Paperboard Capacity (% of Total U.S.)**		7.9			10.1	
Pulp Capacity (% of Total U.S.)**		5.6			3.1	

	NORTH CENTRAL			SOUTH ATLANTIC		
	First Six Months			First Six Months		
SOURCES	1976	1975	1972	1976	1975	1972
Purchased Electricity	6.2	6.3	5.4	2.4	2.3	1.7
Purchased Steam	1.8	1.8	1.7	0.1	0.1	0.1
Coal	32.6	33.1	40.2	10.7	10.1	11.4
Residual Fuel Oil	9.1	6.8	3.9	29.1	29.1	29.9
Distillate Fuel Oil	1.8	1.9	1.6	0.4	0.5	0.3
Liquid Propane Gas	0.1	0.1	0.1	0.1	0.1	–
Natural Gas	31.7	34.8	34.6	3.6	7.0	7.4
Other Purchased Energy				0.4	0.2	
Fossil Fuel & Purchased Energy	83.3	84.8	87.5	46.8	49.4	50.8
Hogged Fuel	0.9	0.7	0.7	3.0	2.1	0.9
Bark	1.9	1.6	1.4	6.5	6.3	7.8
Spent Liquor	12.9	11.9	9.6	43.2	41.6	40.2
Self-Generated Hydro-Electric Power	0.9	0.8	0.7	–	–	–
Other Self-Generated Energy	0.1	0.2	0.1	0.5	0.6	0.3
Total Self-Generated & Waste Fuels	16.7	15.2	12.5	53.2	50.6	49.2
Total Energy	100.0	100.0	100.0	100.0	100.0	100.0
Paper and Paperboard Capacity (% of Total U.S.)**		19.4			22.6	
Pulp Capacity (% of Total U.S.)**		8.9			30.4	

Table 12-4

Energy Use by Region (cont.)

% OF TOTAL ENERGY INPUT* BY SOURCES

SOURCES	SOUTH CENTRAL			MOUNTAIN & PACIFIC		
	First Six Months 1976	1975	1972	First Six Months 1976	1975	1972
Purchased Electricity	4.0	3.9	3.3	9.7	9.4	8.1
Purchased Steam	0.9	0.9	0.9	1.7	1.8	1.9
Coal	3.8	3.4	3.7	1.2		
Residual Fuel Oil	19.1	15.8	10.0	13.1	13.6	12.7
Distillate Fuel Oil	0.9	0.5	2.1	0.3	0.5	0.2
Liquid Propane Gas	0.1	—	0.2	—	—	—
Natural Gas	20.4	25.4	30.2	21.0	23.9	28.9
Fossil Fuel & Purchased Energy	49.2	49.9	50.4	47.0	49.2	51.8
Hogged Fuel	3.3	2.5	1.3	9.5	8.2	7.2
Bark	5.9	6.2	6.7	0.3	0.3	0.3
Spent Liquor	41.2	41.1	41.5	42.9	41.9	40.3
Self-Generated Hydro-Electric Power				0.2	0.3	0.3
Other Self-Generated Energy	0.4	0.3	0.1	0.1	0.1	0.1
Total Self-Generated & Waste Fuels	50.8	50.1	49.6	53.0	50.8	48.2
Total Energy	100.0	100.0	100.0	100.0	100.0	100.0
Paper and Paperboard Capacity (% of Total U.S.)**		26.0			14.0	
Pulp Capacity (% of Total U.S.)**		33.6			18.4	

* Determined by using "Total Energy" + "Energy Sold" as a denominator
** API Capacity Survey 1974 (latest available data)
A dash (—) indicates less than 0.05%

SOURCE: J.M. Duke and M.J. Fudali, **Report on the Pulp and Paper Industry's Energy Savings and Changing Fuel Mix** (New York: American Paper Institute, September 1976), pp. 13-15.

The industry uses various manufacturing processes to make the same product and sometimes to make uniquely different products. The energy requirements associated with each of these major product/process categories are quite dissimiliar. Figure 12-8 indicates the annual and per unit energy requirements for pulping via each of the processes. The figure provides a convenient summary by which to assess the relative importance of each process from an energy usage

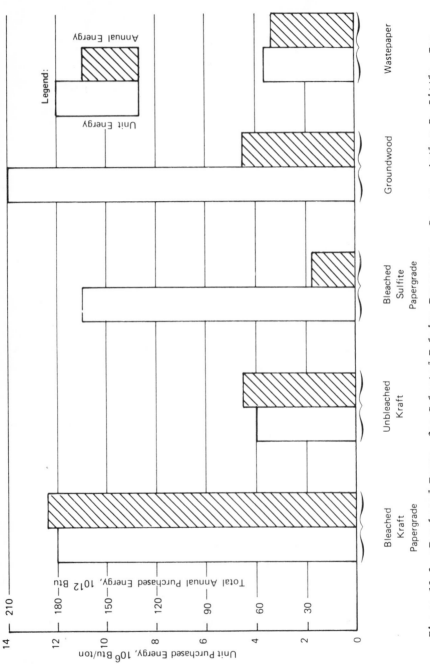

Figure 12-8. Purchased Energy for Selected Pulping Processes. Source: Arthur D. Little, Inc., estimates.

point of view. For example, the annual production of groundwood is comparatively small, but the per unit energy consumption is the highest of the pulping processes. As is also evident, the kraft process is the major energy user from an overall point of view. The pulp and paper industry currently purchases approximately half of its electrical power from utilities. The remaining half is generated within the mills, which use steam turbogenerators in combination with extraction or non-condensing steam for processing and with condensing installations for peak load capacity or to increase the mill's generating capacity.

The unit energy intensiveness of selected process/product combinations for new mills are shown in Fig. 12-9. Several factors are evident from the figure. There is a high purchased energy requirement for non-integrated paper production. This is due primarily to double drying and to the fact that this production process does not have byproduct fuels to use. Bleached kraft products also require high total energy and relatively high purchased energy, and waste paper products require less total energy than any virgin fiber product.

Energy consumption has a seasonal variation. Figure 12-10 shows how fossil fuel and purchased energy use per unit of output declines significantly during the summer months and increases as much as 30% in the winter. Self generated and waste fuels use remain fairly constant. Figure 12-11 shows the impact of this factor from a regional point of view. J. M. Duke found that the decline from winter peak to summer trough in the North Central region was 20.6%, whereas the decline in the South Central region was only 16.7%.

Another important aspect of energy consumption is the impact of capacity utilization. A regression analysis which illustrates the inverse relationship between fossil fuel and purchased energy use per unit of output and capacity utilization is shown in Fig. 12-12. It shows that for every 10% decline in capacity utilization, there will be a 4.9% increase in fossil fuel and purchased energy use per unit of output (5).

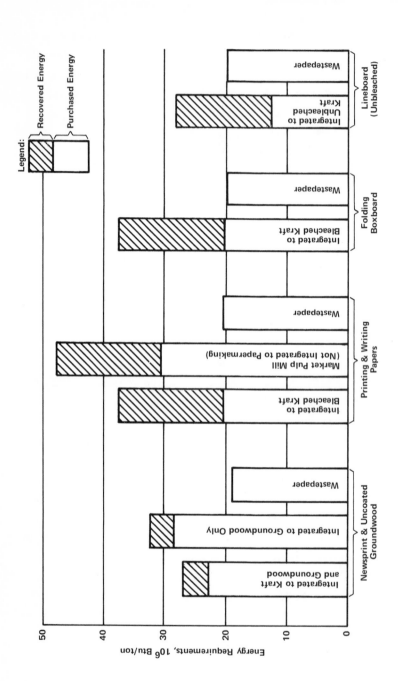

Figure 12-9. Energy Intensivity for Selected Pulp and Paper Process/Product Combinations. Source: Environmental Protection Agency, Environmental Considerations of Selected Energy Conserving Manufacturing Process Options, Vol. V, Pulp and Paper Industry Report (Cincinnati, Ohio: Environmental Protection Agency, December 1976), p. 45.

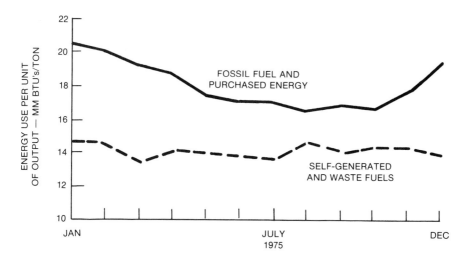

Figure 12-10.   Seasonal Variation in Energy Use.   Source: Duke and Fudali, Report on Pulp and Paper Industry's Energy Savings and Changing Fuel Mix, p. 6.

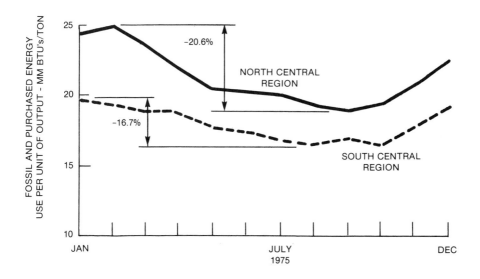

Figure 12-11.   The Seasonal Impact North and South.   Source: Duke and Fudali, Report on Pulp and Paper Industry's Energy Savings and Changing Fuel Mix, p. 6.

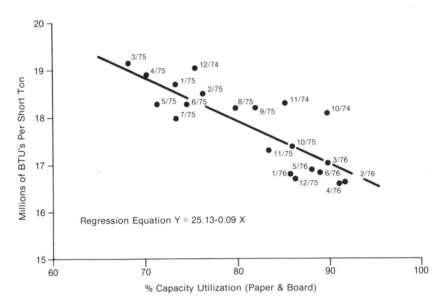

Figure 12-12. Seasonally Adjusted Consumption per Ton of Fossil Fuel and Purchased Energy in Relation to Paper and Board Capacity Utilization (October 1974 through June 1976). Source: Duke and Fukali, Report on Pulp and Paper Industry's Energy Savings and Changing Fuel Mix, p. 8.

Energy usage per unit of output in the pulp and paper industry has been declining. However, since 1972, the cost of energy has tripled and is still rising. Much can be done to further reduce energy consumption in this industry. The remainder of this chapter will discuss those techniques and technologies which can assist in this process.

WOOD PREPARATION

The objective of the wood preparation steps is to convert the heterogeneous raw material into a generally uniform bulk feedstock. Wood arrives at the mill in the form of pulp logs (3-10 feet long), long logs (up to tree length), or chips. Roundwood is debarked either by dry or wet processes, depending upon the size of the logs handled.

Small diameter logs are debarked by friction in a dry state. In the case of larger logs, wet friction debarking is performed in a drum or pocket barker by abrading the bark from the logs. Hydraulic barking, which uses high pressure water jets to separate bark and log, is also commonly practiced.

Energy requirements are different for friction barking and hydraulic barking, using 0.9 HP per day per cord for friction barking and 1.4 HP per day per cord for hydraulic barking (6). About 12-15% of the weight of the roundwood processed is removed by barking (7). The bark removed is burned in bark boilers to produce steam and power used in mill operations. The combustible bark produces 10.5 million BTU of gross heating value for every ton of bark burned, with the average efficiency of bark boilers at 70% (8).

As prime virgin fiber grows short in supply, alternate sources of pulpable material will be needed. One of these is whole tree chipping, in which tops and branches trimmed from pulpwood during logging operations are chipped at the forest site. This material is then shipped to mills, where it is blended with virgin wood chips and pulped (9). A second source receiving increasing attention is sawdust and wood waste from lumber and veneer mills (10). In the future, other agricultural products, such as cornstalks, wheat straw, and bagasse, may receive more common use (11). The economic importance of these practices is large in terms of lower feed costs and reduced labor and woodyard requirements, but their energy conservation potential is probably minor.

PULP MAKING

The second step in the conversion of wood to paper is the manufacture of pulp, the principal objective of which is to decompose the wood structure and recover the fibers in usable form. Wood pulp is manufactured by a variety of processes which are determined to some degree by the use to which it will be put. Table 12-5 provides

Table 12-5

Characteristics of Major Pulpmaking Processes

	Stone groundwood	Refiner groundwood	Thermomechanical	Chemimechanical	Neutral sulfite semichemical (NSSC)	Sulfite	Kraft
Raw materials	Spruce, pine, hemlock and poplars	Softwoods, hardwoods or mixed; resinous woods if pre-pressed; non-wood vegetable materials	Same as refiner groundwood	Hardwoods (with cold soda process); light hardwoods and less-resinous softwoods (with hot sulfite process)	Hardwoods, single or mixed; softwoods usable but require more refining.	Hardwoods, low-resin softwoods	Virtually any woody material
Products	Unbleached: board, partial newsprint furnish, cheap papers. Bleached: tissue, book, commercial and writing papers	Same as stone groundwood but stronger and higher quality due to longer fibers	Same as refiner groundwood	Partial furnish for newsprint and publication papers	Unbleached: corrugating medium, various boards; Bleached: book and specialty papers	Unbleached: news print Bleached: tissue, towelling, glassine, food packaging board; book, bond, other fine papers; α-cellulose pulp for rayon manufacturing	Unbleached: bag and wrapping paper, linerboard, box-board, container-board. Semibleached: news-print, boxboard Bleached: bond, book, commercial and specialty papers; food container board; tissue
Pulp characteristics	Short fibers, hence limited strength	Improved fiber length and tensile strength	Like refiner groundwood	Similar to re-finer ground-wood; hot sulfite pretreat-ment gives good brightness	Long-fibered strength approaches kraft, especially in bleached state	Light color; easy bleaching; can produce very low lignin product	Longest fibers. Very strong, especially in un-bleached state
Typical yield	To 95%	To 95%	To 90+%	85-94%	60-80%	45-55%	45-55%

Source: Kaplan, Energy Use and Distribution, p. 25.

information on the feed materials and product characteristics of the
major pulping processes used in the U.S. today.

Pulping has been performed by using two generic types of processes: chemical and mechanical. There are four primary chemical
processes--kraft, sulfite, neutral sulfite semichemical (NSSC), and
chemi-mechanical. A fifth process, soda, is declining in use. The
mechanical process can be divided into groundwood and thermo-mechani-
cal pulping.

In the chemical processes, the cellulose fibers of the wood are
separated from the noncellulose components by chemical action in a
hot (270-375°F) sulfur solution. The logs to be pulped are first
debarked, made into chips, and then fed into digesters, where they
are cooked for 3-6 hours under steam pressure in the presence of a
cooking liquor. The lignin and other non-cellulose components are
dissolved, leaving the cellulose or pulp.

The digestion step in chemical pulping is carried out in either
batch or continuous digesters. Batch digesters are the older of the
two types, and they are gradually being replaced by continuous digesters which offer advantages in greater compactness, increased
throughput, and reduced labor and energy utilization. The energy
associated benefits derive from the lower and more uniform steam consumption and from the ease with which the continuous digester can be
adapted to computerized automated control (12). Heating of the charge
in batch digesters has been done via direct steam injection, but indirect heating via heat exchangers and continuous circulation has
recently been used. Pulp yields are generally in the 50% range, because nearly all of the lignins are removed during full chemical
pulping. Table 12-6 shows the relative energy requirements for the
chemical pulping processes.

Kraft Pulping

Figure 12-13 shows the principal operations of the kraft process.
Approximately 70% of the paper and paperboard manufactured in the
U.S. is made using the kraft process, which requires about 4-9 $\times$ 10^{6}

Table 12-6

Energy Requirements for Chemical Pulping

Energy end use	Process			
	Chemimechanical pulping 10^6 BTU/ton	Neutral Sulfite Semichemical (NSSC) 10^6 BTU/ton	Acid sulfite 10^6 BTU/ton	Kraft 10^6 BTU/ton
Digestion energy				
Steam	0-1.6	2.2-3.0	3.0-5.8	2.2-4.8
Refining energy				
Mechanical	2.8-4.8	0.6-2.0	0.4-0.6	0.4-0.6
Pumping, screening and washing				
Steam	0.4-0.6	0.4-0.6	0.4-0.6	0.4-0.6
Mechanical	0.5-0.7	0.5-0.7	0.5-0.7	0.5-0.7
Bleaching energy				
Steam	n.a.	1.0	0.7-0.85	1.0-2.0
Mechanical	n.a.	0.2	0.2	0.2-0.25
Approximate average	6.0	4.5-5.3[a] 5.7-6.7[b]	5.2-6.5[a] 6.0-8.5[b]	4.2-5.5[a] 7.0-9.0[b] 5.4-7.5[c]

[a]Unbleached.

[b]Bleached.

[c]Semibleached

SOURCE: Kaplan, Energy Use and Distribution, p. 32.

BTU per ton of pulp. The primary energy consumption operations are digestion of wood chips, evaporation of water from the cooking liquor, and calcining of wet $CaCO_3$ to lime. The major energy sources for the energy intensive heating operations are process wastes, natural gas, and fuel oil. Most of this fuel produce high pressure steam, which is used to produce electricity and lower pressure steam for the process operations. The electricity is used in a number of operations, which include barking, chipping, pumping, screening, draining, pressing, and drying.

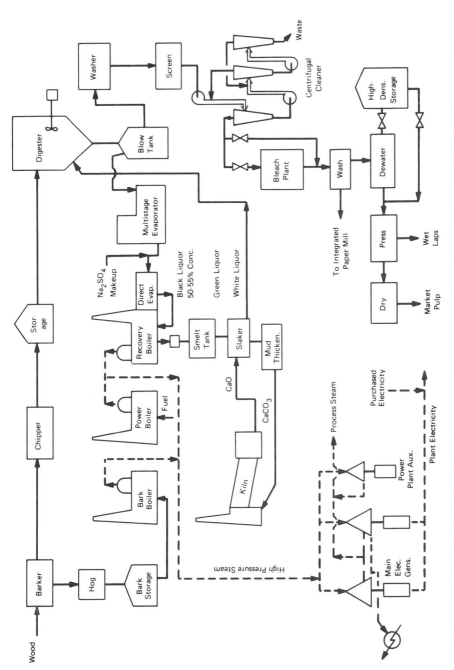

Figure 12-13. Kraft Pulpmaking Process. Source: Kaplan, Energy Use and Distribution, p. 35.

The kraft process has become the predominant process route in the U.S. for the following reasons:

- It can use a wide variety of species of wood, including southern yellow pine, which has a high resin content.

- It is capable of recovering from the black liquor much of the process chemicals, plus a large fraction of the thermal energy needed for the process.

- It provides a higher yield of pulp for a given degree of brightness, which results in a lower cost per ton than the other chemical pulps.

- It produces important byproducts, such as tall oil and turpentine.

- It has greater strength than other chemical pulps.

- It is compatible with mild steel process equipment for most of the operations.

- It can be substituted in white papers and board and in other uses previously dominated by sulfite pulp because of multi-stage bleaching.

The two disadvantages of the kraft process are: (1) the greater difficulty in bleaching than in sulfite pulps and (2) the disagreeable odor produced.

Sulfite Process

Figure 12-14 shows the principal operation of the sulfite process. Sulfite pulp is the second most important of the chemical pulps produced in the U.S., producing about 5% of the total pulp production. Prior to 1937, the production of sulfite pulp exceeded that of kraft pulp, but since then, the output of sulfite has increased only slightly, whereas the production of kraft has increased so that it now far exceeds that of sulfite.

Sulfite pulp is made from non-resinous softwoods, principally western hemlock, spruce, and fir. The cooking liquor employed in the sulfite mills is calcium sulfite or bisulfite. While there are many differences involved in the preparation of the pulping solution and

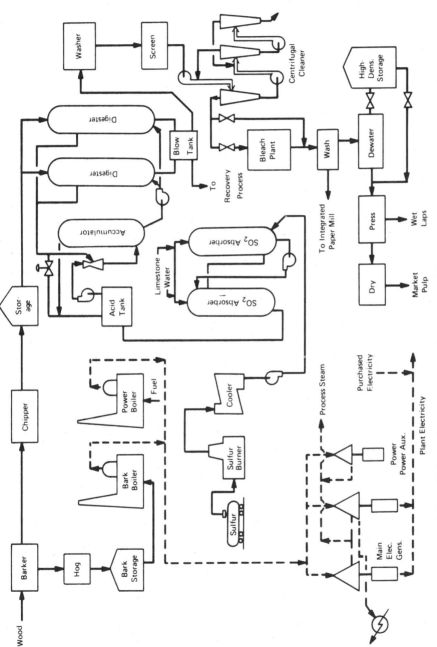

Figure 12-14. Sulfite Pulpmaking Process. Source: Kaplan, Energy Use and Distribution, p. 36.

in the treatment of spent pulping liquor for the different chemical
pulping processes, there are also basic similarities. Wood preparation and post digestion operations are nearly interchangeable. The
digestion step consists basically of heating the chips in a liquid
solvent, and while average cooking times for sulfite pulp are somewhat longer, the average energy expended to make a ton of chemical
pulp by any of the processes is not very likely.

Semichemical Pulping

Semichemical pulping involves digestion in a hot $(320^{\circ}F\text{-}375^{\circ}F)$
sulfite or bisulfite ion containing solution for 1/2-3 hours, depending upon whether continuous or batch digestion is practiced, again
followed by mechanical refining. In commercial practice, nearly all
semichemical pulping is conducted with sodium sulfite-bisulfite
solution at pH near 7 and is hence referred to as Neutral Sulfite
Semichemical Pulping, or NSSC. The NSSC pulping process is illustrated in Fig. 12-15. NSSC pulping accounts for about 8% of the total
pulp production. It consumes somewhat more electricity than kraft
pulping but somewhat less fuel in other forms. NSSC pulping requires
from 4.5 x 6.7 x 10^{6} BTU per ton and yields about 77% pulp.

There are three main features in the NSSC process (13):

- impregnation of hardwood chips with cooking liquor

- cooking at high temperature

- mechanical fiberizing

Some mills buy the cooking chemical, although most mills prepare it
on the premises by burning sulfur and absorbing it in soda ash or
ammonia. Newer mills employ continuous digesters, even though a large
percentage of NSSC pulping still occurs in batch digesters.

In some mills, the softened chips are compressed in one or more
stages of screw presses as they come from the digester. This helps
to obtain maximum recovery of spent liquor and partial washing with
minimum dilution (14). Either from this stage or directly from the

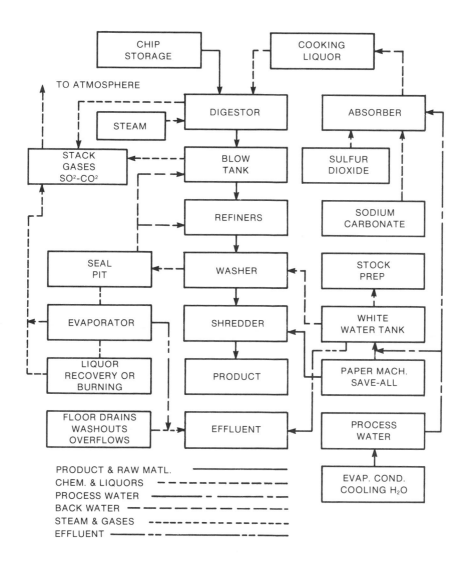

Figure 12-15. Neutral Sulfite Semichemical Pulp Process Diagram. Source: Environmental Protection Agency, Unbleached Kraft and Semi- chemical Pulp, (Washington, D.C.: Government Printing Office, May 1974), p. 29.

digester, the chips are sent to a disk mill for fiberizing. The chips
then undergo washing, screening, and cleaning.

The pulp is diluted with white water from the paper mill to the
consistency of feed for the secondary refiners, which service the papermaking operation. Either hardwoods or softwoods may be processed
by NSSC, but hardwood pulping predominates. The major commodity produced by this method is corrugating medium for packaging.

Chemimechanical Pulping

The chemimechanical process differs from the semichemical in that
comparatively little dissolving of the wood occurs during pulping.
Two variations of this process are the cold caustic and the chemi-
groundwood processes. Chemimechanical pulping employs a mild chemical
treatment, such as one-half hour soaking in $85^{o}F$-$105^{o}F$ soda solution
or approximately $300^{o}F$-$320^{o}F$ sulfite solution, to soften the chips.
In the cold caustic process, wood chips are treated in a mild chemical
and are then fiberized in disc refiners; in the chemi-groundwood process, whole logs are subjected to a mild chemical pretreatment and
then are ground in a conventional grinder. Energy requirements for
chemimechanical·pulping is about 6×10^{6} BTU per ton. Pulp yields
range from about 85 to over 90%, depending upon wood species and processing conditions. The method is used almost exclusively with hardwoods.

Mechanical Process

Wood pulp produced by mechanical means is known as groundwood.
Groundwood processes can work with cut logs (stone groundwood) or with
wood chips (refiner groundwood). In refiner groundwood, pretreatment
such as steaming or chemical softening is sometimes used to reduce the
mechanical energy required for pulping. As the pretreatment becomes
more intricate, the distinction between mechanical and chemical pulping becomes blurred. The mechanical pulp process are shown in Fig.
12-16.

In stone groundwood, barked logs precut to a uniform length are
mechanically pressed against a revolving grindstone. The pulp is

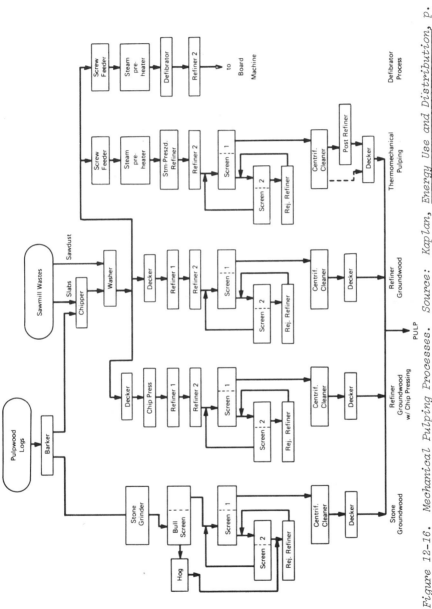

Figure 12-16. Mechanical Pulping Processes. Source: Kaplan, Energy Use and Distribution, p. 24.

suspended in the wash water and continuously piped from the grinders to reject recovery and fiber classifying stages. The pulp suspension leaving the stone grinders is passed through a bull screen to remove large material or unground residue from log scraps and is next pumped to finer screens that separate acceptable fibers from oversize splinter and fiber bundles. The large material constituting approximately 1% of the grinder output are normally fed into a hammer mill. The hogged fragments are then combined with the rejects from the fine screens and reground in a refiner.

Refiner groundwood uses, as feed material, chips which are suspended in water as a thick slurry and fed to refiners that are high speed disk mills. The proportion of water to wood feed is an important parameter in chip refining and must be controlled to maintain pulp quality and minimize energy use. Refining is performed in multiple stages, after which the refiner output is fine screened to remove oversize material, which is sent to reject refiners. The acceptable pulp flow to storage or to the papermill.

Chips lend themselves well to thermal or chemical pretreatment, as the pretreatments lessen the mechanical energy required to refine them. Thermomechanical pulping is now coming into use for making newsprint and other paper pulps. The chips are first sent to a preheating chamber, where they are subject to 2-3 minutes of steam under pressure. They are then discharged into a screw conveyor that charges the heated chips into a disk refiner. The ligneous component of the wood is softened by the heat, which permits the refining to be carried out with less energy than in raw chip refining. The energy requirements and disposition for the mechanical pulping processes are summarized in Table 12-7.

Groundwood pulp accounts for 9-10% of the total pulp production and consumes large quantities of electricity. Because of their high electricity consumption, groundwood mills tend to be located in regions of low electricity cost or of abundant hydropower.

Table 12-7

Energy Requirements for Mechanical Pulping Processes

Pulping process	Wood form used	Wood preparation energy 10^6 BTU/ton	Mechanical energy required 10^6 BTU/ton	Thermal energy required 10^6 BTU/ton	Total pulping energy required 10^6 BTU/ton
Stone groundwood	Logs	0.10	4.5-5.2	a	4.5-5.2
Refiner groundwood	Chips or sawdust	0-0.13[b]	4.5-5.8	a	4.5-5.8
Refiner groundwood with pressing	Chips	0-0.13[b]	4.8-5.8	a	4.8-5.8
Defibrator (hardboard)	Chips	b	0.5	1.1	1.6
Defibrator (insulating board)	Chips or sawdust	b	1.4-1.7	0.9-1.1	2.2-2.8
Thermomechanical	Chips	b	3.7-5.5	0.2-0.4	3.9-5.9

[a]Assumes shower water heated by steam from refiners.

[b]Part of the chips, etc., are obtained from lumber mill waste, where only a minor portion of the original cutting energy can be allocated to this material.

SOURCE: Kaplan, Energy Use and Distribution, p. 31.

Rapson Effluent Free Kraft Process

A completely effluent free bleached kraft pulp mill has been developed by Professor W. H. Rapson of the University of Toronto. Using this technology, the Great Lakes Paper Company has entered the startup phases of its new, $150 million, 250,000 tons per year bleached kraft pulp mill at Thunder Bay, Ontario. This is the first venture in building an effluent free kraft mill by incorporating a closed cycle process into a standard kraft mill (15).

In the Rapson process, a number of changes in the conventional
kraft process have been made. The major ones are the following:

- About 70% of the chlorine normally used in the first
  bleaching step has been replaced with chlorine dioxide.
  This stage is followed by the conventional sequence of
  caustic extraction, chlorine dioxide, etc.

- The R-3 process is used for chlorine dioxide generation.
  In this process, sodium sulfate is crystallized from the
  aqueous sulfuric acid used as a reaction medium, and the
  sulfuric acid is recycled to the chlorine dioxide generator. The only byproduct output from the chlorine dioxide
  generator is solid sodium sulfate, which is used as makeup chemical in the kraft pulp mill.

- Countercurrent washing through the entire bleaching sequence
  makes it possible to reduce the total amount of fresh process water into the pulp mill from 25,000-4,000 gallons per
  ton.

- All bleach plant effluent is reused in the countercurrent brown
  stock washers. In this manner, all of the bleach plant chemicals and dissolved organics eventually go to the recovery furnace.

- Various mill process changes were made to close the screening
  and cleaning systems. Because of this, a somewhat larger recovery furnace is required to accommodate the organics recycled
  from the bleach plant, and a larger black liquor evaporation
  facility is needed.

- Salt is removed from the recovered cooking liquor by evaporating
  the liquor and filtering off the crystallized salt.

A number of advantages are expected by using the Rapson process.
These include (16):

Lower Steam Consumption. Steam usage is reduced, because higher
white liquor concentration permits higher wood to liquor ratio; the
hot water requirement in the bleach plant is drastically reduced,
due to countercurrent washing; and very little steam is required
to heat up pulp prior to its entering a bleaching stage after washing. The combined steam use reduction amounts to about 116,000
lb/hr. Since 62,000 lb/hr of steam is required to concentrate the
mill white liquor in the salt recovery process in a quadruple effect
evaporator, the net steam reduction is 54,000 lb/hr.

Increased Steam Production. More dissolved organic substance is
sent to the furnace with the black liquor by recycling bleach plant
filtrate and closing up the unbleached pulp mill. There is, however,
a net reduction in the input of organics, since the salt input from
the filtrates is more than counteracted by the salt recovery process.
The increased organics are 14,500 tons per year.

Fiber Gain. Fiber losses are reduced by 1% with the recycling
of filtrates and virtual elimination of discharges from the bleaching,
screening, and washing department. This fiber pickup increases pulp
production output and decreases specific chemical usage or makes it
possible to decrease wood input with the attendent lower load on the
recovery boiler.

Higher Pulp Yield. There is a bleached pulp yield increase of
at least one percentage point when compared to a sequence that uses
only chlorine in the first stage, due to the 70% substitution of
chlorine dioxide for chlorine. This increase is due mainly to the
protection of the hemicelluloses.

Lower Consumption of Bleaching Chemicals. A comparison of the
standard kraft bleaching process with the recommended closed cycle
bleaching shows a considerable cost reduction in the closed cycle sequence. If the standard kraft bleaching sequence is practiced and
there is no reclamation of bleach filtrate, there is no need within
the mill to add extra balancing caustic. Then it would be necessary
to install and operate external primary and secondary effluent treatment and, in some areas, tertiary treatment.

Saving External Treatment Cost. The steam stripping of contaminated condensates from the black liquor and subsequent incineration
of the non-condensibles in the lime kiln reduces kraft mill odor,
generates reusable heated water, and reduces BOD by 85-90%.

Lower Water Consumption. Average treated water consumption in
a kraft pulp bleachery amounts to about 18,000 gallons per ton. In
closed cycle mill operation, the quantity is basically reduced to
zero. However, some fresh water is used during startups. Heated
cooling water can be discharged without effluent treatment or can be

used in some plants for associated facilities. At mill sites where
temperature levels of reused water are of concern, cooling towers may
be installed.

Salt Recovery. One hundred and twenty lb/ton of salt is removed
and purified in the salt recovery process. The sodium chloride may be
used for generation of bleachery agents or for roads, etc.

When compared with conventional kraft pulping and bleaching
methods, the technical/economic evaluation of the Rapson process indicates it would provide significant energy savings (7 million versus
2 million BTU/air dried ton, respectively) and cost savings ($290
versus $259 per air dried ton, respectively) (17). In addition, since
much of the overall water effluent associated with bleached kraft pulp
manufacture originates in the bleach plant, the elimination of efflu-
ents from this source would cause a major reduction in the overall
water effluent from an integrated kraft pulp and paper mill.

This comparison clearly indicates the theoretical superiority
of the Rapson process on all points. If operating experience at
Thunder Bay confirms that the process performs as designed, it is
expected that the Rapson process would become the chosen technology
for new kraft pulping capacity.

Oxygen Pulping

Over the last few years, extensive efforts have been made in
oxygen pulping research. Oxygen pulping would control all sulfur
emissions from kraft mills and greatly alleviate the bleach plant
effluent problem. The low solubility of oxygen and the fact that
wood chips do not readily absorb it require a two stage pulping process. See Fig. 12-17 for a generalized material and energy balance
for the oxygen pulping and bleaching process. The wood chips are
first treated in a similar manner to those of the kraft process. The
softened chips are discharged to disc refiners, disintegrated mechanically while still under pressure, and washed and separated from the
black liquor. A second digestion occurs in which the pulp is treated
with oxygen under alkaline conditions at about 250°F and about ten

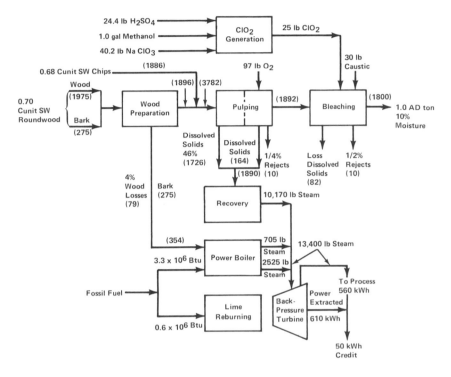

( ) Indicates bone-dry (BD) lb; 1.0 air-dry (AD) ton = 1800 BD lb
1 cunit = 100 ft³ of solid wood; weight per BD cunit ranges from about 2200 lb to 3400 lb, depending on wood species.
*Heat-to-power conversion in this type of turbine is typically about 4000 Btu/kWh.

*Figure 12-17. Material and Energy Balance for A-O Slush Pulp.
Source: Arthur D. Little, Inc., estimates.*

atmospheres pressure. At the end of the oxygen reaction time, the
pulp is washed and sent to the remaining three bleach stages (18).

Modifications to the basic two stage oxygen pulping process have
also been investigated. In one, wood chips are pretreated with a
soda pulping process that is followed by a defibration, but the second
pulping stage is done in the presence of sodium carbonate, instead of
caustic with pH controlled in the range of 7-8. The pulp yield is
higher than kraft, but strength is weaker.

Another alternative in oxygen pulping is the use of a sulfide first stage, before the oxygen treatment. In this variation, wood is first treated with about 15% sodium sulfide to approximately 75% yield. Following the defibration, a standard type of oxygen pulping treatment with caustic and magnesium carbonate is employed. Pulps of higher yield, equivalent to kraft pulp in burst, tensile, and ring crush, can be obtained. They have much lower tear strength than kraft pulp and are much brighter in color.

Table 12-8 compares the process economics for the oxygen pulping process with standard kraft and with the Rapson processes. The table indicates that there would be a substantial reduction in initial plant

Table 12-8

Comparison of A-O and Rapson Process with Standard Kraft Slush Pulp (New mill basis)

Item	A/O	Standard Kraft	Rapson Process
Plant Investment, incl. pollution control ($ million)	146	154	140
Operating Cost, incl. pollution control ($/ADT)	256	290	259
Purchased Energy (10^6 BTU/ADT)	4.3	7.4	2.1
Pollution Loads			
Water—Volume (10^3 gal/ton)	31	31	20
BOD_5 (lb/ton)	33	66	None
TSS (lb/ton)	66	66	None
Color (lb/ton)	150	300	None
Air Emissions			
Particulates (lb/ton)	146	200	200
TRS (lb/ton)	0	24	24

Source: Environmental Protection Agency, **Environmental Considerations,**
Vol. V, **Pulp and Paper,** p. 9.

investment and operating costs with the oxygen process, although it
is not as efficient as the Rapson process. The comparison is not
entirely correct, because oxygen pulping is not a direct substitute
for kraft pulping. The major uncertainty is the quality of the pulp.
At high brightness levels, quality appears to be lower than that of
kraft.

The first alkaline oxygen kraft pulp mill in the U.S. has been
installed by Weyerhaeuser in Everett, Washington. The mill was built
on an experimental basis, with kraft pulping capabilities and with
provisions for converting to alkaline oxygen pulping.

Solvent Pulping (19)

Virtually every commonly available chemical has been evaluated
at one time or another for potential use as a pulping chemical. Solvent pulping is an approach using organic chemicals to dissolve and
extract lignin from wood physically, rather than by chemically attacking it, as in the other methods. Since there is little chemical
damage to lignin, solvent pulping gives a relatively high yield pulp.
Since many solvents boil at lower temperatures than water, energy
costs can be considerably lower. In most cases, quality is equal to
kraft.

There are no solvent pulp mills in the world because of the difficulty anticipated in recovering the solvent to the extent necessary
to make the process economical. If a solvent, such as ethanol, costing
$0.30 per pound, is used, only about 40 lbs/ton could be lost before
the process becomes uneconomical. Factors such as solvent toxicity
and flammability must also be considered in the evaluation of a new
solvent pulping process.

Biological Pulping

Over 1000 species of wood attacking fungi are known, but most
of these have not yet been screened for pulping properties (20).
However, researchers in Sweden have isolated a strain of fungus that

digests lignins in wood while leaving the cellulose content intact.
Other possibilities for biological pulping agents include bacteria
and yeasts. Once such factors as adequate speed, process reproduce-
ability, control conditions, and the physical properties of the product have been standardized, biopulping methods might consume significantly less energy than conventional pulping. Except for initial
warming, this type of process might not require heat addition; in fact,
low grade heat could conceivably be a byproduct of such a reaction.
Energy savings of 1.7-4.3 BTU/ton might be realized through biological
pulping (21).

PAPERMAKING

The term papermaking generally includes all of the equipment used
in handling and forming the refined water suspension of mixed fibers
into the finished paper product. Regardless of type, papermaking
machines have three basic units--the web-forming unit, the press section, and the dryer.

The paper machine furnish, which consists of a combination of
refined pulps together with additives, such as fillers and whiteners,
starts in the stock chest at a consistency of 2-4% solids. It then
passes through the stages of cleaning, screening, and deaerating.
Following this the feed is diluted by recycled white water and enters
the headbox of the machine at a consistency of about 0.5% solids.
Coming from the headbox, the feed is formed into a sheet by either
Fourdrinier or cylindrical machines. Table drainage increases the
sheet dryness to 2.5%, and suction boxes increase it further to about
12%. The suction couch roll at the end of the Fourdrinier wire increases dryness to 18-23%. Presses remove more water and increase
the dryness to 35-50%. The remaining moisture is removed by evaporation in the dryer, although the degree of dryness depends on the grade
of paper being produced. Coating and finishing operations follow the
drying section, and finally, the sheet is rolled for cutting and
shipment.

The primary energy requirement for the papermaking process is thermal energy for water removal and drying. Various methods, including improved drying and waste heat recovery, are available for reducing the energy load. Several experimental paper-forming techniques, for example, high consistency paper-forming, are being investigated.

Drying

Drying is by far the single largest consumer of purchased energy in the entire pulp and paper manufacturing process. Paper and paperboard are produced in dozens of varieties, employing various types of dryers and additives. Broadly speaking, the paper is compacted and dried by its passage over and between an array of heated, felt-covered steel rolls that are usually surrounded by a hood or a complete housing.

The dryer section consists of a number of hollow steel cylinders over which the paper web passes. The cylinders are rotated in synchronization, and heat is supplied by low pressure steam condensing inside the cylinders. The water evaporated from the felt is carried away by a current of warm air passing through the housing and exhausting to the atmosphere. For certain types of paper, drying is augmented by external air impingement as the sheet passes over a special type of heated roll. The air blast is heated by direct firing, by medium pressure steam, or by electric coils.

Because of the high energy requirement for drying, there is a large incentive for decreasing the evaporative work required in drying paper. To accomplish this, several systems can be used: waste heat recovery, a Yankee dryer hood, radio frequency, efficient infrared process heaters, and improved pressing.

Waste Heat Recovery. There is little or no recovery of waste heat in U.S. mills. Old mills have open hoods and low temperature exhaust air, although it would be desirable to have closed hoods for such operations. Most of the technology for utilizing as much as 50% of the heat that is exhausted from dryer stacks are presently

available. Economizers and heat pumps, as well as the instrumentation
to control air flows, are on the market.

Figure 12-18 shows a diagram of a typical dryer section exhaust
air heat recovery system which can achieve heat savings of approximately 0.2×10^6 BTU/ton of water, or 8-10%, in comparison with a
system with an open hood. In addition, a closed dryer hood allows

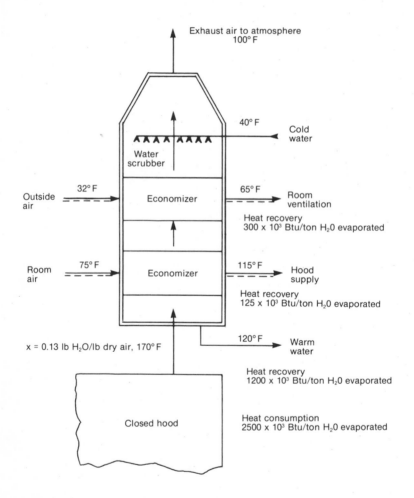

*Figure 12-18. Exhaust Air Heat Recovery System. Source: Raimo
Asantila et al., "Design for High Cost Energy—The Scandanavian Approach," TAPPI, 57 (October 1974), 120.*

a reduction in air flow by as much as 60% of what is required for a
canopy type hood (22), permits controlled recirculation of the dryer
air, and permits better heat recovery. The arrangement also facili-
tates control of the machine room space ventilation, independent of
the dryer requirements.

Heat exchangers are justifiable on a conservation basis, even
for the relatively low temperature air ($150°F-200°F$) in a steam roll
dryer hood. A well designed recuperator section can recover 10% of
the dryer input heat, and by adding waterheating economizer sections
upstream and downstream from the recuperator, another 10-20% can be
saved (23).

Yankee Dryer Hoods. Air impingement dryers are used in special
product drying applications in which steam heated roll dryers are
either inadequate if used alone or are not well suited to the requirements. The Yankee dryer roll is internally steam heated and has a
very highly polished surface. The damp sheet is rolled tightly against
it, with essentially no intervening air film. As the roll rotates,
the exposed side of the sheet is dried by a stream of hot, high velocity air from a dryer hood that surrounds two-thirds to three-fourths
of the periphery.

Heat input to the sheet is high because of the good conductive
bond between the sheet and roll on one side, and the very thin diffu-
sion boundary layer on the air side. Water removal rates as high as
150 $lb/hr/ft^2$ of roll surface have been achieved in this way, compared
to 6.5-13 $lb/hr/ft^2$ for a conventional felted roll (24). Commonly,
the air blast is heated to $570°F-850°F$ by using electric resistance
heaters, through indirect heating via finned tubes, or through direct
firing with gas or oil.

Impingement drying requirements depend upon the moisture perme-
ability and retentivity of the sheet. A modern Yankee dryer with a
roll heated by steam will use about $5.5-7 \times 10^6$ BTU per ton of sheet,
assuming 65% moisture by weight in the entering material and 5% moisture at the removal point. About 55% of the drying heat will be supplied by the air side. This compares to $6.5-12.5 \times 10^6$ BTU per ton

of sheet, using steam roll drying. Dryer hoods are equipped with heat
recovery devices, so that the supply air enters the heater section at
a temperature around 550°F.

The annual production of paper products involving Yankee dryers
is around 8% of total production (25). Air blast heaters of lower
rating, as well as radiant heaters and, occasionally, dielectric units,
are used where extra capacity is needed and space requirements are
restrictive.

Radio Frequency Drying. Use of radio frequency drying at the
final stage has been successful and has resulted in significant savings
in steam used for each ton of paper produced. It has been calculated
that adding 60 KWH R. F. energy to a ton of paper saves about 560 KWH
equivalent of steam. This is equal to about 5.76×10^6 BTU per ton of
paper (26).

Infrared (IR) Process Heater. Infrared energy can be generated
in a flat quartz infrared heater in a relatively narrow wave band.
In comparison, natural gas heaters provide a mixture of radiation and
convection heat in a wide band, which is to a large extent exhausted
out the stack instead of applied to the product being processed. To
be efficient, a radiant heater should concentrate its output within
the spectrum of the peak absorption characteristics of the matter being
heated, e.g., between 2.7 and 2.9 microns of infrared rays for water.
A flat quartz infrared heater can do this.

Some of the potential applications of IR radiant heat to the
papermaking process are as follows (27):

- At the beginning of the drying section (predryers). The
  incoming web usually has about 50% water content at approximately 130°F. If the temperature is boosted to 190°F-200°F,
  the drying process can be speeded up by beginning the evaporation earlier.

- At the size press. A flat quartz heater can control uniformity of drying. A twelve inch IR heater is usually sufficient,
  depending on the required production speed.

- At the dry end. A heater installation here can help boost
  production speed and insure a streak-free finish product.

• In the coating or converting sections. Supplementary flat
quartz infrared drying can aid in efficient drying and in
curing any type of coating.

Recent application at Federal Paper Board Company and Weyerhauser's Longview, Washington, mill have shown the cost and operating
expenses of radiant energy to be attractive. The Weyerhauser installation is the largest IR installation in the U.S.

Improved Pressing. More concentrated effort is needed in the
area of pressing. The paper industry has been moving to heavier
loaded presses and are squeezing out more water from the paper before
drying, since whenever water can be squeezed out instead of evaporated,
much less energy is used. Depending upon the type of sheet being
produced, modern roll pressing conditions commonly range from about
7800 pounds per lineal inch for paper grades to about 12,000 pounds
per lineal inch for the heavier board grades (28). The upper limit
is governed by the ability to press without crushing the fibers in the
sheet. At 12,000 pounds per lineal inch, exit moisture conditions
for 42 pounds linerboard have been measured at 57-58% water by weight
(29). Pressing technology continues to improve, and one U.S. mill
reported adding a third press with 16,000 pounds per lineal inch capacity into a linerboard machine, which added 15% to existing production
capacity (30). This represents approximately 150 pounds additional
moisture per 1000 pounds removed mechanically rather than by heating,
an energy savings of 0.36 x 10^6 BTU per ton. A 1% drier sheet means
4-5% less energy required in the dryer.

High Consistency Paper-Forming

Present paper machines use a very low consistency aqueous suspension of fibers, about 0.5%, during forming to produce a sheet in which
the fibers are uniformly distributed. Large amounts of energy must
therefore be used to remove the water. On the other hand, high consistency forming uses a slurry containing over 1-2% solids, which reduces energy requirements. Because of fiber entanglement as the sheet
is forming, the resulting sheet is more open and tends to drain better in the press. While this reduces the energy load to the dryer,

the sheet has more bulk and lower strength, limiting its application to tissues and composite sheets.

The major operating concerns with this high consistency technique are the difficulties of cleaning the pulp slurry in conventional cyclone cleaners, the lack of basic knowledge regarding control of flocing at high consistency, and the absence of demonstrated proof that high consistency forming can produce acceptable sheet properties under typical industry conditions. However, recent pioneering efforts in paper-forming at high consistencies greater than 1% indicate that potential savings in capital equipment costs, reduced energy requirements, and increased production are possible. These experiments include the Wiggins Teape RAD-Foam process, the Swedish Cellulose Research Institute's headers, and the Lodding K Former. The Lodding K Former claims a reduction in water throughput by as much as a factor of 4 and a reduction in fuel demand of the papermaking phase of production by 55% (31). Although these prototype high consistency paper-forming techniques are available, considerable developmental work will be needed before the high consistency headbox can match the formation and uniformity obtainable from conventional headboxes.

GREATER USE OF RESIDUE FUELS

The fibers required for papermaking are primarily in the woody portion of trees. Very little useful fiber is contained in the bark. The large volume of bark generated requires some means of disposal. Approximately 50% is used currently by the industry as fuel for steam raising. About 300-400 pounds of bark are produced for each standard cord of pulpwood, of which small amounts are processed and sold as byproducts. The average heating value of the dry bark is 8700 BTU/lb; the fuel value of the bark is 2.6×10^6-3.5×10^6 BTU per cord. The industry uses approximately 70×10^6 cords of pulpwood a year, with a total potential bark energy of 180×10^{12} BTU/year (32). Fuel consumption reports issued by the American Paper Institute show that

bark fuels yield about 100 x 10^{12} BTU/year. The percentage use of the
available bark is, therefore, approximately 50%.

Wood is a composite of three generic constituents--resins, lignin,
and cellulose--each with its own BTU value, according to carbon and
hydrogen content. Resins represent 2.8% of the wood and contains
17,000 BTU per pound; lignin represents 18-30% of the wood and contains 13,000 BTU per pound; and cellulose represents 62-80% of the
wood and contains only 7500 BTU per pound (33). In the chemical pulping process, the resins and lignin are dissolved, constituting about
15-20% of the black liquor leaving the digester. In order for the
fuel value of the black liquor to be used, it must be concentrated.
Water is evaporated from the liquor in a multi-effect evaporation system until the solids concentration is approximately 50%. This step
is necessary to allow the recovery of caustic and sulfide contained
in the liquor and to allow the use of organics in the black liquor as
fuel. In addition, soaps that are used to make tall oil are obtained
in this step. Steam at a pressure of 35-80 psi is used to provide
heat for this operation.

Figure 12-19 shows black liquor being concentrated to approximately 65% in a direct heat evaporator. The black liquor is then
burned in a furnace, and caustic and sulfide are recovered in a dissolving tank. The burning of the black liquor supplies heat which is
used to make high pressure steam, which is in turn used to produce
electricity and provide process steam. Flue gases from the furnace
supply heat to the direct heat evaporator. Approximately 10% of the
heat produced in the furnace is used to reduce makeup sodium sulfate
to sulfide in the bottom of the furnace.

The energy required to concentrate and burn the black liquor
negates, to a large extent, the energy available from the solids.
New and improved methods of water removal or of water reduction would
increase the energy recovery from black liquor. New and improved
evaporation systems, such as vapor recompression, should be investigated. It is estimated that 25% of black liquor could be conserved
if new techniques were used.

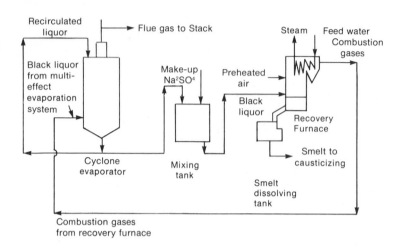

Figure 12-19. Direct Heat Evaporation and Recovery Furnace.
Source: John Reding and B. Shepard, Energy Consumption: Paper,
Stone/Clay/Glass/Concrete, and Food Industries (Midland, Michigan:
Dow Chemical Company, April 1975), p. 12.

About twelve million tons of waste paper are now being recovered
annually for use in making paper. This is presently used mostly in
non-integrated mills. The value of waste paper as a source of fiber,
compared with virgin fiber, depends on the particular grade of waste
paper involved and on the kind of virgin fiber it replaces. The fiber
value of waste paper is obtained by calculating the cost of virgin
fiber in a ton of paper or board and then multiplying that figure by
the yield.

Table 12-9 shows the investment and operating costs for converting
fiber into pulp and then into newsprint. It is much more economical
to make pulp from recycled newsprint than from virgin fiber, primarily
because of lower energy and maintenance costs. On the other hand,
the conversion of pulp to newsprint is more expensive because the
smaller scale of operation typically used with secondary fiber offsets much of the cost advantage of the pulping operation. Alternative-
ly, papermakers can use a blend of virgin and deinked fibers in a large

Table 12-9

Comparative Cost of Newsprint Manufacture
(1977 dollars; f.o.b. Factory Costs)

	Case 1 (New Mill)	Case 2 (New Mill)	Case 3A (Existing Mill) 20% Recycled 80% Virgin	Case 3B (Existing Mill)
Basis: Grade	Virgin	Recycled		Virgin
Production: tpd	550	330	550	550
tpy	187,000	112,000	187,000	187,000
Investment: $ Million	126	80	Assume $45 Million Book Value + New Deinking	37
$000/daily ton	230	240	Not Applicable	N.A.
Cost Item ($/ton)				
Fiber	40	38**	39	40
Conversion	80	30	76	88
Capital Related*	41	39	22	13
Total Slush Pulp	161	107	137	141
Conversion	47	60	47	41
Capital Related*	66	78	17	19
Total Paper (Excludes Slush Pulp)	113	138	64	66
TOTAL PULP & PAPER	274	245	201	207

* Includes depreciation and cost of capital at 5.5% and 10.5% of investment respectively.
** Derived cost of waste paper at its fiber value.
SOURCE: Arthur D. Little, Inc., estimates.

scale, integrated operation. This strategy permits greater output from a machine that is limited by insufficient pulp availability and it enables a manufacturer to divert virgin fiber from one machine to another.

Although paper manufacture from recycled fiber consumes less energy overall than equivalent production from virgin fiber, the average consumption of purchased energy tends to be higher per unit of product. This is because the processing of virgin fiber allows energy recovery from chemical pulping and wood preparation wastes. This is valid when comparing recycled fiber processing with kraft

processing, but it is not true universally. The net energy recoverable
from spent chemical pulping liquor combustion varies inversely with
the process yield and approaches zero for NSSC under certain conditions. Since the overall energy required to produce paper or board
from secondary fiber is less than from virgin pulp via NSSC, purchased
energy for the secondary fiber must also be less than for the virgin
pulp. Similarly, mechanical pulping relies greatly on purchased energy, except for bark burning, so here, too, recycled fiber should enjoy an energy advantage over mechanically pulped fiber.

     Energy is conserved when waste paper is burned in the sense that
wood is used as a fuel. Except for highly filled or heavily coated
papers that contain a large amount of ash, all grades of waste paper
have essentially the same heat of combustion, about 8000 BTU per pound.
Figure 12-20 compares the energy values with the fiber values of waste
paper. OCC is clearly more valuable as a source of fiber for linerboard than as a fuel. The case for old news is not quite as clear
cut, since its fiber value is generally higher than its energy value
at average costs, but the reverse is true if we substitute old news
for the highest cost coal or oil. Figure 12-20 indicates that mixed
waste is more valuable as a substitute for any of the fossil fuels
than as a fiber source.

     From these evaluations, we cannot generalize about the relative
merit of recovering waste paper for its energy or its fiber values.
The most economical use for waste paper depends on which fossil fuel
it would replace and on what kind of recycled paper would be made
from it. Furthermore, we cannot say whether industry can reduce its
energy usage by recycling more waste paper, because there are as many
instances in which the reuse of waste paper would actually increase
the fossil fuel requirement as there are those in which the fuel requirement is decreased. We do know, however, that secondary fiber
recovered from waste paper is competitive with virgin fiber, and that
the characteristics of waste paper pulping operations make the addition
of small increments of pulping capacity at an existing mill site
economically attractive.

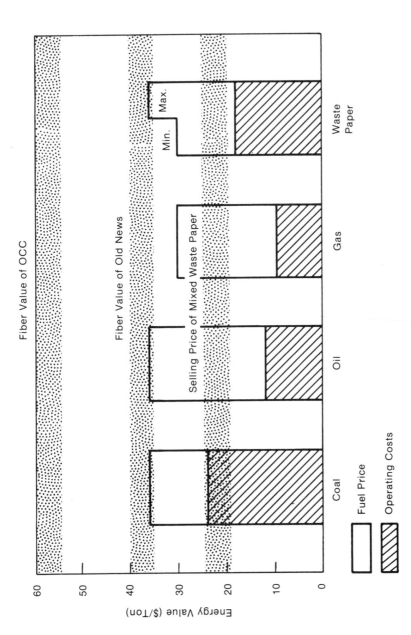

Figure 12-20. Comparison of Fiber and Energy Values of Waste Paper. Source: Fred Iannazzi, "Comparison of Fiber Values of Waste Paper," Paper presented at TAPPI Pulping/Secondary Fibers Conference, Washington, D.C. November 6-9, 1977, p. 13.

CO-GENERATION

The pulp and paper industry purchases about half of the total electrical needs of the industry. The remaining half is generated internally by means of combined turbogenerator-process steam systems (co-generation). The relative efficiency of in-plant electricity generation under paper industry conditions versus the generation by an electric utility has been pointed out in Chapter 6.

Most of the byproduct electricity is generated in new, large, integrated mills with capacities on the order of 500-1000 tons per day. Electricity generation in integrated mills varies widely, ranging from total internal generation to zero generation. Non-integrated mills tend to purchase all of their required electricity. These mills are small in size and have low capital cost installations. As a rule, they produce process steam directly at the low pressures and temperatures required for papermaking and purchase electricity.

Some problems do exist in co-generation in the pulp and paper industry:

- The electricity that could be generated is far in excess of the needs of the industry, so that part of it must be sold to surrounding plants or to utilities.

- The exclusive use of backpressure turbine generator sets for power in a paper mill requires a degree of interaction between mill and utility not commonly encountered in the U.S., plus exceptional stability of operation if the system is to realize the economies of which it is capable.

- Other considerations include reliability of electric supply, especially for relatively small power plant installations; maintenance cost of equipment; relative assurance of fuel supply for the utility's generating sets, as compared with those of the mill; the local cost of fuel and electricity; and the efficiency of existing equipment.

ENERGY CONSERVATION POTENTIAL

The theoretical minimum amount of purchased energy required to produce paper is zero, because the industry is in essence working with a fuel. We know, however, that the pulp and paper industry uses a great deal of purchased energy. Gains in energy efficiency in the past have been through greater use of wood wastes, greater use of spent black liquor, improved waste heat recovery, greater use of waste paper, and advances in process control. Energy consumption per unit of output has been steadily declining. The energy requirements for purchased energy will decrease faster than the total energy needed for the production of pulp and paper. The greater decline of purchased energy results from the greater utilization of residue fuels, from the decline of products made in non-integrated paper mills, and from more on-site power generation.

There is still much to be done in improving energy efficiency in the pulp and paper industry. Improved management techniques, as discussed in Chapter 2, can result in significant energy savings. In addition, there are a number of new process modifications and improvements on the horizon, which can result in significant energy efficient improvements. Table 12-10 reveals the potential for energy conservation for various alternatives in the pulp and paper industry.

Current operating practices by the pulp and paper industry have developed over many years as a result of extensive development by individual paper companies and equipment manufacturers. Because of its low cost, fuel consumption played a secondary role in the establishment of these practices. Fuel costs for the production of paper have typically been about 5% of the total; however, this has recently increased to about 15%. With further fuel cost increases, the industry can be expected to pay increased attention to modifications in equipment and procedures which can reduce fuel consumption.

Major process changes being implemented by the industry will reduce energy consumption as well as alleviate many of its air, water,

Table 12-10

Summary of Energy Conservation Potential
(over conventional practices)

Conservation Practice or Technique	Potential Savings[*] 10^6 BTU/ton
Continuous rather than batch digesters	1.3
Co-Generation	1.5
Recycle waste paper	1.0
Increase wood waste use	2.0
Increase waste heat recovery	3.5
Rapson Process	5.3
Oxygen Pulping	2.9
Yankee dryer	2.0
Dryer hood heat recovery	0.7
RF Drying	5.76
Improved pressing (for each 1% increase in water removal)	0.25
Energy Recovery from black liquor	2.0
Computer application to energy management in mills	1.5
Improved kraft chemical recovery	0.7

[*]These savings are not additive and are based on efficiency improvements over 1974 conventional practices.

and solid waste emission problems. Thus, major process changes, such as the Rapson effluent free kraft process and oxygen pulping, which are in the commercial development stages, will begin to increase their market penetration. However, the magnitude of capital costs required in the pulp and paper industry, plus difficulties associated with perfecting new processes, tend to preclude rapid introduction of new or radically different techniques of production.

An exception is computer control of individual processes, which is becoming more commonplace. Since it was first introduced in paper-making operations, the computer is now being used for pulping, bleaching, and pollution control processes. Further expansion to other areas is necessary. In all of these areas, the computer aids in optimizing operating conditions and in balancing process flows. Thus, both cost and fuel consumption are reduced.

Overall, use of the techniques discussed which are economical and are presently available can result in an energy savings of 20-40% over conventional practices, as well as in substantial productivity gains.

NOTES AND REFERENCES

1. A. Kouvalis, Industrial Practice and Technology Review: The Pulp and Paper Industry (Argonne, Illinois: Argonne National Laboratory, April 1977), p. 3.

2. Environmental Protection Agency, Environmental Considerations of Selected Energy Conserving Manufacturing Process Options, Vol. V, Pulp and Paper Industry Report (Cincinnati, Ohio: Environmental Protection Agency, December 1976), p. 20.

3. Joint Textbook Committee of the Paper Industry, Pulp and Paper Manufacture, 2nd ed. (New York: McGraw Hill, 1950), p. 105.

4. J.M. Duke and N.J. Fudali, Report on the Pulp and Paper Industry's Energy Savings and Changing Fuel Mix (New York: American Paper Institute, September 1976), p. 5.

5. Ibid., p. 8.

6. American Gas Association, A Study of Process Energy Requirements in the Paper and Pulp Industry (New York: American Gas Association, 1965), p. 20.

7. R.G. MacDonald, Pulp and Paper Manufacture, Vol. 1, The Pulping of Wood, 2nd ed. (New York: McGraw Hill, 1969), pp. 18-19.

8. J.M. Duke, Patterns of Fuel and Energy Consumption in the U.S. Pulp and Paper Industry (New York: American Paper Institute, March 1974), Appendix IV.

9. M.J. Osborne, "The APPA-TAPPI Whole Tree Utilization Committee," TAPPI, 57 (December 1974), pp. 5-7.

10. H. Wahlgren, "Forest Residues--The Timely Bonanza," TAPPI, 57 (October 1974), p. 117.

11. C.A. Moore, et al., "Economic Potential of Kraft Product," TAPPI, 59 (January 1976), p. 117.

12. D.B. Brewster and W.I. Robinson, "How Computers are Controlling Functions in the Pulping Process," Pulp and Paper, 48 (May 1974), p. 88.

13. S.A. Rydholm, Pulping Processes (New York: Interscience Publishers, 1965), p. 423.

14. MacDonald, Pulp and Paper Manufacture, Vol. 1, The Pulping of Wood, p. 238.

15. "Great Lakes Paper Launches First Closed Cycle Kraft Pulp Mill," Paper Trade Journal, 161 (March 15, 1977), pp. 29-34.

16. A. Kouvalis, Some Qualitative and Quantitative Information on Variations of the Kraft Process (Chicago, Illinois: Argonne National Laboratory, September 1977).

17. Environmental Protection Agency, Environmental Considerations, Vol. 1, Industry Summary Report (Cincinnati, Ohio: Environmental Protection Agency, December 1976), p. 41.

18. Ibid., p. 85.

19. Esther Dorfman, Pulp and Papermaking Technology (Tuxedo Park, New York: International Paper Company, January 1976), p. 60.

20. T.K. Kirk and J.M. Harkin, "Lignin Biodegradation and the Bio conversion of Wood," AICHE Symposium Series, 133 (1973).

21. Kouvalis, Some Qualitative and Quantitative Information, p. 81.

22. Joint Textbook Committee of the Paper Industry, Pulp and Paper Manufacture, 2nd ed. (New York: McGraw Hill, 1950), p. 538.

23. Ibid., p. 539.

24. Ibid., p. 469-71.

25. S.I. Kaplan, _Energy Use and Distribution in the Pulp, Paper, and Boardmaking Industries_ (Oak Ridge, Tennessee: Oak Ridge National Laboratory, August 1977), p. 53.

26. Kouvalis, _Industrial Practice and Technology Review_, p. 108.

27. Ibid., p. 110.

28. R.A. Deane, "Papermaking and Finishing," _Chemical Process Technology Encyclopedia_, ed. D.M. Considine (New York: McGraw Hill, 1975), pp. 806-813.

29. A.J. Schmitt, "Sheet Quality and Modern Press Section Arrangement," _TAPPI_, 56 (October 1973), p. 56.

30. "Union Camp Adds Big Third Press to Board Machine," _TAPPI_, 57 (June 1974), p. 15.

31. D.J. Kalmes, _A Review of the Lodding Paper Making Project_ (Cambridge, Massachusetts: M/K Systems, January 1973).

32. Sander Nydick and John Dunlay, _Recommendations for Future Government Sponsored R & D in the Paper and Steel Industries_ (Waltham, Massachusetts: ThermoElectron Corporation, August 1976), p. 1-1.

33. Robert T. Clark, "Energy," _Pulp and Papermaking Technology_, ed. Esther Dorfman, p. 107.

34. U.S. Department of Commerce, _1972 Census of Manufactures: Papermills, SIC 2611_ (Washington, D.C.: Department of Commerce, February 1974), p. 3.

Chapter 13

PETROLEUM REFINING

Petroleum refineries use over 3 x 10^{15} BTU of energy per year, which represents about 4% of the annual U.S. energy consumption and about 15% of annual industrial energy consumption. The petroleum refining industry is primarily engaged in producing gasoline, kerosene, distillate fuel oils, residual fuel oils, lubricants, and other products from crude petroleum. The industry in the U.S. processed 12,457 x 10^3 barrels per day in 1975, which includes 4,107 x 10^3 barrels per day of imported crude oil. Table 13-1 shows the pattern of products since 1940. As is evident, the major product is motor gasoline, of which 6,516 x 10^3 barrels per day was produced in 1975 [1]. The most obvious trend in the table is a decrease in the production of residual fuel oil and the increase in middle distillate products (jet fuel and distillate fuel oil). Since 1960, gasoline production has remained roughly constant as a percentage of total crude runs.

U.S. refinery capacity has been increasing steadily since 1932. On the other hand, the number of refineries reached a peak in the late 1930s and has been decreasing steadily. Figure 13-1 shows the changes since 1930. In 1975, the industry consisted of 135 companies operating 266 refineries. Table 13-2 shows the impact of large companies which operate nearly 80% of the U.S. capacity. Table 13-3 provides industry breakdowns by size of refinery. It shows that the forty-seven largest

Table 13-1

U.S. Refinery Yield Pattern
(percentage of production, by volume)

	Motor Gasoline	Kerosene	Jet Fuel	Distillate Fuel Oil	Residual Fuel Oil	Other Products
1940	43.2	5.7	—	14.9	23.7	12.5
1945	40.6	4.7	—	15.6	26.2	12.9
1950	43.0	5.6	—	19.8	19.4	12.2
1955	44.0	4.3	2.1	22.6	14.7	12.3
1960	45.2	4.6	3.0	23.0	10.6	13.6
1965	44.0	2.8	5.8	23.5	7.6	16.3
1970	45.5	2.4	7.5	22.7	6.1	15.8
1971	46.3	2.1	7.4	22.4	6.3	15.5
1972	46.5	1.8	7.2	22.2	6.8	14.6
1973	45.8	1.7	6.8	22.5	7.7	15.5
1974	46.2	1.3	6.8	21.8	8.7	15.2

Source: Bureau of Mines, **Minerals Industry Surveys** (Washington, D.C.:
              Government Printing Office, 1940-1974).

refineries account for 62% of the petroleum produced, and that more
than half (54%) of the nation's refineries have capacities of less than
30,000 barrels per day, accounting for only 9.7% of the total capacity.

Products and the specific processes used to produce them vary
considerably among refineries, causing the kinds and quantities of
energy consumed also to vary significantly. The most energy intensive
companies use more than ten times as many BTUs per barrel of oil processed as the least energy intensive companies. This reflects differences in refinery complexity and product mix. Table 13-4 summarizes
general refinery process and production capacities related to refinery
size as of 1974.

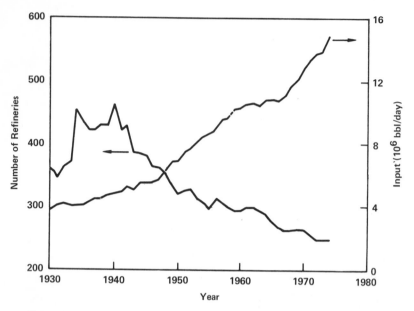

Figure 13-1. U.S. Refinery Capacity and Number. Source: Virgil Haynes, Energy Use in Petroleum Refineries (Oak Ridge, Tennessee: Oak Ridge National Laboratory, September 1976), p. 27.

MANUFACTURING PROCESSES

Petroleum refining is a very diverse industry because of wide variations in process methods, feedstocks, and product mix. Figure 13-2 shows a typical refinery flow chart. Initially, crude oil is distilled at nearly atmospheric pressure. Heavy distillates are removed at the bottom of the distillation tower, and lighter distillates are removed at the top. The lighter distillates are further separated into crude natural gas, gasoline, and naptha. Some of the products are further distilled in a vacuum to separate the heavier molecular weight components from those of lower molecular weight. The fractions having high boiling points are used as feedstock to thermal or catalytic crackers, where hydrocarbon chains are decomposed to increase

Table 13-2

Industry Breakdown by Company Processing Capacity

Capacity Range MBPD	Number of Companies	Number of Refineries	Total Capacity MBPD	Total Capacity % of Total
over 200	19	111	12,435	78.2
100 - 200	9	19	1,338	8.4
50 - 100	11	25	778	4 9
30 - 50	10	11	609	2.6
10 - 30	34	47	700	4.4
under 10	52	53	233	1.5
Total	135	266	15,693	100.0

Source: Gordian Associates, **An Energy Conservation Target for Industry SIC 29**
(New York: Gordian, June 1976), p. 24.

Table 13-3

Industry Breakdown by Refinery Size

Refinery Size, MBPD	Number of Refineries	Total Capacity MBPD	% of Total
over 500	1	700	4.4
300 - 500	7	2,591	16.3
200 - 300	8	2,033	12.8
100 - 200	31	4,499	28.3
50 - 100	40	2,975	18.7
30 - 50	36	1,561	9.8
10 - 30	57	1,122	7.1
under 10	86	412	2.6
	266	15,893	100.0

Source: Gordian Associates, **An Energy Conservation Target**, p. 24.

Table 13-4

Distribution of U.S. Refinery Process and Product Capacities
Relative to Refinery Size—January 1974

Process	Percent of U.S. refineries[a]/percent of U.S. crude-oil capacity[a] for refinery size ranges (10^3 BTU/bbl) of:					
	0-25	25-50	0-50	50-100	100-450	0-450
Vacuum distillation	46.1/23.0	76.1/28.8	54.7/26.6	95.2/32.0	100.0/39.1	70.0/35.2
Thermal processes[b]	15.7/ 4.6	34.7/ 5.7	21.1/ 5.3	57.1/10.6	75.0/10.0	36.8/ 9.3
Catalytic cracking	20.9/12.4	71.7/28.2	35.4/22.1	97.6/31.4	93.2/30.1	56.3/28.9
Catalytic reforming	41.7/10.3	84.8/18.6	54.0/15.3	100.0/22.4	100.0/22.5	70.0/21.2
Catalytic hydrocracking	2.6/ 0.6	13.0/ 2.4	5.6/ 1.7	14.3/ 3.4	56.8/ 7.3	16.2/ 5.4
Catalytic hydrorefining	4.3/ 1.4	21.7/ 4.4	9.3/ 3.3	35.7/12.8	40.9/ 7.3	19.4/ 6.4
Catalytic hydrotreating	31.3/10.0	87.0/24.4	47.2/18.9	97.6/31.7	100.0/32.8	65.2/29.9
Alkylation	13.0/ 1.8	69.6/ 5.7	29.2/ 4.3	95.2/ 6.1	90.9/ 5.4	36.8/ 5.3
Aromatics/isomerization[c]	2.6/ 0.2	15.2/ 0.7	6.2/ 0.5	45.2/ 3.0	47.7/ 2.3	20.2/ 2.2
Lube production	13.0/ 2.8	8.7/ 0.3	11.8/ 1.2	7.1/ 0.4	38.6/ 1.7	16.2/ 1.3
Asphalt production	36.5/10.6	58.7/ 7.2	42.9/ 8.5	38.1/ 3.6	59.1/ 3.3	44.9/ 4.3
Coke production	4.3/ 0	21.7/ 0	9.3/ 0	30.9/ 0	52.2/ 0	20.6/ 0

[a]Percent by size range — not percent of all refineries.
[b]Cracking; visbreaking, coking.
[c]BTX; hydroalkylation;cyclohexane; C_4 feed; C_5 feed; C_5 and C_6 feed. Production capacity.

Source: Haynes, **Energy Use in Petroleum Refineries**, p. 33.

the yield of gasoline and light oils. The residue that exits the bottom of the distillation tower is processed by visbreaking, catalytic cracking, hydrocracking, lube oil manufacture, or asphalt manufacture.

The requirement for increasing yields of light products over the years has led to a continuous development of conversion process technology whereby residual fuel fractions are converted to distillates and gasoline blending components. These processes included thermal cracking and coking in the early years and, more recently, include catalytic cracking and hydrocracking. Catalytic cracking continues to be the single most important refinery conversion process.

Improved product quality has also led to major technological advances. The need for higher octane gasolines had been satisfied primarily by the thermal reforming processes until the late 1940s, when catalytic reforming became dominant because more favorable yields could be gained. The catalytic reforming process itself has seen many improvements over the years, such as the use of bi- and multi-metallic

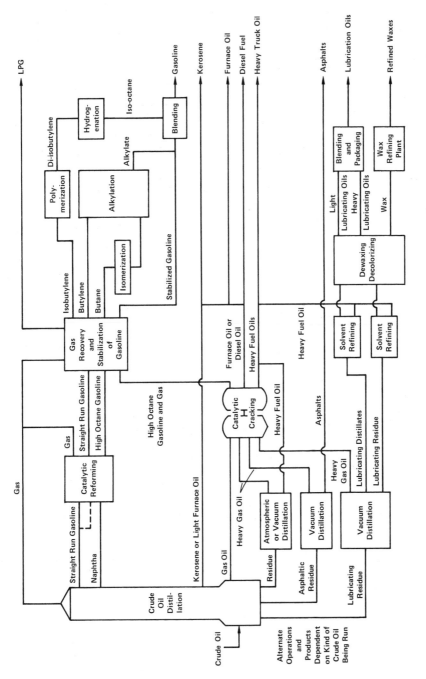

Figure 13-2. Typical Refinery Process Flow.

catalysts. Current developments include a "continuous reforming"
process, in which a portion of catalyst is continuously regenerated
in a separate section of the plant, allowing fresh catalyst activity
to be continued without shutting down the complete unit.

The growth of catalytic reforming has made large quantities of
hydrogen available. Processes to improve naptha and distillates based
on catalytic hydrogenation have subsequently been developed and are
being used universally. Currently, hydrotreating processes are being
employed on heavy fractions, even residual fuel oils in some refineries.

The following sections discuss the various process options in a
typical refinery.

Crude Distillation

The first step in the manufacture of petroleum products is the
separation of the crude oil into light distillates (gases, naptha,
kerosene), gas oil, and residue. See Fig. 13-3. Each of these distillation products contains many compounds and boils within a limited
range. Separation is accomplished by pumping crude oil through heat
exchangers into a distillation column at a temperature high enough
to separate the crude oil into its various fractions. Because the
required temperature at atmospheric pressure is often high enough to
decompose the crude oil, steam is injected into the tower to lower the
required feed temperature to about $650^\circ F$-$700^\circ F$. Gases are removed in
side streams and are partially recycled through a stripper that may use
steam. The heavier fraction is transferred for further processing,
such as vacuum distillation.

Energy requirements for atmospheric distillation consist of fuel
to heat the crude oil, steam to inject into the atmospheric tower and
stripper, and energy to pump the fluids. Energy required to heat the
crude oil depends on the extent of heat exchange, percent of crude
vaporized, and composition of the feed. Energy requirements for atmospheric distillation of crude oil in a plant making liberal use
of heat exchange and air preheat are approximately as follows (2):

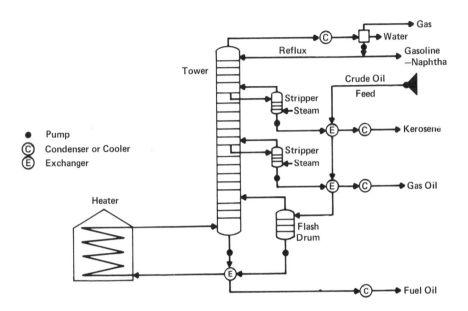

Figure 13-3. Crude Oil Distillation Unit. Source: Federal Energy Administration, The Potential for Energy Conservation in Nine Selected Industries, Vol. 2, Petroleum Refining (Washington, D.C.: Government Printing Office, 1975), p. 219.

	Energy Requirement (Btu/bbl)	Process Temperature (°F)
Electricity[a]	6,000	
Fuel	80,000	675
Steam	10,000	500
TOTAL	96,000	

[a]Electricity converted to energy requirement based on 10,000 Btu/kWhr.

Energy is lost from the atmospheric distillation system as heat in the flue gases from the tube still; convection-radiation heat from the system; heat rejected by air- and water-cooled heat exchangers; and heat in blowdown water.

Vacuum Distillation

In 1974, vacuum distillation capacity was 35% of crude oil capacity (3). The second stage of processing of the residual from the
atmospheric distillation tower obtains heavy distillates suitable for
cracking and for manufacturing lube oils. Feed to the vacuum tower
is heated to about $750^{\circ}F-800^{\circ}F$ and is maintained at 100-150 psig
with a condenser steam ejector system. The bottoms from the vacuum
tower may be used as asphalt, may be coked, or may be further cracked.
Figure 13-4 is a schematic diagram of a vacuum distillation system.

Energy requirements for vacuum distillation consist of fuel to
heat the feed, reboil heat, motive steam for steam jet ejectors, and
energy to pump the fluids. The heat required depends on the composition of the residual, the degree of vaporization desired, and the
vacuum desired, and the vacuum employed. Approximate energy requirements for vacuum distillation of the residual from atmospheric columns
are as follows (4):

	Energy Requirement (Btu/bbl feed)	Process Temperature ($^{\circ}F$)
Electricity[a]	3,000	
Fuel	75,000	775
Steam		
Jets	9,000	500
Process	13,000	500
TOTAL[a]	100,000	

[a] Electricity converted to energy requirement based on
10,000 Btu/kWhr

As in the case of the atmospheric distillation system, energy is lost
from the vacuum distillation system in the form of heat in the flue
gases from the tube still; convection-radiation heat from the system;
heat rejected by air- and water-cooled heat exchangers; and heat in
blowdown water.

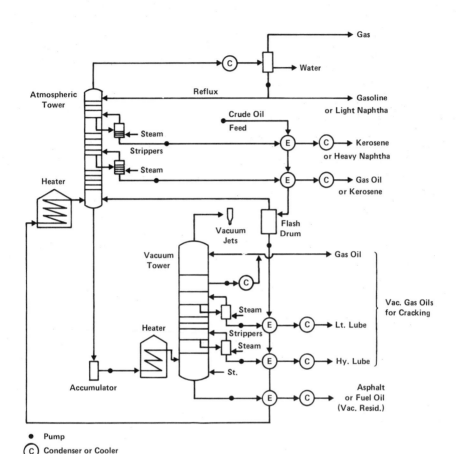

Figure 13-4. Vacuum Distillation Unit. Source: Federal Energy Administration, The Potential for Energy Conservation, Vol. 2, Petroleum Refining, p. 220.

Thermal Cracking Processes

Thermal cracking processes include thermal cracking, visbreaking, delayed coking, and fluid coking. In each of these operations, heavy gas oil fractions are broken down into lower molecular weight fractions, such as domestic heating oils and catalytic cracking stock,

through the use of heat but without using a catalyst. Typical thermal
cracking conditions are $900°F$-$1100°F$ and 600-1000 psig pressure. The
high pressures result from the formation of light hydrocarbons in the
cracking reaction. There is always a certain amount of heavy fuel
oil and coke formed by polymerization and condensation reactions.

Thermal processes are used by 37% of U.S. refineries. Coking is
by far the dominant thermal process, with delayed coking using seven
times more than fluid coking. Most of the remainder of thermal cracking capacity is equally divided between gas oil cracking and visbreaking. Visbreaking is a mild form of thermal cracking used to maximize
furnace oil and minimize gasoline production. Gas oil cracking using
steam dilution is primarily employed to produce light olefins.

Thermal Cracking and Visbreaking. Thermal cracking and visbreaking are similar processes which differ mainly in degree (i.e.,
in time, temperature, and pressure). The many variations of these
processes depend on the characteristics of the feed and the desired
products.

For visbreaking, feed is heated to about $900°F$, and at a pressure
of 50-300 psig, it is mixed with a recycle stream of heated heavy
gas oil (5). See Fig. 13-5. After quenching by a recycle stream of
product heating oil, the gas, gasoline, heating oil, and residuum
are separated in a fractionating section. The residuum is then passed
through a vacuum tower, the overhead returning to the fractionator
and the bottoms becoming a fuel oil.

Coking. Coking is a form of thermal cracking in which heavy
residues are converted into gas, naptha, heating oil, and coke. In
coking, heavy oil is maintained at a high temperature for a longer
period of time than for thermal cracking. The delayed coking process,
for example, similar to thermal cracking in other respects, feeds
only heavy oil to the furnace and uses soaking drums to prolong the
reaction time. The accumulation of coke in the soaking drums must
be removed every one or two days; hence, multiple soaking drums are
provided to allow continuous operation. Furnace outlet temperature
is about $930°F$, and coke drum pressure and temperature are 10-70 psig

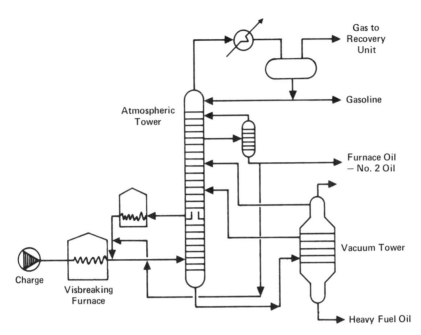

*Figure 13-5. Visbreaking Unit. Source: Federal Energy Admin-
istration, The Potential for Energy Conservation, Vol. 2, Petroleum
Refining, p. 221.*

and about 810°F, respectively. Figure 13-6 shows schematic diagrams
of the delayed coking system and the fluid coking systems.

Fluid coking is a continuous process in which coke is deposited
on coke fines as they emerge from a heater chamber that receives coke
from the reactor vessel. Heavy residue is fed to the reactor vessel
in which there is a fluidized bed of coke particles. The top of the
reactor vessel has a scrubber fractionator to separate the particulate
material and hydrocarbon fractions from the heavier oils being recycled. Because some of the coke is burned, the fluid coking process
produces less coke than the delayed coking process. Operating temperatures and pressures are 900°F-1,050°F and 0-15 psig.

Energy requirements for delayed coking consist of steam for
driving the water-jet pumps (used to remove coke from the soaking
drums and sometimes to drive feed pumps), fuel for the heater, and

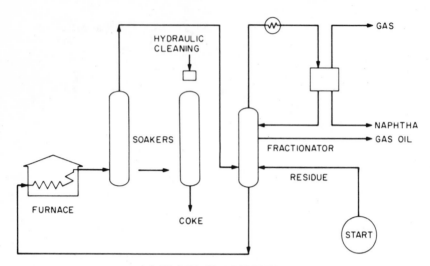

(a) DELAYED COKING PROCESS

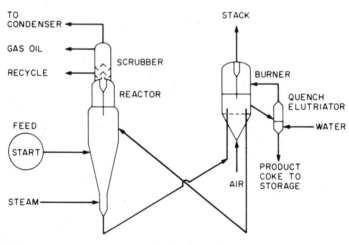

(b) FLUID COKING PROCESS

Figure 13-6. Coking Processes. Source: G. D. Hobson, Modern Petroleum Technology, 4th ed. (New York: John Wiley and Sons, 1973), pp. 283, 285.

electricity for other pumps. Sometimes steam is also used for re-boiling in the fractionator section. Energy requirements for fluid coking consist of steam for fluidizing the reactor bed (and for

gasification in the alternate version) and sometimes for driving
feed pumps. Electricity is also used for pump drives. No external
fuel supply is required for process heating; in fact, a net production
of steam can result.

The substantially lower energy consumption shown for the fluid
process in the table below is misleading, because energy is supplied
by burning coke that is produced in the process. Overall differences
in energy consumption probably are not significant, and although
delayed coking is used much more than fluid coking, the choice is
influenced more by the kind of coke desired and by considerations
such as maintenance and operating continuity. Approximate energy
requirements for coking are as follows (7):

	Energy Requirement (Btu/bbl feed)			Process Temperature	
				Delayed	Fluid
	Delayed	Fluid[a]	Fluid[b]	°F	°F
Electricity[c]	70,000	15,000	160,000		
Fuel	80,000	0	0	930	1000
Steam	45,000	100,000	-134,000	750	500
TOTAL	195,000	115,000	26,000		

[a]Without gasification.

[b]With gasification.

[c]Electricity converted to energy requirement based on 10,000 Btu/kWhr

Catalytic Cracking

Catalytic cracking breaks heavy fractions, principally gas oils,
into lower molecular weight fractions. The use of a catalyst permits
operations at lower temperatures and pressures than those in thermal
cracking and inhibits the formation of undesirable polymerized
products. Fluidized bed catalytic processes, in which the finely
powdered catalyst is handled as a fluid, have largely replaced the
fixed bed and moving bed processes which use a beaded or pelleted
catalyst. Figure 13-7 shows a schematic flow diagram of fluid catalytic cracking.

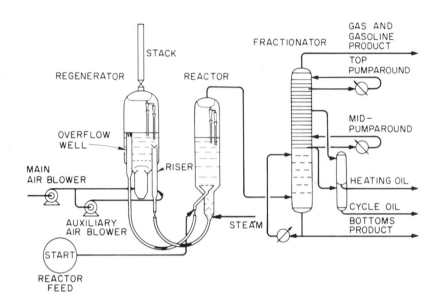

Figure 13-7.   Fluid Bed Catalytic Cracker Units.   Source: "1974 Refining Process Handbook," Hydrocarbon Processing, 53 (September 1974), 118.

The catalytic cracking process involves thermal decomposition, with primary catalytic reactions occurring between the primary products.   The catalyst is usually heated and then lifted into the reactor area by an incoming oil feed that is immediately vaporized upon contact.   Vapors from the reactors pass upward through a separator, which removes most of the entrained catalyst.   These vapors then enter the fractionator, where the desired products are removed and heavier fractions are recycled.

In the fluid bed process, the hot, regenerated catalyst supplies some and often all of the energy to heat the feed.   The catalyst is regenerated at a temperature of $1,050°F$-$1,400°F$ and at pressures of 10-30 psig.   Steam is used to strip the hydrocarbon products from the catalyst and is sometimes used in the fractionator system.   Flue

gas from the regenerator is pressurized, hot, and combustible--con-
ditions which provide opportunities for recovering energy and which,
if fully exploited, could make the fluidized bed catalytic cracker
net energy generators. Approximate energy requirements for catalytic
cracking are as follows (8):

	Energy Requirement (Btu/bbl feed)	Process Temperature (^{O}F)
FCUU		
Electricity[a]	50,000	
Steam	35,000	500
Total	85,000	
FCUU[b]		
Electricity[a]	57,000	
Fuel	50,700	900
Steam	- 177,000	500
Total	- 69,000	

[a]Electricity converted to energy requirement based on 10,000 Btu/kWhr

[b]Industry source estimate for a new, large refinery.

Hydrocracking

Hydrocracking is a pressurized hydrogen catalytic process used
to convert a wide range of hydrocarbon feedstocks to high octane
gasoline, reformer feed, and high grade fuels by cracking, hydrogenation, isomerization, and hydrotreating reactions. See Fig. 13-8.
Most hydrocracker designs are of the fixed bed type. The temperature
for hydrocracking is 644^OF-790^OF and the pressure, 1200-2000 psig.
Energy requirements can vary considerably with the feed and product
requirements.

Approximate energy requirements for hydrocracking, with and without energy required for the hydrogen consumed, are as follows (9):

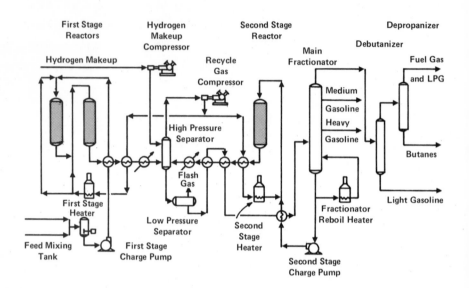

Figure 13-8. Fixed Bed Hydrocracker Unit. Source: Hobson, Modern Petroleum Technology, p. 313.

	Energy Requirement (Btu/bbl feed)	Process Temperature (°C)	(°F)
Electricity[a]			
Hydrocracking	100,000		
Hydrogen	19,000		
Fuel			
Hydrocracking	200,000	380	716
Hydrogen	388,000		
Steam			
Hydrocracking	13,000	260	500
Subtotal			
Hydrocracking	313,000		
Hydrogen	407,000		
TOTAL	720,000		

[a]Pumps and compressors electrically driven. Electricity converted to energy requirement based on 10,000 Btu/kWhr.

Catalytic Hydrorefining

Catalytic hydrorefining is used to treat middle distillates, catalytic cracker feeds, and stock feeds, and to desulfurize residual and gas oils. Many variations of catalytic hydrorefining exist; however, they are all basically very similar. Hydrogen and feed are heated and passed over a fixed bed catalyst. The materials are then cooled, and the gases, after H_2S removal, are recycled as the liquids are further separated into the desired products. Temperature and pressure are in the range of $559°F$-$851°F$ and 200-1000 psig. Figure 13-9 is a schematic diagram of a catalytic hydrorefining unit.

Energy for catalytic hydrorefining is needed to heat the mixed feed and the gas stream to the necessary reaction temperature and to pump and compress the fluids. Approximate energy requirements with and without energy required for the manufacture of hydrogen are as follows (10):

	Energy Requirement (Btu/bbl feed)	Process Temperature (°F)
Electricity[a]		
Hydrorefining	22,000	
Hydrogen	8,000	
Fuel		
Hydrorefining	73,000	725
Hydrogen	155,000	1,650
Subtotal		
Hydrorefining	95,000	
Hydrogen	160,000	
TOTAL	260,000	

[a]Electricity converted to energy requirement based on 10,000 Btu/kWhr.

The electricity requirement is for a relatively low pressure and low makeup hydrogen rate; high pressure and high makeup hydrogen could

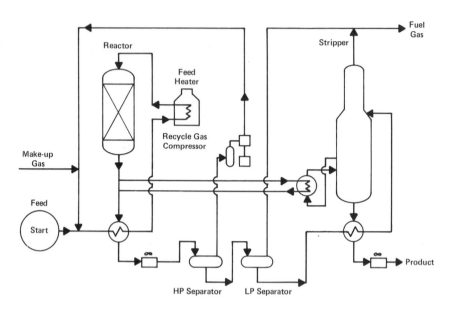

Figure 13-9.   Catalytic Hydrorefining Unit.   Source: "1974 Refining Process Handbook," Hydrocarbon Processing, p. 148.

easily triple the requirement. Thus, to conserve energy, the operation should be controlled both at the lowest practical pressure and with hydrogen conditions commensurate with a satisfactory product.

Catalytic Hydrotreating

Catalytic hydrotreating processes are very similar to the catalytic hydrorefining processes. The main purpose is to desulfurize the feed, which are generally the lighter lower and middle boiling components. Over half of the hydrotreating capacity is for desulfurizing reformer feed, and most of the remainder is for desulfurizing various light refinery liquid products.

The schematic diagram shown for catalytic hydrorefining is also applicable to catalytic hydrotreating, but operating temperatures for the latter are somewhat lower at $347^{\circ}F$-$806^{\circ}F$, and pressures are in the lower part of the hydrorefining range (<600 psig). Energy

is required to heat the mixed feed and the gas stream to the necessary reaction temperature and to pump and compress the fluids. Approximate energy requirements for catalytic hydrotreating with and
without energy required for the manufacture of hydrogen are as
follows (11):

	Energy Requirement (Btu/bbl feed)		Process Temperature (°F)
	a	b	
Electricity[c]			
Hydrotreating	14,000	11,000	
Hydrogen	3,000		
Fuel			
Hydrotreating	39,000	37,000	500
Hydrogen	67,000		1,650
Steam			
Hydrotreating	34,000	48,000	500
Subtotal			
Hydrotreating	87,000	96,000	
Hydrogen	70,000		
TOTAL	160,000	96,000	

[a] Processes that require makeup hydrogen.
[b] Processes that do not require makeup hydrogen.
[c] Electricity converted to energy requirement based on 10,000 Btu/kWhr.

The electricity requirement is for a relatively low pressure and low
makeup hydrogen rate. High pressure and high makeup hydrogen rate
could increase this requirement substantially. As in hydrorefining,
to conserve energy, the catalytic hydrotreating operation should be
controlled both at the lowest practical pressure and with hydrogen
conditions commensurate with a satisfactory product.

Catalytic Reforming

Catalytic reforming is used primarily to upgrade low octane naptha to high octane gasoline by blending components containing significant quantities of aromatic hydrocarbons. A catalytic reforming unit is composed of reaction, separation, and fractionation sections. In the reaction section, the charge is contacted with a catalyst under the proper conditions for the desired reactions to occur. The principal chemical reactions involved are dehydrogenation of napthenes to aromatics, dehydrocyclization of paraffins, hydrocracking of high molecular weight paraffins, isomerization of paraffins and of napthenes, and desulfurization of organic sulfur compounds to form hydrogen sulfide.

In the separation section, the reaction product mixture is cooled and separated into liquid and gas streams. Some of the hydrogen-rich gas is compressed and recycled to the reactors, and the net gas produced is withdrawn from the system. The separator liquid is then sent to fractionation facilities.

Catalytic reforming is used by about 70% of U.S. refineries. There are about fifteen variations of reformers, comprising two basic types--cyclic and semi-regenerative--each of which uses either a conventional platinum on alumina catalyst or a bimetallic catalyst. See Fig. 13-10. Use of the semi-regenerative type predominates, with the use of conventional and bimetallic catalysts being equally divided. Most of the remaining reforming is by the cyclic process, using conventional catalysts (12).

Energy requirements for catalytic reforming consist of fuel to heat the feed and of recycled hydrogen and energy to pump the feed, recycled hydrogen, and coolants. Feed and hydrogen entering the reactors must be at temperatures between $842^{\circ}F$-$1004^{\circ}F$, and, because the predominant reactions in the first two reactors are endothermic, heat must be added to the process stream between reactors. Approximate energy requirements for catalytic reforming are as follows:

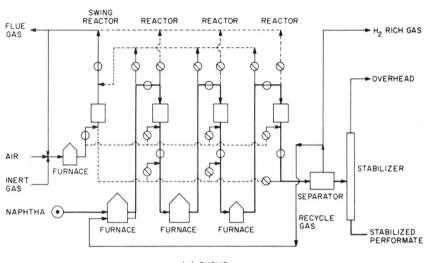

(a) CYCLIC

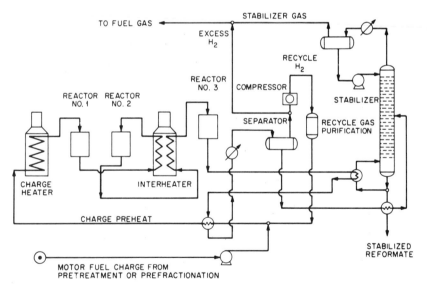

(b) SEMIREGENERATIVE

Figure 13-10. Catalytic Reformer Units. Source: Hobson, Modern Petroleum Technology, pp. 338, 342.

	Energy Requirement (Btu/bbl feed) for:	
	Cyclic	Semi-regenerative
Electricity[a]	50,000	75,000
Fuel	280,000	355,000
TOTAL	330,000	430,000

[a]Electricity converted to energy requirement based on 10,000 Btu/kWhr.

Two features of catalytic reformers affect energy consumption. First, higher pressure and higher hydrogen-to-oil ratios of the semi-regenerative units result in more energy consumption and less hydrogen production. Secondly, bimetallic catalysts allow operation at lower pressure, temperature, and hydrogen-to-oil ratios, resulting in an energy savings.

Alkylation

The alkylation process combines propylene, butylenes, and amylenes with isobutane in the presence of a catalyst to produce high octane branched chain hydrocarbons (alkylate) for use in aviation gasoline and motor fuel. Although alkylation can be achieved thermally, essentially all existing U.S. units use a catalyst--either H_2SO_4 or HF.

The H_2SO_4 and HF processes are very similar, operating at temperatures of $36°F$-$59°F$ with 150 psig pressure in the H_2SO_4 process, and at $59°F$-$122°F$ for the HF process. Removing heat from the reactions requires a refrigeration system for the H_2SO_4 units; HF units can use water cooling, because they can operate at higher, less controlled temperatures. For both processes the isobutane, olefin, and acid are fed to a reactor that is provided with a cooling system. The outlet stream from the reactor enters a settler tank and proceeds to a fractionator where isobutane is stripped for recirculation to the

reactor and alkylate is separated from the butane and propane. Intimate contact of the acid catalyst with the hydrocarbons is provided
by various means, such as stirring or high velocity circulation.
Figure 13-11 shows schematic diagrams of H_2SO_4 and HF alkylation
units.

Energy required for alkylation units is used to pump the feeds,
recycled feedstock, products, and coolant water; to run air-cooled
heat exchanger fans; to operate the refrigerant compressor for H_2SO_4
units; and to heat the reboilers of the fractionation columns. Most
of the energy used by the alkylation process is for reboilers. The
quantity of energy required for alkylation are given in the following
tabulation (14):

	Energy Requirement (Btu/bbl product)		Process Temperature	
	H_2SO_4	HF	H_2SO_4 (°F)	HF (°F)
Electricity	48,000	15,000		
Fuel	160,000	13,000	500	500
Steam	520,000	370,000	750	500
TOTAL	730,000	400,000		

Polymerization is an alternate process for upgrading the octane
rating of fuels. However, alkylation has some basic advantages in
that its product yield is higher, because the isobutane as well as
the olefins are converted and because the alkylation is free of gum-
forming materials. Alkylation will therefore continue as an important
refining process, as long as there is a demand for high octane fuel.

Isomerization

Isomerization is a process by which hydrocarbon structures are
rearranged without adding or removing molecules. It is used to form
branched molecules from straight chain molecules. One important

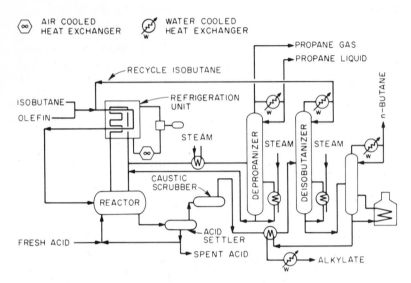

(a) H$_2$SO$_4$ ALKYLATION PROCESS

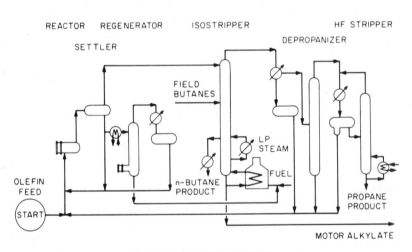

(b) HF ALKYLATION PROCESS

Figure 13-11. Alkylation Units.   Source: "1974 Refining Process Handbook," Hydrocarbon Processing, p. 118.

process, C_4 isomerization, converts normal butane to isobutane (a feed for alkylation units). C_5 and C_6 isomerizations produce iso-paraffins that can be blended into motor gasoline to improve the octane rating. Figure 13-12 shows the butane and pentane/hexane isomerization processes.

The isomerization process consists of contacting the hydrocarbon with a catalyst under proper conditions of temperature and pressure. Aluminum chloride with hydrochloric acid or platinum-containing material are used as catalysts. Operating conditions are essentially in the same range for butane and pentane/hexane isomerization (i.e., $250°F-550°F$ and 200-500 psig, and a hydrogen-to-oil mole ratio of 0.1-0.5:1), with butane isomerization normally performed in the lower portion of the ranges.

Energy required for isomerization is used to pump the feed, recycled feedstock, products, and coolant water; to run the air-cooled heat exchanger fans; and to heat the feed and reboiler streams. Approximate energy requirements for isomerization are as follows (15):

| | Energy Requirement (Btu/bbl feed) | | Process Temperature | |
	C_4 isom.	C_5/C_6 isom.	C_4 isom. (°F)	C_5/C_6 isom. (°F)
Electricity[a]	12,000	15,000		
Fuel	32,000	42,000	350	500
Steam	23,000	45,300	500	500
TOTAL	67,000	102,000		

[a]Electricity converted to energy requirement based on 10,000 Btu/kWhr.

Polymerization

Polymerization is used to produce a high octane gasoline blending component from olefins, typically from propylene, butylene, or mixtures of the two. Phosphoric acid is the most common catalyst used. Solid and liquid phosphoric acid polymerization processes are

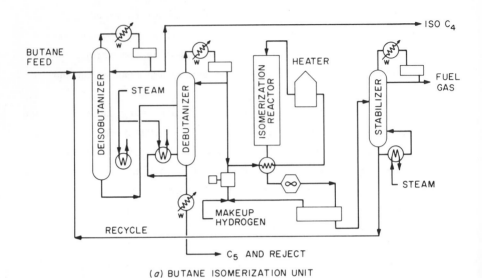

(a) BUTANE ISOMERIZATION UNIT

(b) PENTANE/HEXANE ISOMERIZATION UNIT

AIR COOLED HEAT EXCHANGER   WATER COOLED HEAT EXCHANGER   PROCESS STREAM HEAT EXCHANGER

Figure 13-12.   Isomerization Units.   Source: Haynes, Energy Use in Petroleum Refineries, p. 126.

shown in Fig. 13-13. For the solid polymerization process, the
temperature in the reactor is about 392°F, and the pressure is from
400-1200 psig. The reactions are exothermic, and quenching is required to control the temperature. Control is accomplished by introducing C_3 recycle at several levels in the reactor vessel.

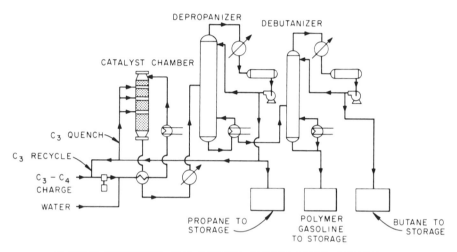

(a) SOLID PHOSPHORIC ACID CATALYST POLYMERIZATION PROCESS

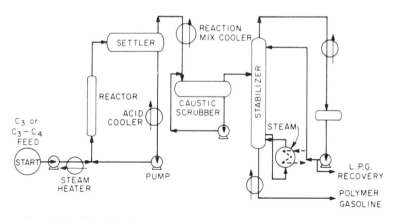

(b) LIQUID PHOSPHORIC ACID CATALYST POLYMERIZATION PROCESS

*Figure 13-13. Polymerization Units. Sources: (a) Hobson,
Modern Petroleum Technology, p. 360. (b) "1966 Refining Process
Handbook," Hydrocarbon Processing, 45 (September 1966), p. 211.*

Energy required for polymerization units is used to pump the feeds, products, recycled feedstock, and coolant water; to run air-cooled heat exchanger fans, and to heat the reboilers of the depropanizer and debutanizer columns. Energy requirements for polymerization are as follows (16):

	Energy Requirement (Btu/bbl product)	Process Temperature (°F)
Electricity[a]	48,000	
Steam	800,000	500
TOTAL	850,000	

[a]Electricity converted to energy requirement based on 10,000 Btu/kWhr.

ENERGY CONSUMPTION IN THE INDUSTRY

In 1974, petroleum refineries used approximately 3.1×10^{15} BTU of energy annually, or about 4% of the yearly U.S. energy consumption. Plant size, product, and location affect refinery energy consumption, as do crude oil composition and the types of processes and equipment used. Light crude oils require somewhat less energy in refining than do heavy crude oils, in order to yield similar products. Similarly, low sulfur crude oils require less energy in refining than do high sulfur crude oils.

The kinds of fuel consumed by U.S. petroleum refineries are given in Table 13-5. Energy consumption per unit of output has decreased slightly at a rate of 0.8% per year since 1960, and in 1974 was about 707,000 BTU per barrel of crude oil. As is evident, most of the energy used is from refinery products: refinery gas, oil, LPG, and coke. About a third of the energy comes from natural gas.

Large variations exist in both the amount of energy consumed per unit of crude oil processed and the kinds of fuel used. Table 13-6 shows both the average energy consumption of various fuels per unit

Table 13-5

Energy Consumed at U.S. Petroleum Refineries

Fuel	Energy (10^{12} BTU/year) consumed for the years:								Percent of energy for 1974
	1962	1964	1966	1968	1970	1972	1973	1974	
Fuel oil	262.3	266.9	251.1	263.2	267.1	276.3[a]	309.1	316.3	10.1
Acid sludge	1.6	1.2	0.9	0.5	0.3	b	b	b	b
Gas	1701.0	1685.0	1783.2	1942.5	2070.5	2169.8	2220.2	2154.9	68.9
LP	23.7	45.0	9.5	23.7	16.2	53.8	40.7	39.5	1.3
Refinery	847.9	785.6	841.4	913.6	993.6	1043.0	1072.5	1043.1	33.3
Natural	829.4	854.4	932.3	1005.1	1060.7	1073.0	1107.0	1072.3	34.3
Coke	309.1	321.5	322.2	325.7	313.6	338.3	400.0	374.0	11.9
Coal	19.1	21.2	25.7	19.2	12.4	8.1	7.9	5.2	0.2
Purchased electricity[c]	114.2	126.7	126.5	163.7	194.7	226.1	233.8	231.0	7.4
Purchased steam[d]	25.0	27.0	25.8	29.6	33.0	45.1	45.2	47.6	1.5
Total	2432.3	2449.5	2535.4	2744.4	2891.6	3063.7	3216.2	3129.0	
(BTU/bbl)[e]	792	756	735	727	729	716	709	707	

[a]Includes 2.6×10^{12} of crude-oil consumption.
[b]Included in fuel oil.
[c]Assumes 10,000 Btu/kWhr.
[d]Assumes 1333 Btu/lb steam.
[e]Values are about 5% higher than referenced data because of adjustment of purchased energy by efficiency factors. Only barrels of crude input used; total barrels input is about 8% higher because of unfinished rerun, natural gas liquids, and other hydrocarbons.

Source: U.S. Department of the Interior, Bureau of Mines, **Mineral Industry Surveys, Crude Petroleum, Petroleum Products, and Natural-Gas-Liquids** (Washington, D.C.: Government Printing Office, 1962-1973 [final summaries] and April 1974 [monthly statement]).

of crude oil and the large regional variation of energy consumption in 1974. As can be seen, District 5 used only 54% of the energy used in District 8 per unit of output. Use of natural gas is less in the East Coast, Appalachian, and North Midwest districts, where the gas is being replaced primarily by fuel oil. Texas refineries use considerably more natural gas than do other refineries and rank highest in the amount of energy use per barrel.

The major petroleum refining processes and their requirements for fuel, steam, and electricity are summarized in Table 13-7. The total energy required is estimated to be 379,000 BTU per barrel of crude oil (17). Not included in the total is energy for lubricating oil, wax, and asphalt processing, a factor which might raise.the total energy estimate to 455,000 BTU per barrel of crude oil. Other energy uses not included are those for catalyst and additive manufacture, product blending and storage, treatment of water wastes, and terminal

Table 13-6

Energy Consumption Per Unit of Crude Oil in U.S. Petroleum Refineries
by Bureau of Mines Districts — 1974

Energy consumption (10^3 BTU/bbl/day) of:

District	Crude Oil	Distillate	Residual	LPG[a]	Natural Gas	Refinery Gas	Coke	Coal	Purchased Electricity	Purchased Steam	Total
1		7.3	190	2.3	62	228	103		46	32	671
2		50	206	29	73	195	39	80	80	1.5	754
3			115	13	64	163	97	20	58		530
4		4.4	131	9.3	55	235	103		52		590
5			110	0.9	26	183	104		49		473
6	0.3	1.6	27	5.9	274	242	75		45		671
7		5.0	10	4.4	379	220	121		60	0.2	800
8		5.4	1	2.2	524	223	83		39		878
9		23	14	12	211	259	49		50	31	649
10	0.9	31	38	16	261	151	35		28		561
11			15	35	246	159	56		29		540
12		3.5	89	8.2	171	224	109		48		653
13		11	41	21	189	265	81		83	21	712

[a]Liquified petroleum gas.

Source: Haynes, **Energy Use in Petroleum Refineries**, p. 37.

Table 13-7

Energy Requirements for Petroleum-Refining Processes

Process	Process feed (% crude)	Energy requirement (10³ BTU/bbl feed to process)				(10³ BTU/bbl crude oil)
		Fuel	Steam	Elect.	Total	
Desalting	80	0.05	neg	0.6	0.65	0.5
Atm. distillation	100	80	10	6	96	96
Vac. distillation	35.2	75	22	3	100	35.2
Gas separation	10	0	62	1.8	63.8	6.4
Cracking						
Thermal	1.4[a]	700	162	15	877	12.3
Visbreaking	1.4[a]	160	30	10	200	2.8
Coking	6.5					
Delayed	5.7	80	45	70	195	11.2
Fluid	0.8	0	100	15	115	0.9
Catalytic	28.9					
Fluid[b]	26.5	0/51	35/-177	50/57	85/-69	22.5/-18.3
Gas lift[c]	2.4	140	−67	22	95	2.3
Hydrocracking	5.4					
w/o Hydrogen		200	13	100	313	16.9
w Hydrogen		588	13	119	720	38.9
Reforming	21.2					
Cyclic	6.0[d]					
w/o Hydrogen		280	0	50	330	19.8
w Hydrogen		2	0	36	38	2.3
Semiregenerative	15.2[d]					
w/o Hydrogen		355	0	75	430	65.4
w Hydrogen		189	0	67	256	38.9
Hydrorefining	6.4					
w/o Hydrogen		73	0	22	95	6.1
w Hydrogen		228	0	30	258	16.5
Hydrotreating	29.9					
w/o Hydrogen		39	34	14	87	26
w Hydrogen		106	34	17	157	46.9
Alkylation	5.4					
H_2SO_4	3.4	100	323	48	471	16.0
HF	2.0	13	373	15	401	8.0
Isomerization	0.7					
C_4 feed	0.5	32	23	12	67	0.4
C_5/C_6 feed	0.2	42	45	15	102	0.2
Miscellaneous				20.9	20.9	20.9
Total (with hydrogen charge and credit)						379.1

[a]0.1 added to account for "other."

[b]Numbers to right of / are for a new, grass-root plant.

[c]Includes 0.4 of Houdriflow.

[d]0.3 added to account for "other."

SOURCE: Haynes, Energy Use in Petroleum Refining, p. 13.

operations. The estimated unit energy usage is much below the U.S.
average reported by the Bureau of Mines, suggesting that the estimates may be optimistically low, relative to actual use. Reasons
for major differences between these estimates and those of other
studies include the much lower energy requirements for catalytic
cracking and alkylation and a credit assignment for hydrogen produced
by the reforming process. These factors together could account for
about 190,000 BTU per barrel of crude oil.

The most energy intensive petroleum refining processes are
thermal cracking, hydrocracking, reforming, alkylation, and polymerization. However, based on extent of use, the less energy intensive
processes of distillation and hydrotreating are also important energy
consumers, while polymerization is relatively unimportant. The distribution of energy consumption among the various processes is shown
in Table 13-8.

Petroleum refining has become more complex as plant size and the
demand for gasoline have increased (see Fig. 13-14). This should
require more energy per unit of output. However, over the past
thirty years, the use of catalytic processes as distinct from
straight thermal processes have enabled energy consumption to remain
approximately constant, despite the increase in complexity.

PROCESS IMPROVEMENTS

Table 13-9 provides a summary of the major energy consumers in
the petroleum refining industry. Most of the energy conservation
techniques discussed in Chapters 2, 4, 5, and 6 are applicable for
implementation in the petroleum refining industry in both existing
and proposed plants. The techniques depend not only on the plant's
applying the correct equipment but also on its having strong management commitments to energy conservation. The following sections will
describe some of the process improvements which may assist in enhancing energy efficiency in the petroleum refining industry.

Table 13-8

Energy Distribution Among Processes

Process			Energy Distribution %
Cracking			18.1%
Thermal		3.2	
Visbreaking		0.7	
Coking		3.2	
Delayed	3.0		
	0.2		
Catalytic		6.5	
FCC	5.9		
Gas Lift	0.6		
Hydrocracking		4.5	
Crude Separation			36.4
Desalting		0.1	
Atmospheric Distillation		25.3	
Vacuum Distillation		9.3	
Gas Separation		1.7	
Reforming			22.5
Semi-Regenerative		17.3	
Cyclic		5.2	
Hydrogen Treating			8.5
Hydrotreating		6.9	
Hydrorefining		1.6	
Alkylation			6.3
HF		2.1	
H_2SO_4		4.2	
Hydrogen			2.5
Isomerization			0.2
Miscellaneous			5.5
		Total	100.0

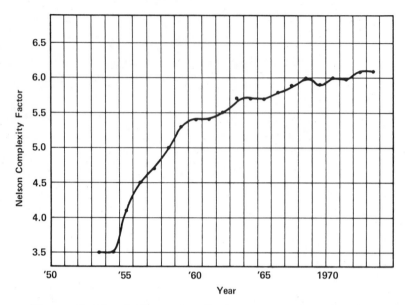

Figure 13-14. Refinery Complexity. Source: Federal Energy Administration, The Potential for Energy Conservation, Vol. 2, Petroleum Refining, p. 374.

Table 13-9

Energy Consumption Patterns

Directly fired fuel in process heaters	78%
Electrical energy	11%
Steam	10%
Cooling water system	1%
Total	100%

Improvements in Steam Generation and Usage

The three major functions that use steam in a refinery are (1) process heating, (2) power generation, such as for steam driven turbines, compressors, and pumps associated with the process, and (3) steam jet ejectors, for dilutants, stripping mediums, or sources of vacuum. Steam is provided to the different operations throughout the plant either via natural circulation, via vapor phase systems, or via forced circulation liquid heat transfer systems.

Fuel used for steam generation is about 10% of total refinery energy input. Although various techniques for improving boiler operation and steam energy conservation were discussed in Chapters 2 and 5, additional conservation potential should include the following:

Optimization of Steam System. Steam is required to perform mechanical work and heating duties. The design of a steam system that utilizes to the maximum all the available energy at the different pressure and temperature levels required in a plant is very complicated. It is quite common for steam within a refinery to be distributed and used at several pressure levels and to have reducing stations between the levels. Any shortfall in low pressure steam availability is made up by reducing some of the next higher pressure steam, and so on. The steam is produced by the boiler at the highest pressure and is "cascaded" down to the use point. Unless each steam pressure level is in good supply/demand balance, there can be considerable waste of energy.

In general, wherever steam pressure reduction occurs, consideration should be given to determine whether the pressure reduction can be used to perform useful work through a turbine that drives a process pump, through a turbine that drives an electric generator, or through a thermocompressor that recompresses low pressure waste steam to some intermediate level. Optimization of the steam system can result in savings of up to 40% (18).

Condensate Recovery. Condensate from turbines, reboilers, and tank heating coils contain energy which can be usefully recovered and returned to the boiler feed water system. Recovery of condensate reduces the quantity of new boiler feed water needed, which, in turn, reduces the energy and materials used to prepare boiler feed water. In some cases, condensate can be removed at sufficiently high pressure to warrant the installation of a flash drum from which steam can be used in another part of the plant. It is estimated that a 5% savings of steam used can be achieved through condensate recovery (19).

Combustion Control Instrumentation. The proper control of combustion is important for refinery energy efficiency. Control of excess air may be achieved by using air blowers and stack dampers, and the quantity of excess air can be monitored by observing the oxygen content of the stack gases. For accomplishing the latter, a number of stack gas analyzers are commercially available. This technology is considered well-proven in the industry and can usually be justified on all sizes of heaters. Rates of return on combustion control instrumentation are generally quite high but vary, depending on initial energy efficiency.

Conversion of Natural Draft to Forced Draft Operation. A major difficulty with natural draft burners is the pressure drop across the burner and the relatively low velocity air flow which inhibits good fuel air mixing. A forced draft burner system overcomes this pressure drop limitation, making intimate mixing of the air with the fuel possible. With forced draft systems, increases in the heat release per burner are possible, and air flow control is easier to achieve.

In the past decade, EXXON affiliates have resorted to almost exclusive use of forced draft combustion systems in liquid or gas liquid fueled furnaces. They have developed burners which permit clean combustion of even the most difficult fuels at economical, low excess air levels.

Advantages of large forced draft burners over natural draft systems include the relatively larger capacity per burner, which permits fewer burners per furnace. A typical 400×10^6 BTU per hour furnace requires only 10-14 burners, compared with the 40 or 50 burners required by the natural draft system (20). Fewer burners make control easier. Whereas natural draft systems require a multitude of burner primary and secondary air registers that must be manipulated for proper control, combustion air in the forced draft system is controlled from a single point. This feature makes adaption of the forced draft system to remote, automated control very convenient and practical. Fewer burners also allow a more effective burner cleaning and maintenance program, which uses less manpower. Finally, forced draft is easily adaptable to air preheat systems for added economy at lower incremental investment.

Air Preheaters to Recover Stack Gas Waste Heat

By far the largest part of the energy consumed in refining is in the form of fuel to process heaters or boilers. The efficiency of a fired heater is determined by the amount of excess air required to burn the fuel and by the stack temperature. Stack gases from the combustion of gaseous and liquid fuels will leave the heaters at temperatures varying from 400°F to above 1000°F. These gases represent the major area of energy loss in the combustion process.

There are basically two ways to recover heat from stack gases: (1) install a "convection section" of tubes in the stack area, and pass process liquid or water through the tubes to absorb heat or raise steam, and (2) use the hot gases to preheat cold combustion air. An air preheat system saves furnace fuel by transferring heat from the flue gas to the combustion air. This reduces the furnace flue gas temperature and correspondingly improves the operating efficiency. But before an air preheater system is applied to an existing furnace, operating data which include process heat duty, heat fired, stack temperature, and flue gas excess air levels should be obtained. With this data and a runaround chart (see Fig. 13-15), the fuel savings can be estimated. For example, potential heat recovery with a stack temperature of 850°F and a process duty of 150×10^6 BTU per hour amounts to 30×10^6 BTU per hour if the flue gas temperature were reduced to 300°F. At a fuel cost of $2.00 per 10^6 BTU, this fuel savings amounts to $500,000 per year. At today's price levels, such a system may cost on the order of $350,000-$500,000 to install.

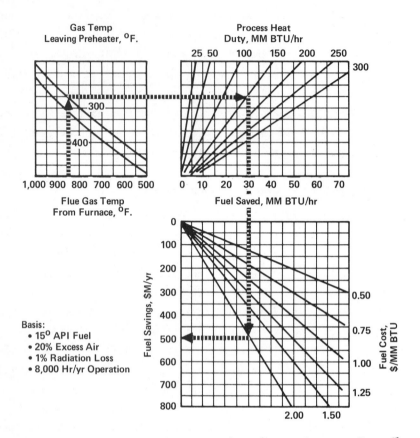

Figure 13-15. Air Preheater Fuel Savings. Source: Dean Cher-
rington and Herb Michelson, "How to Save Refinery Furnace Fuel,"
Oil and Gas Journal, 72 (September 2, 1974), p. 59.

Steam generation, or waste heat boiler systems, can also be used
to recover heat that would normally be lost by high temperature furnace stacks. These systems are generally more expensive and offer
less operating credits than an air preheater system. To evaluate
the addition of a waste heat boiler system to a process furnace,
operating data which include heat fired, stack temperature, and
excess air, should be obtained. With these data and the runaround
chart (see Fig. 13-16), potential heat recovery rate and steam credits
can be estimated. For example, potential heat recovery from a furnace

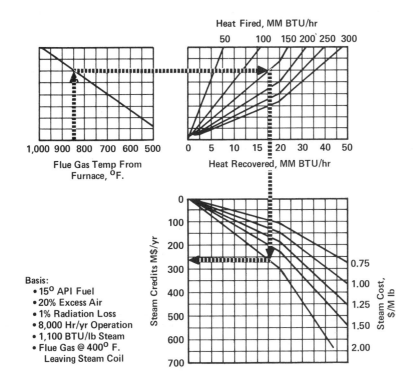

Figure 13-16. Steam Generation Heat Recovery. Source: Cherrington and Michelson, "How to Save Refinery Furnace Fuel," Oil and Gas Journal, p. 61.

firing 150×10^6 BTU per hour with a stack temperature of 850°F is 18×10^6 BTU per hour. At a steam cost of $2.00 per thousand pounds, operating credits are approximately $250,000 per year. At current prices, such an installation may cost $450,000-$600,000 to install.

Air preheaters are becoming increasingly common on new furnaces, although retrofitting a preheater may be difficult, due to lack of space for ducting and for stack gas and air bypasses, and because of the preheater itself. If installed, a preheater may be expected to improve efficiency by 10-15%.

Carbon Monoxide Boilers for Fluid Catalytic Crackers

The operating temperature of the typical fluid catalytic cracker regenerator runs about $1200^{\circ}F$. Its waste gases contain about 9% CO by volume, so that energy can be recovered both from the high temperature of the gases and by burning the carbon monoxide. The latter can be accomplished directly or after passing the gases through a turbine to lower the gas pressure and recover energy. The hot gases can be discharged into a boiler firebox where combustion air is added. The energy recovered by a CO boiler typically amounts to about 90,000 BTU per barrel of FCC charge. In 1972, about 30% of FCC capacity in the U.S. did not have CO boiler installations (21).

Heat Exchanger Applications

A typical refinery contains large numbers of heat exchangers to transfer heat from one process stream to another. As fuel costs rise, it is probable that existing plants will profit from the installation of additional heat exchanger surface that were previously considered only marginally economic. Installation of heat exchangers, a well-accepted method of saving energy, can, in many cases, be carried out simply and quickly, affording almost immediate energy conservation benefits. But heat exchanger applications and benefits are quite variable, as can be seen from the representative cases shown in Table 13-10. The amount of energy that can be saved in a refinery, or by modifying heat exchange trains, is highly specific to the refinery in question.

Process Heat Integration

The heat available from a process stream in one unit can be put to use in another process unit to preheat feedstock or boiler feed water. By "coupling" the two units, energy efficiency can often be improved significantly. In a large refinery, there will exist many opportunities for heat integration, each of which must be judged on

Table 13-10

Representative Heat Exchanger Investments

Application	Cost $	Energy Savings (billion BTU/yr)	Return On Investment, %
Fractionator Preheater	64,600	268	663
Crude-Naphtha exchanger	60,500	8.07	27.2
Waste Heat Recovery for Feed Preheat	452,000	93.2	23.2
Reactor Feed Effluent Exchanger	122,000	66.6	96.5

Source: Gordian Associates, **An Energy Conservation Target for Industry SIC 29**, p. 10.

its own merits for profitability and for less tangible factors, such as reduced flexibility of operation.

A form of heat integration between two units, "hot feed" systems, can be applied in the following manner. Crude distillation units commonly produce materials which are to be processed further in downstream conversion units. One such material is vacuum gas oil for use as catalytic cracker feed. In many refineries, the vacuum gas oil is cooled and stored in a tank before being pumped to the catalytic cracker. In the cracker, the cold feed is heated prior to contact with the circulating catalyst. It is often possible to avoid the cooling and subsequent reheating of gas oil by bypassing the storage tank and feeding hot gas oil directly from the crude unit to the catalytic cracker. In a representative industry example, the installation of such a direct hot feed system cost $678,000 but saved 238 billion BTU per year, for a rate of return of 64.7% (22).

Power Recovery Turbines on Gaseous Streams

There are many possibilities for gas turbine units in process plants. One process unit in which energy recovery has been successfully applied is the fluid catalytic cracker (FCC). Regeneration

of catalyst produces large volumes of heat gases that are available
at pressures ranging from around 10-40 psig and at temperatures up
to about $1,250^{\circ}F$. Figure 13-17 shows that the quantity of gases
available even at low pressures is such that enormous energy savings
are potentially available for use in driving other plant equipment,
such as the main blower. A power recovery turbine investment at a
cost of \$93,000 will result in energy savings of 1.04×10^9 BTU per
year, which represents a 23.1% rate of return (23).

Hydraulic Turbines on High Pressure Liquid Streams

Energy from previously lost high pressure liquid process streams
may be recovered with hydraulic turbines. These turbines may be
either single or multistage, depending on the pressure energy available from the process streams. A representative investment of
\$298,000 for a hydraulic turbine will save 4.95×10^6 KWH per year,
providing a 22.9% rate of return (24).

Recovering energy from high pressure process streams by decreasing the pressure through the use of turboexpanders or hydraulic
turbines can be done whenever the process streams are being reduced
at pressure ratios greater than a value of about 2. This practice
would save over half of the energy required to provide the high
pressure stream. It is applicable to fluid catalytic cracking, hydrocracking, some reforming, hydrorefining, and hydrotreating processes.
The estimated savings, assuming application to one-third of the 1974
hydrocracking, hydrorefining, and hydrotreating capacities, is about
1.2×10^{13} BTU per year (25).

Heat Exchanger Cleaning

Heat exchanger systems are usually very effective when first
installed; however, these systems become dirty in use, and as a result, heat transfer rates suffer significantly. Heat exchanger
cleaning schedules based on observations of the actual efficiency
deterioration of the exchangers should be established. A representative energy saving is 120×10^6 BTU per day for each on-stream

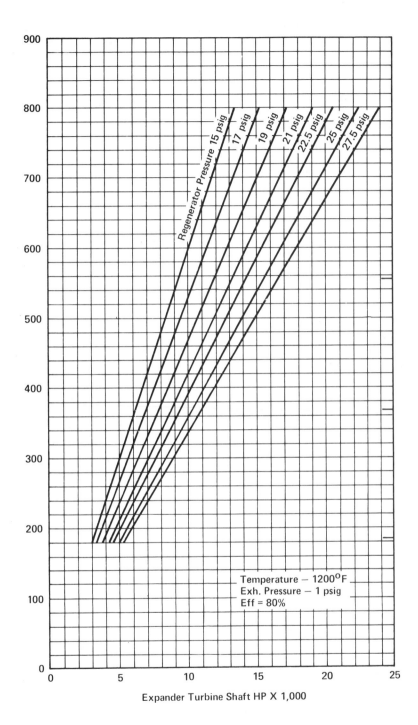

Figure 13-17. Expansible Horsepower Available from Flue Gas.
Source: Federal Energy Administration, The Potential for Energy Conservation, Vol. 2, Petroleum Refining, p. 188.

cleaning system installed in a refinery processing on the order of
100,000 BPD crude oil. The kinds of systems considered as candidates
for the installation of on-stream cleaning facilities include (26):

- Crude oil/atmospheric residuum exchange

- Crude oil/vacuum residuum exchange

- Catalytic reformer steam driver surface condenser

- Catalytic cracker air blower steam driver surface condenser

- Catalytic cracker gas compressor steam driver surface condenser

- Crude distillation unit overhead condenser

Installation of these systems will result in energy savings of
7.2×10^6 BTU per barrel of crude oil.

Replacement of Steam Jets by Vacuum Pumps

Although vacuum pumps of all types are used in chemical process
plants, the steam jet type has been widely used because of its low
initial cost, its operating simplicity, lack of moving parts, and
supposed reliability. A recent review of vacuum producing equipment
suggesting potential advantages of other types of vacuum pumps concludes that mechanical pumps offer significant energy savings over
steam jets, in spite of higher initial investment costs (27). Producing vacuum with mechanical pumps rather than with steam jet
ejectors has the potential for reducing energy requirements by
75-90%.

Improvement of Fractionation Efficiency (28)

There are many examples of new tray or tower packing designs
which can reduce the energy required to operate fractionation towers.
In some cases, new tower intervals installed to increase tower
throughputs have realized energy savings as a side benefit.

Intermediate Reboilers on Towers. More efficient operation can
often be achieved by installing an intermediate reboiler at some

point higher up the tower. A lower temperature is usually required
at that point, which often allows reboiler heat to be supplied by a
lower grade heat source than what is used for the main bottom reboiler. The tower internal hydraulics may well be improved, allowing
an increase in capacity.

Optimum Balance Between Number of Trays and Reflux Rate. The
degree of separation achieved in a fractionating tower operating for
specific product purities and at a specific pressure is a function
of the number of trays and of the reflux ratio. Adding trays, which
results in greater initial investment, will allow lower reflux ratios
to be used, saving on energy input to the system. The economic balance will undoubtedly shift towards the use of lower reflux ratios
as the cost of fuel increases, but this is likely to impact only on
new plants.

Control of Fractionating Towers. Excessive reflux and reboil
on a tower are wasteful of energy. The purity of a fractionation
product should be monitored carefully to ensure that it meets specifications but is not excessively pure. Hence, towers represent major
areas for "housekeeping" for good energy savings.

Reduction of Tower Operating Pressure. Since low pressure improves the relative volatility of components being separated in a
fractionating tower, a reduction in tower operating pressure should
allow a lower reflux ratio while achieving the same degree of separation. In practice, the lowering of pressure may be limited by the
capacity of the tower overhead condensing system, since the condensing temperature drops as the pressure is lowered.

Operation at Optimum Tray Loading and Efficiency. The trays in
a fractionating tower generally operate at maximum efficiency at high
loading. To achieve a given separation with a fixed number of trays,
the usual situation in an existing tower, and to minimize reflux, the
trays should be run at maximum efficiency. The loading of trays may
often be adjusted to the maximum efficiency point by such methods as:

• changing tower throughput

• changing tower pressure

• changing the percentage of feed vaporization

In some rare cases, retraying a tower or blocking off parts of each
tray may be advantageous.

Use of Appropriate Feed Tray Location and Feed Condition. For
efficient operation, the feed to a tower should be put in at the
correct point according to its composition, and should be added in
the form of a vapor liquid mixture in which the proportion of vapor
is about the same as the proportion of overhead product to be recovered from the feed. This is important, since it is often more
economical to add heat to the feed than to the tower bottom. The
feed point is at a lower temperature than the tower bottom, which
allows lower grade heat sources to be utilized for preheating the feed
before it enters the tower.

Insulation

Although insulation can be used to minimize heat loss from equipment, the choices of material and thickness are governed by economics.
In the case of oil storage tanks, many refinery products have to be
stored above ambient temperatures to maintain the ease with which it
can be pumped. Until the recent dramatic energy price rises, tanks
were only insulated at temperatures of about $200^{\circ}F$ or higher; but
present economics justify insulation of tanks at $125^{\circ}F$. The return
on investment in tank insulation depends on the tank size and temperature. Insulation should be installed up to the point at which the
marginal return on investment is equal to the hurdle rate.

If a standard tank forty feet high and sixty feet in diameter,
with a capacity of about 20,140 barrels, is used as an example, aggregate savings of insulated over non-insulated tanks for the different classes of materials are as follows (29):

Crude 3233×10^6 BTU per hour (basis $100°F$)

Distillate 20366×10^6 BTU per hour (basis $200°F$)

Residual 954×10^6 BTU per hour (basis $225°F$)

Instrumentation and Computer Control

Attractive economic benefits have been shown to result from closed loop computer control of complex units. A computer system will guide operators on all product qualities, yields, and energy used. For each feed product package, an optimum mode of operation can be established and the process unit run against some physical limitation within that operating mode to maximize profits. Direct digital control can be used to maximize profit and minimize energy consumption. Gordian Associates reports that, based on refining industry contacts, literature references, and discussions with design contractors, about 1% of energy use can be attributable to savings by direct digital control installation on the following typical units (30):

Unit	Some Critical Parameters
FCC	Air rate, coke balance, catalyst circulation rate, feed to catalyst ratio, regenerator pressure.
Cat. reformer	Octane control, reactor temperature, hydrogen to hydrocarbon ratio (hydrogen recycle).
Desulfurization units	Hydrogen to hydrocarbon ratio, reactor temperature.
Alkylation units	Isobutylene to butylene ratio, contractor temperatures, refrigeration system.
Hydrocrackers	Hydrogen recycle, reactor temperature.
Luboil plant	Various
Crude distillation	Various

Refinery Loss Control (31)

In refinery operations many losses occur. Among the more common are:

- Flare losses. Relief valves often leak during normal operation, and valuable products will escape to flare and be burned,
  with no useful heat recovery. In some plants, minor gas
  streams are vented to flare as a routine disposal operation.

- Relief valve leaks. Some relief valves may discharge directly
  to atmosphere in an emergency. If these valves leak during
  normal operation, valuable products are lost.

- Filling losses from tanks. When a tank is filled with liquid,
  the vapor space in the tank is displaced through pressure/
  vacuum valves, resulting in hydrocarbon losses.

- Evaporation losses from tanks. During the period that a tank
  contains liquid, the vapor space above the tank contains a
  hydrocarbon-saturated vapor which expands and contracts according to temperature. At night, the tank vapor space will
  cool down, and the vapor will contract. Air will be drawn in
  through the pressure/vacuum valve, and hydrocarbon molecules
  will evaporate to maintain saturation. During the day, as
  the vapor space warms up, the vapor will expand, and saturated
  vapor will escape through the pressure/vacuum valve and be
  lost to the atmosphere. When tank hatch covers are left open,
  losses will be exaggerated.

- Evaporation from oil/water separators. Oily water from plant
  areas are collected in separators. Light components will
  evaporate from separators under normal operation, although
  some designs of separators include a closed hood to trap
  vapors for recovery of the valuable hydrocarbons.

- Leaks in pipes, flange, pump glands, tanks, water coolers, etc.

- Loading operation spillages. Vapor losses and direct liquid
  losses through overfilling of tanks, trucks, railcars, or
  ships can represent significant losses.

The magnitude of each of the above losses will vary from day to
day in any refinery. Given good operating practices and well-main-
tained equipment, it is estimated that losses will be in the range
of 0.5-1% of refinery throughput. With poor control practices, losses
could be much higher. Tighter operation of processes and good maintenance of equipment should lead to a loss recovery of about 0.1% of

throughput. Since energy use is about 8-10% of refinery throughput
for a representative refinery, 0.1% of throughput represents a 1%
savings in energy use (32).

PRODUCT CHANGES

A number of factors impact on the petroleum refining industry
and on its energy usage patterns. The most important of these include:

- The large rise in crude oil prices and the government controls
  which resulted in its aftermath.

- The trends toward producing more residual fuel and toward reducing processing complexity.

- The trend toward poorer quality crudes being made available.

- The need for higher quality products in the marketplace, such
  as increased clear pool octane numbers for motor gasolines
  and lower lead levels and lower sulfur contents of fuel oils.

- Increased environmental restrictions which regulate emissions
  into the air and water and regulate the disposal of solid
  wastes from the refining industry.

The impact of some of these changes is reflected in a computer simu-
lation on gasoline production and gasoline lead levels, performed
by Gordian Associates for the Federal Energy Administration, is
described below (33).

Impact of Ratio of Gasoline to Distillate Fuel Oil Production

For the first series of computer runs performed, the variable
under consideration was the ratio of motor gasoline to distillate
fuel oil production. The combined motor gasoline and distillate
(no. 2) fuel oil volume was held constant at 70,000 BPD. Based on
the crude unit intake used in the runs of 100,000 BPD, this represents a 70% volume yield. The motor gasoline element of the 70% yield
was fixed as follows:

Gasoline Grade	Percent of Total	Research Octane Number
Regular	60	93 with 2cc TEL/gal
Premium	30	99 with 2cc TEL/gal
Unleaded	10	92 unleaded
	100	

A minimum yield of 5% residual fuel oil was set so that the yields of products would correspond reasonably with the overall U.S. market for petroleum products. The quantity of motor gasoline was then varied from 35,000-65,000 BPD, with the corresponding distillate fuel oil yield varying from 35,000-5,000 BPD. For each gasoline yield, the breakdown of the products was as stated above (60% regular, 30% premium, etc.)

Two series of runs were performed, one with a light 32° API crude oil, which represented the overall average crude oil run in the U.S., and the second with a heavy 17° crude most suited for high fuel oil production. For all cases, the model was allowed a free choice of units and throughputs, subject only to the constraints of product yields and product specifications, and to the objective of minimum internal refinery energy consumption.

Details of the results, including the exact refinery configurations developed, are given in Table 13-11, and the energy use is plotted against motor gasoline yield in Fig. 13-18. The pertinent quantitative findings which emerge from an analysis of the results are as follows:

- Depending on the case chosen, the internal refinery energy use represents a loss of 7-12% by volume of the crude oil intake.

- Energy consumption is almost constant for a product slate corresponding to a motor gasoline yield between 40 and 50% volume on crude oil. Current U.S. market demand is around 50%.

- The penalty in terms of internal energy consumption becomes high when about 60% motor gasoline is exceeded.

Table 13-11

Summary of Energy Consumption Data Corresponding to Changes in Motor Gasoline Yield

		Motor gasoline yield, % vol. on crude oil	No. 2 fuel oil yield % vol. on crude oil	Energy Consumption	
Case Number				FOE barrels(c) per 100,000 BPD crude	MMBTU/bbl. crude
Heavy crude	B-1	35	35	9837	0.626
	B-2	40	30	9631	0.613
	B-3	50	20	9641	0.614
	B-4	60	10	10426	0.664
	B-5	65	5	11720	0.746
Light crude	L-1	37 (a)	37 (a)	8059 (b)	0.513
	L-2	42 (a)	31 (a)	7416 (b)	0.472
	L-3	50	20	7684	0.489
	L-4	55	15	8232	0.524
	L-5	60	10	8356	0.532
	L-6	68	2	10116	0.644

Notes: (a) These cases ran approximately 95000 BPD of crude oil and therefore the combined motor gasoline and No. 2 oil yields are greater than 70%. All other cases ran 100,000 BPD of crude.

(b) Based on 100,000 BPD crude to make all cases comparable.

(c) If these values are divided by 1000, the result represents the percentage by volume of crude oil consumed as internal energy.

- The internal energy use does not continue to decline after the motor gasoline yield falls below about 40%. There is actually an increase in energy requirement, since increasing quantities of high boiling range crude oil components are converted into the middle range to produce the distillate fuel oil specified.

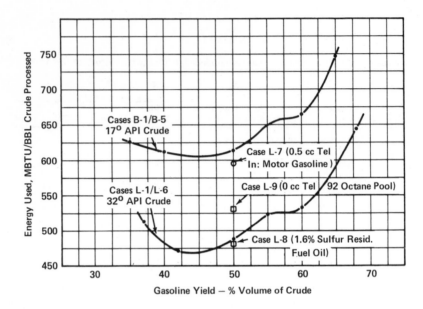

Figure 13-18. Energy Use and Gasoline Yield.

- The refinery internal energy consumption is dependent on the crude type. For the crude oils studied, the 17° API heavy crude required about 25% more than the 32° API light crude for a corresponding product slate.

Impact of Leaded Gasoline on Refinery Energy Consumption

For this part of the study, the motor gasoline yield was fixed at 50% of crude oil intake, with the refinery running the light 32° API crude. Changes in the lead levels allowed in the gasoline pool were made and the model run to determine the corresponding refinery configurations and internal energy consumptions.

In general, reducing the lead level in gasoline, while keeping octane number constant, has a very significant effect on internal energy requirements. For example, dropping lead level from 2.0-0.5cc TEL per gallon at a constant pool research octane of 96.5 requires additional processing complexity which uses an additional 22% internal energy. On an annual basis, this corresponds to between 0.4-0.5 quadrillion BTUs extra energy.

The results of several cases run with varying octane and lead levels may be summarized as follows (see also Fig. 13-19):

Premium and Regular Lead Level (cc TEL/gallon)	Average Pool Research Octane Number	Relative Energy Used
2.0	96.5	1.00
0.5	96.5	1.22
0.0	92.0	1.09
0.5	92.0	0.99

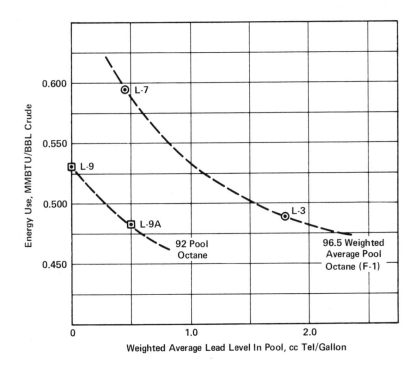

Figure 13-19. Energy Changes with Pool Octane and Lead Levels.

ENERGY CONSERVATION POTENTIAL

While the petroleum refining process is reasonably efficient in
a conventional first law engineering sense, in principle substantial
room for improvement exists. This is so because, theoretically, little, if any, work is required to effect the separation of the crude
into narrower petroleum cuts. Based on second law thermodynamic
availability analysis, the theoretical minimum specific fuel consumption should be 70,000 BTU per barrel of crude oil. Present refineries use about 700,000 BTU per barrel of crude oil.

The reason for the comparatively high loss of available energy
in the refining process is the utilization of direct fired fuels to
provide the energy for crude separation into narrower boiling cuts.
The various reboiler streams are invariably at temperatures considerably less than 1400°F-1500°F. The maximum flame temperature of
the fuels, on the other hand, is on the order of 3500°F-4000°F,
signifying a high capacity which can be transformed into work, a
capacity which is grossly unterutilized.

Table 13-12, based on 1974 capacities and processing rates,
summarizes some of the potential for energy savings. Total savings
equal 61 x 10^{13} BTU per year, or about 19.7% of the energy used to
refine petroleum. Some specific energy conservation measures were
identified in a report to the Federal Energy Administration, shown
in Table 13-13. The overall conclusion that can be drawn from these
studies is that there is substantial potential for energy conservation
in process operations. Management commitment must be obtained to
implement these improvements, and capital must be raised. In addition, long term research and development must be performed, and
significant modifications to the processes are required to raise
operating efficiencies above what they are today.

Additional long term research may prove fruitful for the following:

Table 13-12

Potential Annual Energy Savings for Petroleum Refining

Process	Potential energy savings (10^{13} BTU. year) by:						Percent of refinery energy consumption
	Improved combustion	Recovery of low-grade heat	Improved process	Alternate process	Other	Total	
Atm. distillation	4.0	3.9	2.2[b]		0.2[c]	10.3	3.3
Vac. distillation	1.4	1.5	0.8[b]		1.2[c,d]	4.9	1.6
Gas separation	0.3	0.9	0.2[b]			1.4	0.5
Thermal cracking	0.5	0.1			0.1[b,d]	0.6	0.2
Visbreaking	0.1					0.2	0.1
Coking	0.4	0.1				0.5[e]	0.2
FCC			15[f]			15.0	4.8
Hydrocracking	0.5	2.0			0.6[g]	3.1	1.0
Reforming	2.6			5.0[h]		7.6	2.4
Hydrorefining	0.2	0.3	0.3[i]		0.2[g]	1.0	0.3
Hydrotreating[j]	1.0	1.1	0.3[k]	1.2	0.4[g]	4.0	1.3
Alkylaton[l]	1.0	0.2		0.6		1.8	0.6
Hydrogen[m]	1.1	0.3				1.4	0.4
Other	—	—	—	—	9.3[n]	9.3	3.0
Total	13.1	10.4	18.8	6.8	12.0	61.1	19.7

[a]Values less than 5×10^{11} not included.

[b]Improved fractionation.

[c]Improved insulation.

[d]Mechanical vacuum pumps.

[e]Arbritrary estimate of 10% savings.

[f]Modern, energy-conserving FCC process.

[g]One-fifth of estimated electrical energy (use of hydraulic turbines and more energy-efficient prime movers).

[h]Conversion to cyclic units.

[i]30% of electricity on 2/3 of capacity (lower pressure and less recycle and makeup of hydrogen).

[j]15% of feed processed with no makeup hydrogen.

[k]20% of electricity on 1/2 of capacity (lower pressure and less recycle and makeup of hydrogen).

[l]conversion of 1/2 of the less efficient capacity to the more efficient process.

[m]Based on 1/2 the estimated hydrogen requirements.

[n]Use of back-pressure turbines for direct drives and electricity generation. More efficient pumping systems.

SOURCE: Haynes, **Energy Use in Petroleum Refineries**, p. 18.

Table 13-13

Contribution of Specific Conservation Measures

Measure	%
Air preheater installation (process heaters)	1.0
Convection sections on process heaters	0.5
Sootblowers on process heaters	0.3
Combustion control instrumentation for process heaters	0.4
Improved insulation for process heaters	0.1
Replacement of old heaters with new equipment	0.7
Convection section for boilers	0.1
Combustion control instrumentation for boilers	—*
Boiler blowdown heat recovery	0.1
Replacemnt of old boilers with new equipment	0.1
Sootblowers for boilers	—*
Improved insulation for boilers	—*
Optimization of steam balances	0.4
Condensate recovery (and 3-3, replacement of steam traps)	0.5
Additional heat exchangers, hot oil systems Interchange of heat between different process units }	2.3
Optimization of heat exchanger cleaning cycles	0.5
Insulation of storage tanks	1.0
Power recovery turbines (FCC systems)	—*
Instrumentation and computer control of complex units (including 7-7 and 11-2, fractionating tower controls)	0.3
Direct hot rundown/hot feed systems	1.1
CO boilers for FCC units High temperature regeneration on FCC units }	1.3
Refinery loss control	1.0
Housekeeping	7.1
Other small items not included as separate contributions	0.2
TOTAL	19.0

* Less than 0.05%, see "Other small items."

SOURCE: Gordian Associates, **An Energy Conservation Target for Industry SIC 29**, p. 171.

- Replacement of direct fired heaters by other methods for heating below 1400°F. High, theoretically available energy from fuel combustion processes is seriously wasted when used in the relatively low temperature process heating typical of petroleum refineries. One method to resolve this would be to combine chemical and metallurgical processes, so that direct fired furnaces can be used to melt metals or ores which then give up heat to thermal fluids for transfer to chemical units for the accomplishment of chemical work.

- Development of low entropy change heat transfer fluids as replacements for steam which can practically transfer only about 50% of the enthalpy of combustion as useful work.

- Further determination of the effect of process parameters on energy consumption in the production of an equivalent or suitable product.

- Development of reliable instrumentation systems to measure important process characteristics and control the processes at optimum conditions.

- Development of improved catalysts.

- Improvement in overall fuel utilization by considering both the efficiency of internal combustion engines and the petroleum refining necessary to produce fuel for the engine.

- Development of equipment and techniques for the beneficial use of low grade heat.

NOTES AND REFERENCES

1. Bureau of Mines, "Crude Petroleum, Petroleum Products, and Natural Gas Liquids," Annual Petroleum Statement (Washington, D.C.: Government Printing Office, 1975).

2. Virgil Haynes, Energy Use in Petroleum Refineries (Oak Ridge, Tennessee, Oak Ridge National Laboratory, September 1976), p. 43.

3. A. Cantrell, "Annual Refining Survey, Oil and Gas Journal, 72 (April 1, 1974), p. 99.

4. Haynes, Energy Use in Petroleum Refineries, p. 47.

5. G.D. Hobson, Modern Petroleum Technology, 4th ed. (New York: John Wiley and Sons, 1973).

6. Haynes, Energy Use in Petroleum Refineries, p. 103.

7. Ibid., p. 55.

8. Ibid., p. 56.

9. Ibid., p. 58.

10. Ibid., p. 63.

11. Ibid., p. 66.

12. Ibid., p. 109.

13. Ibid., p. 61.

14. Ibid., p. 69.

15. Ibid., p. 71.

16. Ibid., p. 73.

17. Ibid., p. 12.

18. Gordian Associates, An Energy Conservation Target for Industry SIC 29 (New York: Gordian, June 1976), p. IV-10.

19. Ibid., p. IV-10.

20. Dean Cherrington and Herb Michelson, "How to Save Refinery Furnace Fuel," Oil and Gas Journal, 72 (September 2, 1974), p. 69.

21. Environmental Protection Agency, Cost of Clean Air (Washington, D.C.: Government Printing Office, April 1974), Appendix A.

22. Gordian Associates, An Energy Conservation Target for Industry SIC 29, p. 145.

23. Ibid., p. 140.

24. Ibid., p. 141.

25. Haynes, Energy Use in Petroleum Refineries, p. 22.

26. Gordian Associates, An Energy Conservation Target for Industry SIC 29, p. IV-12.

27. E.S. Monroe, "Energy Conservation and Vacuum Pumps," Chemical Engineering Progress, 71 (October 1975), p. 69.

28. Gordian Associates, An Energy Conservation Target for Industry
    SIC 29, pp. 28-32.

29. Ibid., p. IV-13.

30. Ibid., p. IV-16.

31. Ibid., p. II-43 - II-44.

32. Ibid., p. IV-19.

33. Federal Energy Administration, The Potential for Energy Conser-
    vation in Nine Selected Industries, Vol. 2, Petroleum Refining
    (Washington, D.C.: Government Printing Office, June 1974),
    pp. 118-24.

Chapter 14

IMPLICATIONS FOR THE FUTURE

Energy conservation requires implementation.

Due to the relatively low fuel costs and the relatively high
capital costs in the past, industry has had few economic incentives
for implementing energy efficient processes or equipment. However,
recent sharp increases in energy prices, together with present and
future interruptions in energy supplies, have resulted in an ac-
celerated program for energy conservation in manufacturing--a program
in which technology is playing a significant role. Until recently,
national concern has primarily concentrated on energy supply, with
research and development directed toward more productive modes of
supply and toward discovery of new sources. While in the long term
new sources must be developed to provide a stable supply, the immediate problem of the inadequacy of domestic petroleum and natural gas
can only be solved by concentrating on increasing efficiency in
energy use. Moreover, the rate at which potential fuel savings can
be accomplished and the rapidity with which energy conservation equipment will be installed are subject to such economic considerations
as the intricacies of the money market, international relations,
institutional arrangements, and social attitudes.

Several factors set the pace for the growth of industrial conservation. In the very short term, the pacing factor is the

development by government of the proper combination of incentives
and regulations to induce conservation action and investment in more
efficient plants and processes. A second factor is the enormous
investment in existing plants and equipment, much of which use energy
inefficiently. The rate at which existing structures, equipment,
and processes can be retired and replaced sets the pace for growth
of conservation benefits. The economic impacts of changes in energy
use patterns are complex, and the added problems of possible early
phase-out of plant facilities for energy efficient replacements can
cause severe strains. Also, the energy investment in place replacements must be weighed against operational savings of the new processes. Technology is a longer range pacing factor which assumes
greater importance as major system elements are retired and replaced
with wholly new designs. Long term technology is needed to bridge
the gap between more efficient use of present primary energy sources
and the future stable sources expected to emerge from energy supply
research and development.

  Industry and government must work together to solve our energy
dilemma. In addition to the process changes and housekeeping alternatives that industry can implement, as described in previous chapters, industry and government must move toward increased secondary
materials usage, enhanced international technology transfer, and
conversion to more available fuels. Government must also work toward
elimination or minimization of the barriers to energy conservation
and should provide incentives to increase the efficiency of industrial energy use.

SECONDARY MATERIALS USAGE

  It is estimated that in 1971, 5.8 billion tons of materials
were used in the U.S. economy, an equivalent of twenty-eight tons
for each person in the U.S. Of this total, approximately 10% comes
from agriculture, forestry, fishing, and animal husbandry; 34% is

represented by fuels; and 55% comes from the minerals industries in
the form of metals, construction materials, and other minerals.
Materials use is growing at a rate of 4-5% yearly: per capita consumption was 22 tons in 1965, 24.7 tons in 1968, and 28 tons in 1971.

It is also true that nearly all of the materials and energy required in the U.S. comes from virgin or natural resources. Because
of the resource scarcities with which we are faced today, materials
must be increasingly recycled to conserve these resources. Energy
can also be conserved in this way. Secondary materials can be generated at many points during the life cycle of a product. Contrary
to popular belief, most secondary materials that are recycled are
"new"--that is, recycled before being used by consumers--rather than
consumer discarded.

There are four broad categories of recoverable secondary materials:

1. Mixed waste is the secondary material contained in the mixed
solid wastes discarded by residential, commercial, and industrial
generators and collected by municipalities or private contractors.
Very little of this secondary material is presently recovered for
recycling, but the potential is significant.

2. Mill revert is generated in the production process and is
reincorporated by the producer into his production process, without
ever entering the scrap market.

3. Industrial scrap results from the fabrication of materials
into consumer usable products. This is still a high quality material
and can be recycled through conventional scrap channels into production.

4. Obsolete scrap consists of large consumer or industrial products which can be readily separated, such as obsolete ships, rails,
dismantled chemical plants, and shredded auto scrap. It may also
contain impurities, such as non-compatible metals or alloys. Obsolete scrap usually requires significant handling and processing before
it can be reused.

The use of scrap or rejected products of manufacturing processes by industry has been practiced for many years. The recycling of secondary materials from the economy back to industry, on the other hand, has not been widely practiced. Table 14-1 provides an indication of the extent to which recycling occurs. As is evident, copper and lead are extensively recycled, and other materials, to a lesser extent.

In general, the energy required to recover and recycle ferrous metals, aluminum, and copper is considerably less than the energy required to produce these metals from ores. For example, if the metals that are recoverable from urban wastes were recycled in 1975, this would have amounted to 6.9 million tons of ferrous metals, 400,000 tons of aluminum, and 100,000 tons of copper (1). The recovery and recycling of these metals could lead to substantial energy conservation, compared to deriving these metals from virgin raw materials. Table 14-2 indicates the total energy requirements and

Table 14-1

Recycling of Major Materials (1967)

Material	Total Consumption (Million Tons)	Total Recycled (Million Tons)	Recycling as % of Consumption
Paper	53.110	10.124	19.0
Iron & Steel	105.900	33.100	31.2
Aluminum	4.009	.733	18.3
Copper	2.913	1.447	49.7
Lead	1.261	.625	49.6
Zinc	1.592	.201	12.6
Glass	12.820	.600	4.7
Textiles	5.672	.246	4.3
Rubber	3.943	1.032	26.2
TOTAL	191.220	48.108	25.2

Source: A. Darnay and W. E. Franklin, **Salvage Markets for Materials in Solid Wastes** (Washington, D.C.: Government Printing Office, 1972), p. xvii.

Table 14-2

Energy Conservation Potential Resulting from Recycling

Metal	10^6 BTU/ton From Ore	10^6 BTU/ton From Scrap	Energy Savings Per Ton—10^6 BTU (Col 1 - Col 2)	Percent
Steel	49.0	6.75	42.25	86
Aluminum	295.0	9.5	285.5	97
Copper	71.0	6.7	64.3	91

Source: William Franklin **et al., Potential Energy Conservation from Recycling Metals in Urban Wastes** (Kansas City: Midwest Research Institute, May 1974), p. 15.

conservation potential for producing finished metal from virgin ore and scrap. The energy savings is extremely high, in excess of 30 to 1 in the case of aluminum and in all cases exceeding 86%.

Energy savings can also be achieved through the utilization of waste materials as energy. For example, the pulp and paper industry obtains the bulk of its energy from sources such as pulping liquors, bark, and sawdust, and the chemical industry uses waste tars and gas. There is a need for more study to determine the greater use of waste materials, such as municipal wastes, plastics or polystyrenes, biomass, etc., for industrial energy operations, particularly in the generation of steam.

The recycling of materials is by no means easy. A beginning is being made in the metals industry, in particular, by the copper, lead, and aluminum industries. But there are numerous obstacles:

- Virgin materials tend to be more homogeneous in composition than waste materials, and physical separation and upgrading of mixed wastes are costly and require considerable energy.

- Natural resources occur in concentrated form, while scrap and wastes occur in a dispersed manner. Consequently, acquisition for recycling is costly and is particularly sensitive to high transportation costs.

- Natural resources are abundant, and manufacturing industries have directed their operations to exploit these. Plants are generally built near the source of virgin materials (e.g., paper plants near pulpwood supplies). Technology to utilize

virgin materials has been perfected; however, due to adverse
economics, similar technology to exploit wastes has not been
developed.

- Synthetic materials made from hydrocarbons and their combustion with natural materials cause contamination of the latter, limiting their recovery.

- The delivered price of virgin materials to the manufacture is almost as low in many cases as the cost of secondary materials, and virgin materials are usually qualitatively superior to salvages materials.

- There are artificial economic barriers which favor virgin materials over secondary materials use. For example, depletion allowances, favorable capital gains treatments, and favorable freight rates are available to virgin materials processors but not to secondary materials processors. New materials do not include society's cost of disposing of the waste.

New government policies and research are needed to remove these
barriers, real or artificial, to recycling. There must be a trend
away from building things which are used and quickly thrown away.
Durability must be a prime objective of the manufacturing process.

INTERNATIONAL TECHNOLOGY TRANSFER

The transfer of energy efficient technology from the corresponding industries of other nations to the U.S., in conjunction with
wider dissemination of energy saving techniques across industries,
could have substantial benefits to the industrial sector over the
next decade. The adoption of energy conservative technology will not
occur, however, unless economic and social factors are found favorable. International technology transfer has the advantage of potential rapid implementation, and can lead to major savings of time
and money by eliminating unnecessary industrial research and development programs or by giving them a more advanced basis on which to
begin. In view of the national energy situation, it is now more important than ever that opportunities for technology transfer be sought,

that barriers to transfer be identified, and that methods to overcome these barriers be developed.

In each of the industries analyzed, there is a substantial amount of energy efficient technology which is available from abroad. Such technologies include:

Iron and Steel:

• Dry Coke quenching

• Pressurized blast furnace

• BOF off gas recovery

Cement:

• Suspension preheater system

• Precalcining systems

Copper:

• Outukumpu flash smelter

• Mitsubishi smelter

• Momoda blast furnace

• Noranda smelter

Pulp and Paper:

• Rapson effluent free process

• Swedish papermaking technology

Petroleum Refining (2):

• Catalytic naptha reforming. The French Institute Francais du Petrol (IFP) has developed an "aromizing" process for reforming naptha to produce benzene, toluene, and xylenes, for which there is a reduction in energy consumption of about 27% per barrel of aromatics produced.

• Reformate solvent extraction. The German Kopper's company has developed a "morphylane" process for extracting aromatics from catalytic reformate for which a process energy saving over a typical U.S. refinery of about 35% per barrel of reformate feed process can be obtained.

• Polymerization of olefins. The liquid phase "dimersol" process licensed by IFP shows a reduction in process energy consumption of about 70% per barrel of olefin feed polymerized, when compared to the solid phase phosphoric acid polymerization process used in the U.S.

There are obviously many factors which can affect the direction and rate of international technology transfer. Some of the more important of these factors include:

• Market demand for the product or process, including competitive pressures to reduce manufacturing costs or raw materials dependency in the making of the product.

• National policies, laws, and regulations, including taxes, tariffs, quotas, and other restrictions or incentives.

• Corporate policies and attitudes, including the "not invented here" syndrome and the communication gap between those with a knowledge of what is possible and those who may be able to put that knowledge to use.

• Availability of information.

• Availability of venture capital.

It is important to understand that international technology transfer can assist the U.S. in improving energy efficiency in the industrial sector. Although some technologies cannot be applied until new capacity is built in that industry, certain retrofit applications of innovative technology can make useful contributions today. Industrial managers should seek out those opportunities for increased energy efficiency and increased profits.

The fresh perspectives that international technology transfer provides should not be regarded as a threat to U.S. industry but as an opportunity for improving the productivity of the industry and as a stimulus to the development of U.S. technology, which can effectively compete in both domestic and export markets. The Federal government should encourage international technology transfer, because in the short term, foreign technology can improve U.S. industry efficiency, especially with regard to energy use and to reducing the cost of oil imports. Where the foreign technology is to be imported in the form of machinery, there may be the opportunity to manufacture that machinery in the U.S. through a licensing agreement, which provides jobs at home and furthers export opportunities. Finally, the import of efficient foreign technology should encourage the development of still better techniques and technology in the U.S.

If the government does get involved in international technology transfer, any policy carried out must be flexible. The formal and direct government role must be considered carefully, since most of the technology involved are owned by private corporations and are licensed to other private corporations. The direct involvement and responsibility for actual international technology transfer must, therefore, remain in the hands of private companies. However, there are ways in which the government can play a role for enhancing technology transfer. These roles fall into the following categories (3):

- Maintenance of good information services.

- Provision of technology evaluation services to companies with limited resources.

- Provision of translation and interpretation services to improve communications.

- Encouragement of personal contacts with foreign scientists and engineers.

- Sponsorship of demonstration plants to encourage adoption of particular technologies.

- Provision of financial incentives to industry for capital investment in energy efficient technology.

These actions can stimulate international technology transfer and provide benefits of improved energy efficiency for the U.S. industrial sector.

CONVERSION TO ALTERNATIVE FUELS

Fossil fuels are being consumed faster than we can produce them from existing sources. This realization has intensified government

and industry efforts in the areas of fuel conversion and of conversion
to production processes that use natural resources which are in abundant supply. Industry is in the midst of a period of transition in
fuel sources. Natural gas now provides about half of industrial energy, but this resource is becoming scarce and may have to be phased
out as an industrial fuel. Oil is being substituted for natural
gas in process heaters and boilers, and although the increased use
of oil may be contrary to the goal of national self-sufficiency in
energy, industry has few other alternatives at the present time.

Coal and nuclear energy are the only major domestic fuel resources that have reasonable long term resource bases. The technologies required to use these fuels in an economical, environmentally
acceptable way are under development and are, in some instances,
being applied. Domestic coal resources are sufficiently large to
make it a reasonable long term alternative for industrial applications.
Nevertheless, there are major intermediate term problems in exploiting
our coal resources. These problems relate to environmental constraints in mining and utilization, to coal industry capitalization,
and to transportation. When all factors are considered, it appears
that the supply of coal will be hard pressed to meet demand, at least
over the next decade.

However, in the short term, that is, until 1985, coal possesses
the greatest utility as a replacement fuel for oil and gas in the
process industries. Coal based systems which appear capable of meeting environmental standards, especially with respect to sulfur dioxide, include conventional firing that uses either low sulfur coal
or high sulfur coal with stack gas scrubbing, fluidized bed combustion that uses high sulfur coal, low and intermediate BTU gas, high
BTU pipeline quality gas, solvent refined coal (SRC), liquid boiler
fuels, and methanol from coal.

The direct firing of coal in industrial boilers and process
heaters will be more economical than the use of coal derived fuels.
There are three methods for directly using coal to generate steam or
process heat in an environmentally acceptable manner: burning low

sulfur coal, using fluidized bed combustion, and burning high sulfur
coal with stack gas scrubbing. The most realistic coal-based alternative is low sulfur coal fired in a conventional boiler. The most
promising method of using high sulfur coal is the fluidized bed
boiler, which offers flexibility in fuel supply as well as low cost.
Fluidized bed combustion may also hold promise for process heating,
but little development work is being done on fluidized bed process
heaters at this time.

Direct fired coal will be used as a fuel for steam boilers and
will be economically attractive, primarily to those plants producing
more than 100,000 pounds of steam per hour. By 1985, this user group
can potentially replace the equivalent of 2,000,000 barrels per day
of oil or 13 trillion cubic feet per day of natural gas, representing
25% of the U.S. industrial oil and gas requirement. There are, however, two major problems inhibiting the increased use of coal in
direct combustion. The first problem is the absence of simple and
reliable coal handling and combustion systems for the intermediate
to small user. The second problem is the lack of improved processes
for obtaining combustion of coal in an environmentally acceptable
manner by all potential industrial coal users.

Several technologies can be used to produce coal-derived gaseous
and liquid fuels, some of which are also applicable for obtaining
liquid and gaseous products from oil shale and organic wastes. Table
14-3 indicates conversion efficiencies of various processes. Some
of these processes are commercially available, whereas others require
much more development. Costs must also be reduced before widespread
implementation of these processes can be expected. Generally, existing boilers burning one type of fuel do not require modification to
burn synthetic fuels of the same type. For example, problems resulting from burning coal-derived fuel oils in boilers designed for oil
firing are insignificant. However, changing from one type of natural
fuel to another type of synthetic fuel will require modifications to
the boiler and/or to the fuel system.

Table 14-3

Conversion Efficiencies for Making Various Coal-Derived Fuels

Fuel Form	1975 Efficiency Percent	1985 Efficiency Percent
Mechanically cleaned coal	80-90	80-90
Solvent refined coal	85-90	90-95
Liquid Fuel oils	65-70	70-75
High BTU gas	60-65	70-75
Low BTU gas	65-75	85-90

Source: Abstracted with permission of Technical Publishing Company. A Division of Dun-Donnelley Publishing Corp. A Dun and Bradstreet Company, from the article "The Outlook for Coal as an Industrial Fuel," in the February 3, 1977 issue of Plant Engineering Magazine.

Characteristics and estimates of commercial availability for various synthetic fuels are shown in Table 14-4. The principal uncertainties in the commercial utilization of most of these synthetic fuel processes are the capital and operating costs.

Perhaps no area holds as much potential for energy conservation as the burning of waste process gases and residues. As conservation within industry holds, waste fuel utilization can prove to be a key source of energy conservation, as well as a means of reducing environmental problems associated with disposing of process wastes. However, utilization of waste fuel may be hampered in some cases by environmental constraints, and unless provisions are made for variances when burning non-commercial fuels, this useful source of energy may be wasted.

In the longer term, 1985-2000, there should be a significant industrial potential for process heat from nuclear energy. Although nuclear steam can be produced at the lowest cost of any major alternative, the onstream availability of existing reactors, together with the large size required for economical use, restrict its industrial potential.

Table 14-4

Estimated Characteristics and Availability
of Coal-Derived Fuels

Fuel Form	Heating Value	Ash Per Cent	Sulfur Per Cent	Availability
Mechanically cleaned coal	13,400 BTU/lb	7.0	1.0	1977-78
Solvent refined coal	16,000 BTU/lb	0.1	0.8	1983-85
Synthetic No. 6 fuel oil	142,000 BTU/lb	1.0	0.2	1985-87
Low BTU gas	150 BTU/cuft	—	Trace	1981-82
Medium BTU gas	300 BTU/cuft	—	Trace	1985
High BTU gas	950 BTU/cuft	—	Trace	1985

Source: Abstracted with permission of Technical Publishing Company. A Division of Dun-Donnelley Publishing Corp. A Dun and Bradstreet Company, from the article "The Outlook for Coal as an Industrial Fuel," in the February 3, 1977 issue of Plant Engineering Magazine.

In evaluating the feasibility of using a particular fuel in any industrial process, both the physical and chemical composition must be considered. Since no two fuels have identical physical and chemical properties, no two fuels will behave alike when applied to an industrial process. Furthermore, up to the present time, few alternative fuels have been available, so that little information is available about their combustion properties. Substituting a new fuel for conventional fuel depends on the following major criteria:

• Relative rate of heat release

• Combustion temperature profile

• Size of combustion zone relative to furnace volume

• Combustion stability

• Effect of combustion products on process product

• Relative pollution emissions

FEDERAL GOVERNMENT ROLE

There is a large potential for energy conservation within the industrial sector that currently represents about 40% of total U.S. energy demand. But beyond general housekeeping measures and changes in operating procedures, most of the fuel-saving methods require capital investment, either for changes in existing plants or for construction of new plants employing the latest technology. The rate at which the potential fuel savings can be accomplished is also subject to economic considerations. In addition to the financial, there are two other categories of specific potential constraints to industrial energy conservation in which the Federal Government can be of assistance--the technological and the regulatory.

Financial Programs

An industrial firm may face a number of financial barriers as decisions are made regarding energy conservation investments. Availability of capital for discretionary investments for energy conservation is the major constraint. Most of the available capital is consumed by investments related to basic production requirements and by mandatory capital investments, such as those for environmental protection and occupational health and safety.

Relatively low returns on energy conservation investments at today's cost of equipment and the current cost of energy make up a second related constraint. While energy costs have risen dramatically, other production costs over the past ten years have risen at a nearly equal average rate for many industries, resulting in a lower prioritization of energy cost problems than might be expected from observing the recent trend. The capital costs of energy conservation equipment have also increased sharply, offsetting potential advantages of

energy conservation investments. Economic requirements for capital
investments in discretionary conservation equipment are stringent,
with paybacks required of two or less years.

One important factor is the slow rate of growth of several of
the major energy-consuming industries, which results in a minimum of
new plant expenditures for more energy efficient new technologies.
This situation, compounded by the difficult economic requirements for
purchase of energy conservation retrofit equipment, lessens the incentives for industry and equipment manufacturers to produce even
better energy efficient equipment. In addition, the long construction
time required to build new plants and the downtime penalties of plant
modifications caused by lost production can be disastrous to the
financial position of the firm. Thus, a company may elect to forego
energy conservation investments until means can be found for avoiding
the loss in production.

The profitability of energy conservation investments will depend
in large part on the energy cost savings, both present and future,
that result from the investment. Therefore, financial risk may be
related to the uncertainty of future energy prices or to the efficiency of new technology in which an investment might be made. Uncertainty about future energy price trends will make it difficult for
firms to accurately predict future cost savings, thereby increasing
the risk relative to more predictable investments. The perceived risk
of new technology may be caused simply by a firm's lack of complete
information about a process, or by a more fundamental technical
problem requiring further research and development.

One factor affecting the risk/rate of return issue is the accounting methods that firms employ to assess the cost of energy and
the potential profitability of energy conservation. Before a firm
can compute the potential energy cost savings for a particular process change, it must first know the present energy cost of a process.
Companywide energy costs must, therefore, be properly allocated to
the various products and processes. But finding that an energy conservation investment is attractive from a risk/rate of return point

of view does not necessarily ensure a firm's ability to obtain funds
from traditional sources. This situation may be the result of any of
several factors: the condition of the firm's balance sheet (debt/
equity ratio), current interest rates, the ability of the firm to
service the debt over time.

The Federal Government can alleviate or remove some of these
risks and improve the rate of return on these investments through the
use of temporary financial incentives. Table 14-5 lists examples of
such incentives. The fundamental justification for financial incentives arises from the fact that the energy prices paid by the industrial consumers are well below their true resource costs because of
the phased decontrol of oil prices, regulated natural gas prices, and
only partial internalization of energy related pollution and import
vulnerability costs. As a consequence, the consumption of energy is
excessive. Thus, artificially low prices and the market's failure to
overcome the institutional and motivational barriers to energy conservation provide the basic rationale for governmental action. The
best long run approach should be to quickly correct price distortions.
However, if these market distortions cannot be corrected immediately,
an interim option is to provide temporary financial incentives to
energy consumers to partially offset the disincentives created primarily by these artifically low energy prices.

In a study performed for the Energy Research and Development
Administration, Booz, Allen, and Hamilton investigated various economic incentives and financing options to stimulate the commercialization of energy conservation technologies (4). They found that the
majority of industries studied were against the intrusion of the
Federal Government in the capital investment planning processes.
Table 14-6 provides a summary of economic incentives which the various
industrial companies felt were both most helpful and least helpful
in commercializing energy conservation technologies. As is evident,
the investment tax credit is the financial incentive felt most beneficial. The loan guarantee is believed to be of minimal importance.

Table 14-5

Examples of Economic Incentives

Incentive	Example
Cash	
Capital Equipment Grant	Maritime Administration cash payment to U.S. shipbuilders to cover U.S./foreign ship construction cost differential
Operating Grant	Civil Aeronautics Board operating subsidy to air carriers
Demonstration Grant (R, D&D)	EPA solid waste demonstration grants
Tax	
Investment Tax Credit	10% investment tax credit for industrial equipment
Accelerated Depreciation	Five year writeoff for pollution control facilities
Tax-Free Bonds	Industrial revenue bonds for pollution control facilities
Energy Tax	Companies taxed a certain amount for each unit of fuel consumed
Credit	
Direct Government Loans	Small Business Administration loans
Loan Guarantees	Small Business Administration/EDA/ERDA Geothermal loan guarantees
Interest Subsidy	HUD low/moderate income housing support
Purchase	Department of Agriculture Farm price supports

In their study, Booz, Allen, and Hamilton concluded the following, with respect to the effectiveness of the incentives proposed in terms of energy saved:

Table 14-6

Ability of Economic Incentives to Stimulate the Commercialization of Energy Conservation Technologies in 12 Manufacturing Industries

Industry	Major Financing Difficulty	Economic Incentives Deemed Most Helpful	Economic Incentives Deemed Least Helpful
Aluminum	Insufficient cash flows Highly leveraged capital structure	Cost sharing/capital cost expensing Investment tax credits	Loan guarantees and grants Reduction of capital gains tax
Brick manufacturing	Very poor cash flows Little to no access to public or private capital	Loan guarantees and grants Cost sharing/capital cost expensing	Reduction of capital gains tax Elimination of double taxation on corporate profits
Cement	Poor earnings and internal cash flows Moderately leveraged capital structure	Cost sharing/capital cost expensing Investment tax credits	Loan guarantees and grants 60-month amortization
Chemicals	Moderately leveraged capital structure Recent period of poor earnings	Expensing capital costs Investment tax credits	Loan guarantees and grants Reduction of capital gains tax
Copper	No significant financing difficulties	Cost sharing/capital cost expensing Investment tax credits	Loan guarantees and grants Reduction of capital gains tax
Glass	Moderately leveraged capital structure	Cost sharing/capital cost expensing Investment tax credits	Loan guarantees and grants Industrial revenue bond concept
Gray Iron Foundries	Limited access to capital due to insufficient collateral (firm size, net worth)	Loan guarantees and grants Cost sharing/capital cost expensing	Reduction of capital gains tax Elimination of double taxation on corporate profits
Grains and Flour	Largest firms essentially have no significant difficulty Smaller firms have declining cash flows and moderately leveraged capital structure	Smaller firms: investment tax credits Smaller firms: expensing of capital costs	Loan guarantees and direct grants Reduction in capital gains tax
Meat Packing Plants	Minimally constrained by moderately leveraged capital position	Cost sharing/capital cost expensing Investment tax credit	Loan guarantees and grants Reduction in capital gains tax
Paper	Minimally constrained by moderately leveraged capital structures	Cost sharing/capital cost expensing Investment tax credits	Loan guarantees and grants Industrial revenue bond concept
Plastic Materials	Moderately high leveraged capital structure	Elimination of double taxation on corporate profits Cost sharing/capital cost expensing	Loan guarantees and grants Industrial revenue bond concept
Steel	Insufficient cash flows to meet burgeoning capital requirements Moderately leveraged capital structure	Cost sharing/capital cost expensing Investment tax credits	Loan guarantees and grants 60-month amortization

Source: Booz, Allen and Hamilton, **Economic Incentives and Financing Options to Stimulate the Commercialization Of Energy Conservation Technologies** (Bethesda, Maryland: Booz, Allen and Hamilton, May 1976), p. 1-6.

- Loan guarantees and grants will not produce significant energy savings because of their orientation to the fragmented industries

- Economic incentives such as investment tax credits, expensing of capital costs, and rapid writeoffs/amortization can produce significant energy savings if they are energy technology specific

- Cost sharing programs, if they contain a minimum of administration and regulation, can lead to very significant energy savings

- Economic incentives such as the reduction of capital gains tax and the tax deductability of dividends probably will not provide significant short-term energy savings but could provide major benefit in the longer run

  - The debt-to-equity ratios of major industries will recede slowly as equity financing replaces long-term debt

  - Since the incentives are not energy technology specific, there is no guarantee that the reduced cost of capital will induce energy as opposed to non-energy capital investments (5).

There is presently a 10% investment tax credit for any industrial investment. The proposed National Energy Act offers an additional 10% tax credit for energy conserving equipment. One method that provides a larger stimulus to energy savings would be to modify the flat 10% tax credit with a graduated tax credit of up to 40% for improved energy efficient technology. The graduated tax credit should be based on two parameters: (1) the improvement in per unit energy efficiency over conventional practices from a relative sense and (2) the improvement in per unit energy efficiency from an absolute sense (e.g., 10^6 BTU/ton). See Table 14-7. For example, if the Rapson effluent-free papermaking process were installed in place of the conventional kraft process, the energy efficiency gain would be 5.3×10^6 BTU/ton with an overall efficiency gain of 15%. Using Table 14-7, tax credit would thus be:

Energy efficiency gain (5.3×10^6 BTU/ton)	10%
15% efficiency improvement	12%
Total	22%

Table 14-7

Proposed Investment Tax Credit

Energy Efficiency Gain 10^6 BTU/ton	Percent Tax Credit
0-1	4
1-3	6
3-5	8
5-10	10
10-20	16
over 30	20

Percent Improvement over Baseline Process or Equipment	Percent Tax Credit
0-5	4
5-10	8
10-20	12
20-30	16
over 30	20

This proposal must be evaluated must more in depth to determine exactly the nature of the tax credit criteria as well as the impact of this plan on Treasury revenues. There may also be distortions that are created in the marketplace by this differential tax credit. Rough calculations show that the impact on the Treasury would not be significantly more than the impact of the present 10% tax credit plus the proposed 10% tax credit on energy efficient equipment of the National Energy Act. This proposal will focus industrial investments where the greatest energy savings impact can be achieved, although it is recognized that public policy priorities in other areas may make the proposal infeasible. Implementation of the proposal will create a demand for new plant and equipment, which should stimulate the economy by creating new jobs and, in turn, increasing revenue for the Treasury. This should help to offset the losses to the Treasury caused by the tax credit.

Technological Programs

Larger savings in energy may be realized from processes just beyond present horizons of technology. The theoretical and practical
limit of industrial energy efficiency far exceed both the best available technology and the expected actual energy conservation progress.
However, achieving the practical and/or theoretical limits to energy
efficiency requires, for most industrial processes, major research
and development (R & D) and redesign, as well as reconstruction of
all process equipment. As has been discussed previously, for each
of the major key energy intensive industries, there is a variety of
commercially available improved process technologies that can be
economically adopted. With the low rate of new plant additions and
the difficult economic requirements for purchase of energy conservation retrofit equipment, the incentives for industry and equipment
manufacturers to produce even better energy efficient equipment are
not great. Similarly, for the major categories of generic energy
conservation technologies, industry interest is not strong enough to
provide a large enough market for mass production, which is, therefore, inadequate incentive for private research and development.

While industry considers financial constraints to be the key
barrier to more energy conservation, there are a number of important
technological shortfalls. First, there is a clear lack of economical
technology alternatives across all major energy intensive industries
for switching industry off oil and natural gas. Second, there is a
critical shortage of new technology concepts on the drawing boards
for most basic industries. The major sources of new technology for
several of our key industries are foreign.

Priorities on research and development spending vary, to a substantial degree, between industries. At least $16 billion was spent
by U.S. industry on R & D in 1976, an average of 1.9% of sales, or
33.9% of profits, on total R & D. See Table 14-8. In comparison,
the key energy intensive industries spend far less on R & D, only
0.8% of sales. Energy conservation is a primary focus of R & D in
only three of the major energy intensive industries. These industries

Table 14-8

R & D Profiles of Energy Intensive Industries

Industry	Number Of Companies Represented	1976 Total R & D Expenditures ($ Millions)	R & D Expenditures As A Percent Of Sales (1976)	R & D Expenditures As A Percent Of Profits (1976)
Chemicals	42	$1438.8	2.6	39.7
Steel	8	124.4	0.7	17.6
Petroleum	18	752.9	0.4	7.6
Paper	12	111.9	0.8	12.1
Non-Ferrous Metals	17	157.4	1.2	25.0
Building Materials	23	173.0	1.0	17.7
Textiles	9	23.9	0.4	10.1
Subtotals/ Averages for Energy-Intensive Industries	129	$2782.3	0.8	16.4
Total Sample U.S. Industrial Firms	600	$16224.7	1.9	33.9

Source: Abstracted from the June 27, 1977 issue of Business Week (c) 1977 by McGraw Hill, Inc. 1221 Avenue of the Americas, New York, NY 10020. All rights reserved.

are commodity producers. For example, the aluminum industry R & D is highly proprietary and heavily oriented toward energy efficiency. Basic research by aluminum companies has brought out a large number of new concepts for the aluminum manufacturing process. In petroleum refining, R & D is directed toward improving process efficiency and productivity, which includes energy conservation. Approximately 15% of R & D funds are spent on energy conservation. Chemical industry R & D is directed principally toward process technology improvements,

including energy efficiency. Rapid technological changes and the
competitive nature of the industry have contributed to a near term
focus on R & D.

However, a number of key energy intensive industries are not
focusing on R & D for energy efficiency, due to overriding constraints
related to business conditions, capital availability, and perceived
needlessness for using today's best technology. The steel industry's
investment in R & D is spent principally on non-energy related projects in which product quality control and new steel products are the
first priority. Most steelmaking R & D is funded in Japan and Germany, which are the principal sources of new technology. Pulp and
paper industry R & D focuses principally on ways to meet EPA and OSHA
guidelines. Cement industry R & D is directed toward engineering
refinements for energy conservation. Basic process technology is
currently being developed by foreign equipment manufacturers.

The discussion above indicates that while there is some R & D
attention to energy conservation, each industry's R & D priorities
are unique, and only a few industries focus heavily on energy conservation. The purpose of R & D in industry is to maintain and/or
achieve a competitive advantage in the marketplace; thus, the very
nature of industrial R & D is secrecy and assurance of proprietary
information.

Although there is a number of options for applying Federal R & D
funds to improve industrial processes and equipment in the short
and mid terms, this is in the area of industry's own R & D concentration, and the funding required to support a widespread attack on
the general problem of process efficiency is very large. There are,
on the other hand, critical problems which are not being addressed
by industry and which represent major opportunities for Federal
funding. These include:

- Provision of alternative industrial fuels.

- Provision for all new industrial process concepts for radically improved energy efficiency.

- Provision of generic technologies which significantly advance
  the state of the art, e.g., high temperature recuperators
  (over 1800°F), high temperature heat pumps (over 250°F), heat
  recovery from dirty gas streams, co-generation of steam and
  electricity, and use of industrial wastes.

It must be emphasized, however, that R & D is, in most cases,
only an essential initial step to a conservation program and that the
benefits attributed to R & D are contingent on successful implementation. R & D will contribute to the information base on which conservation strategy is devised. To carry out these strategies, proper
institutional structures and authority are needed.

Regulatory Programs

Federal regulatory measures can be used to protect the public
from the adverse effects of monopoly, to regulate industry for the
benefit of industry members, and to allocate responsibilities, like
the tax structure. Regulatory measures are applied by setting prices,
restricting entry, or requiring a specific action.

Presently, there are regulatory programs in effect at both
federal and state levels that may have adverse impacts on industrial
energy conservation. Several examples of such programs include:

- Interstate Commerce Commission (ICC) shipping rates and regulations. Recycling offers significant potential for energy
  and resource conservation in a number of important industries.
  However, shipping rates that discriminate against transport
  of recycled materials in favor of virgin materials constitute
  a significant barrier to increased recycling. ICC regulations
  that preclude the use of the most direct routes by shippers
  or that result in trucks or rail cars returning empty or only
  partially loaded are additional built-in impediments to energy
  conservation.

- Federal Energy Regulatory Commission (formerly Federal Power
  Commission) gas price regulation. The FERC regulation of the
  well-head price of natural gas at an artificially low level
  has encouraged the wasteful use of this valuable resource.
  Price increases would encourage conservation by industrial
  users.

- Environmental Protection Agency (EPA) pollution control requirements. The Clean Air Act of 1970 and the Federal Water Pollution Control Act Amendments of 1972 have imposed requirements on industry to minimize environmental pollution from plants and facilities. The energy required to remove the pollutants before discharge to the environment is an additional burden on industry. It has been estimated that the energy required in 1980 to abate stationary source air pollution will be 267 x 10^{12} BTU, or 0.4% of total U.S. energy demand. Water pollution abatement will require 404 x 10^{12} BTU in 1980, or 0.6% of total energy demand (6).

- Antitrust regulations. The purpose of antitrust laws is to discourage the restraint of trade or competition that might result from interfirm cooperation in price setting, research and development, and other activities. The extent to which these regulations discourage cooperative R & D and information exchange in energy conservation represents a barrier to industrial energy conservation.

Examples of regulations which could enhance energy efficiency or encourage the use of more available fuels include:

- Federal regulation of the amounts of given energy sources--notably of oil and natural gas--that could be utilized by particular industrial sectors and/or plants. This is called the energy budget. In reality, certain industrial managers are already facing a similar situation, as they are advised that natural gas will not be available to them at any price. This energy budget assumes a suspension of conditions that would normally be expected to determine energy price levels.

- The imposition of standards on the manufacture of equipment that determine energy consumption. This option, now being considered by Congress, would involve establishment of standards of energy efficiency for such equipment as electric motors, pumps, boilers, compressors, and grinding mills.

- The assignment of energy efficiency targets to each industry. A voluntary program of this sort is presently on-going with the Department of Energy. Efficiency targets which have been assigned are enumerated in Table 14-9.

Regulation may be the most directly effective of any program in accomplishing a specific goal. Unfortunately, it probably has the highest administrative costs of any program and may result in a low level of efficiency.

Table 14-9

Energy Efficiency Improvement Targets

Industry	1972-1980 FEA Target Improvement
Steel	9.1%
Petroleum Refining	12.0%
Paper	20.0%
Ammonia	15.1%
Aluminum	12.8%
Cement	18.3%
Olefins	24.8%
Textiles	22.0%
Chlor-Alkali	8.8%
Glass	13.5 - 22.6%
Food Processing	12.0%
Major End Products	6.7 - 30.1%

Summary

A number of government policy options are available for encouraging industry to improve production efficiency. These range from financial incentives, to R & D, to government regulation. Maximum energy savings can be realized only if industry examines and pursues all opportunities for conservation. Requiring the development of a conservation program that is approved by corporate management and by the Federal Government may provide the incentive and motivation for identification of savings potential by all levels within each firm. By requiring both the establishment of goals for reduction of energy consumption per unit of output and the reporting of progress made, each level in the firm, from plant worker to corporate officer, may be made aware of the savings required and of the progress achieved.

The largest potential for savings accrues from regulatory actions. However, the cost of such actions, in terms of intangibles such as curtailment of industry growth, is impossible to assess in general terms. Marginal investments made through financial incentives can assist greatly in terms of energy savings. These programs, however, would also defy easy specifications of eligibility and may create "windfalls" to selected portions of industry. Technological programs, if focused on the proper areas, could lead to energy savings without interfering with both the growth of the industry and the selective discrimination within it. At any rate, achievement of energy conservation goals must be a cooperative venture between government and industry, with government providing sufficient incentives to enhance conservation efforts and industry moving ahead vigorously to apply the available economic technologies.

CONCLUSIONS

Energy conservation offers one of the best opportunities for industry to contribute to the lessening of the overall energy situation and offers a positive approach to counter rising energy costs. In general, industrial managers are aware of the energy dilemma, so that there is receptiveness to energy conservation methods. However, there must be greater recognition of the specific possibilities for energy conservation among top management, plant engineers, consultants, and manufacturers representatives.

In the short term, approximately 30% of the energy used in industrial processes could be saved through the application of existing techniques that are economically justifiable. Predicted increases in fuel prices are expected to make energy conservation measures even more attractive in the future. In the long term, through the invention of more efficient devices and more efficient processes, and through the utilization of waste heat on a larger scale, industries can be expected to save more than 30%.

While energy conservation is not the cure-all to our energy dilemma, it does provide us with the opportunity to put good engineering in practice. This, after all, is what we should be doing with or without an energy problem.

NOTES AND REFERENCES

1. William Franklin, David Bendersky, William Park, and Robert Hunt, Potential Energy Conservation from Recycling Metals in Urban Wastes (Kansas City: Midwest Research Institute, May 1974), p. 2.

2. Gordian Associates, International Technology Transfer (Washington, D.C.: Energy Research and Development Administration, 1977), p. 14.

3. Ibid., p. 332.

4. Booz, Allen, and Hamilton, Economic Incentives and Financing Options to Stimulate the Commercialization of Energy Conservation Technologies (Bethesda, Maryland: Booz, Allen, and Hamilton, May 1976), p. I-5.

5. Ibid., p. I-6.

6. Resource Planning Associates, A Brief Analysis of the Impact of Environmental Laws on Energy Demand and Supply (Cambridge, Massachusetts: Resource Planning Associates, October 1974), p. 2-3.

BIBLIOGRAPHY

Advanced Energy and Technology Associates. Energy Conservation in the Pulp and Paper Industry. Durham, New Hampshire: University of New Hampshire, February 1977.

Agawal, J.C. and J.F. Elliott. "High Sulfur Coke for Blast Furnace Use." AIME Procedures of Ironmaking Conference. Vol. 30. New York: The Metallurgical Society of AIME, 1971, pp. 50-67.

Argall, G.O. "Outokumpu Adds Second Catalyzer to Raise Pyrite to Sulfur Conversion to 91%." World Mining, 20 (March 1967), pp. 42-46.

Aldermon, L. and R. Chambers. "Preheating and Charging Coal to Coking Ovens." AIME Proceedings of Ironmaking Conference. Vol. 31. New York: The Metallurgical Society of AIME, 1972, pp. 193-200.

Alexander, Frank. "Why Regenerator Glass Melters." The Glass Industry, 55 (May 1974), pp. 12-13, 26-28, 30.

"Aluminum Faces Power Crisis in Northwest." Metals Week, 47 (May 24, 1976), 1.

American Gas Association. A Study of Process Energy Requirements in the Paper and Pulp Industry. New York: American Gas Association, 1965.

American Iron and Steel Institute. Energy Conservation in the Steel Industry: Handbook on Energy Conservation Technology in the Steel Industry. Washington, D.C.: American Iron and Steel Institute, 1976.

American Paper Institute. Wood Pulp Statistics. 35th ed. New York: American Paper Institute, 1971.

"Are Solvay Plants on the Way Out?" Chemical Week, 112 (March 7, 1973), pp. 40-41.

Asantila, Raimo et al. "Design for High Cost Energy--The Scandanavian Approach." TAPPI, 57 (October 1974), pp. 117-21.

Atkins, P.R. "Recycling Can Cut Energy Demand Dramatically." Engineering and Mining Journal, 174 (May 1973), pp. 69-71.

Atkins, P.R. and C.N. Cochran. "Future Energy Needs in the U.S. Light Metals Industry." Efficient Use of Fuels in the Metallurgical Industries. Chicago, Illinois: Institute of Gas Technology, December 1974, pp. 721-40.

Babcock and Wilcox. Steam: Its Generation and Use. New York: Babcock and Wilcox, 1972.

Barbu, I. and I. Stefanescu. "Use of ICEM Formed Coke in the Blast Furnace and in Other Applications." Journal of the Iron and Steel Institute, 211 (October 1973), pp. 685-88.

Barnes, R.W. "Energy and Industrial Processes: A Look at the Future." Paper presented at the Symposium on Advances in Energy Storage and Conversion, American Chemical Society National Meeting, San Francisco, California, September 1, 1976.

Basiulis, A. and M. Plost. "Waste Heat Utilization Through the Use of Heat Pipes." Paper presented at the Annual Winter Meeting of the Heat Transfer Division of the American Society of Mechanical Engineers, Houston, Texas, November 30 - December 4, 1975.

Battelle Columbus Laboratory. Energy Efficiency Improvement Targets for Primary Metals Industries, SIC 33. Vol. 1. Columbus, Ohio: Battelle, August 1976.

Battelle Columbus Laboratory. Energy Use by the Steel Industry in North America. Columbus, Ohio: Battelle, July 1971.

Battelle Columbus Laboratory. Energy Use Patterns in Metallurgical and Non-Metallic Mineral Processing. Phase 4. Columbus, Ohio: Battelle, June 1975.

Battelle Columbus Laboratory. Energy Use Patterns in Metallurgical and Non-Metallic Mineral Processing. Phase 9. Columbus, Ohio: Battelle, August 1976.

Battelle Columbus Laboratory. Final Report on Industrial Energy Study of the Glass Industry. Columbus, Ohio: Battelle, December 1974.

Battelle Columbus Laboratory. Potential for Energy Conservation in the Steel Industry. Columbus, Ohio: Battelle, May 1975.

Battelle Columbus Laboratory. Study of the Energy and Fuel Use Patterns in the Non Ferrous Metals Industries. Columbus, Ohio: Battelle, December 1974.

Beatson, Cedric. "Recapture the Heat that is Escaping from Your Factory." The Engineer, 222 (October 10, 1974), p. 58.

Beck, K. et al. "A New Technique for Preheating Coking Coal Blends for Carbonization in Slot Type Recovery Ovens." AIME Proceedings of Ironmaking Conference. Vol. 31. New York: The Metallurgical Society of AIME, 1972, pp. 185-92.

Berg, Charles A. "A Technical Basis for Energy Conservation." Technology Review, 76 (February 1974), pp. 15-23.

Berg, Charles A. Energy Conservation Through Effective Utilization. Washington, D.C.: National Bureau of Standards, February 1973.

Belding, John A. and William Burnett. From Oil and Gas to Alternate Fuels: The Transition in Conversion Equipment. Washington, D.C.: Energy Research and Development Administration, November 1976.

Booz, Allen and Hamilton. Economic Incentives and Financing Options to Stimulate the Commercialization of Energy Conservation Technologies. Bethesda, Maryland: Booz, Allen and Hamilton, May 1976.

Boyce, H.P. "The OG Gas Clearing System." Operation of Large BOFs. London: The Iron and Steel Institute, 1972, pp. 1-9.

Brevard, J.C. et al. Energy Expenditures Associated with the Production and Recycle of Metals. Oak Ridge, Tennessee: Oak Ridge National Laboratory, November 1972.

Brewster, D.B. and W.I. Robinson. "How Computers are Controlling Functions in the Pulping Process." Pulp and Paper, 48 (May 1974), pp. 88-92.

Brooks, C.L. "Energy Conservation in the Aluminum Industry." Efficient Use of Fuels in the Metallurgical Industries. Chicago, Illinois: Institute of Gas Technology, 1974, pp. 705-20.

Brown, J.W. and R.L. Reddy. "Direct Reduction--What Does It Mean to the Steelmaker." Iron and Steel Engineer, 53 (June 1976), pp. 37-46.

Bruce, J. and W. Staneforth. "Some Aspects of Experience on the Brookhouse Project." Developments in Ironmaking Practice. London: The Iron and Steel Institute, 1973, pp. 63-77.

Brunger, R. "Sulfur Control by Furnace and External Means." AIME
    Proceedings of Ironmaking Conference. Vol. 31. New York: The
    Metallurgical Society of AIME, 1972, pp. 169-87.

Bryk, P., J. Ryselin, J. Honkasalo, and R. Malmström. "Flash Smelting
    of Copper Concentrates." Journal of Metals, 10 (October 1958),
    pp. 395-400.

Bureau of Mines. "Crude Petroleum, Petroleum Products, and Natural
    Gas Liquids." Annual Petroleum Statement. Washington, D.C.:
    Government Printing Office, 1975.

Bureau of Mines. Minerals Industry Surveys. Washington, D.C.: Gov-
    ernment Printing Office, 1940-1974.

Burroughs, James R. The Technical Aspects of the Conservation of Ener-
    gy for Industrial Processes. Midland, Michigan: Dow Chemical
    Company, May 1973.

Cantrell, A. "Annual Refining Survey." Oil and Gas Journal, 72
    (April 1, 1974), pp. 82-106.

Cantrell, A. "Annual Refining Survey." Oil and Gas Journal, 71
    (April 2, 1973), pp. 99-125.

Carnahan, Walter et al. Efficient Use of Energy: A Physics Perspec-
    tive. Princeton, New Jersey: The American Physical Society,
    January 1975.

Cato, G.A. et al. "Field Testing: Applications of Combustion Modifi-
    cations to Control Pollutant Emissions from Industrial Boilers--
    Phase II." KVB, Inc. Report to Environmental Protection Agency.
    Washington, D.C.: Government Printing Office, April 1976.

Chapman, P.F. "The Energy Costs of Producing Copper and Aluminum from
    Primary Sources." Metals and Materials, 8 (February 1974), pp.
    107-11.

"Chementator." Chemical Engineering, 80 (June 11, 1973), p. 61.

Cherrington, Dean and Herb Michelson. "How to Save Refinery Furnace
    Fuel." Oil and Gas Journal, 72 (September 2, 1974), pp. 59-68.

Clark, Robert T. "Energy." Pulp and Papermaking Technology. Ed.
    Esther Dorfman. Tuxedo Park, New York: International Paper
    Company, January 1976, pp. 107-12.

Cochs, L. "Why Injection?--Some Facts and Figures." Blast Furnace
    Injection: Proceedings of the Symposium on Blast Furnace Injec-
    tion. Sydney: "The Australian Institute of Mining and Metallurgy,
    1972.

"Continuous Casting--More for Less." Commentary by Institute for Iron
     and Steel Studies, No. III-3. Washington, D.C., March 1974.

Cook, E. "Energy Flow Through the United States Economy." College
     Station, Texas: Texas A & M University, December 1975.

Copper Development Association, Inc. Copper Supply and Distribution.
     New York: Copper Development Association, 1973.

"Copper Smelting Today: The State of the Art." Special Edition Joint
     Issue of Chemical Engineering and Engineering and Mining Journal.
     Special Section, March 1973, pp. p-z.

Council on Environmental Quality. The Economic Impact of Pollution
     Control. Washington, D.C.: Government Printing Office, March
     1972.

Deane, R.A. "Papermaking and Finishing." Chemical Process Technology
     Encyclopedia. Ed. D.M. Considine. New York: McGraw Hill, 1975,
     pp. 806-13.

Denton, Jesse, Stephen Webber, and John Moriarty. Energy Conservation
     Through Effective Energy Utilization. Washington, D.C.: Govern-
     ment Printing Office, 1976.

Donnelley, R.G. et al. Industrial Thermal Insulation: An Assessment.
     Washington, D.C.: Energy Research and Development Administration,
     August, 1976.

Dorfman, Esther, ed. Pulp and Papermaking Technology. Tuxedo Park,
     New York: International Paper Company, January 1976.

Dow Chemical Company. Energy Industrial Center Study. Midland, Michi-
     gan: Dow Chemical Company, June 1975.

Duke, J.M. Patterns of Fuel and Energy Consumption in the U.S. Pulp
     and Paper Industry. New York: American Paper Institute, March
     1974.

Duke, J.M. and N.J. Fudali. Report on the Pulp and Paper Industry's
     Energy Savings and Changing Fuel Mix. New York: American Paper
     Institute, September 1976.

Dukelow, S.G. "Charting Improved Boiler Efficiency." Factory, 7
     (April 1974), pp. 31-34.

ECON-I, How to Determine Economic Thickness of Thermal Insulation.
     Mt. Kisco, New York: Thermal Insulation Manufacturers Associa-
     tion, 1973.

Elliott, Thomas C. "Demand Control of Industry Power Cuts Utility Bills, Points to Energy Savings." Power, 120 (June 1976), pp. 19-26.

Energy and Environmental Analysis, Inc. Energy Management in Manufacturing: 1967-1990. Vol. I, Summary Report. Washington, D.C.: Government Printing Office, April 1974.

Environmental Protection Agency. Cost of Clean Air. Washington, D.C.: Government Printing Office, April 1974.

Environmental Protection Agency. Development Document for Effluent Limitations and New Source Performance Standards for the Petroleum Refining Point Source Category. Washington, D.C.: Government Printing Office, April 1974.

Environmental Protection Agency. Development Document for Effluent Limitation Guidelines and New Source Performance Standards for the Unbleached Kraft and Semi-chemical Pulp Point Source Category. Washington, D.C.: Government Printing Office, May 1974.

Environmental Protection Agency. Environmental Considerations of Selected Energy Conserving Manufacturing Process Options. Vol. 1, Industry Summary Report. Vol. 2, Industry Priority Report. Vol. 3, Iron and Steel. Vol. 4, Petroleum Refining. Vol. 5, Pulp and Paper. Vol. 8, Alumina/Aluminum. Vol. 10, Cement. Vol. 11, Glass Industry. Vol. 14, Primary Copper. Cincinnati, Ohio: Environmental Protection Agency, December 1976.

Evans, D.J. "Treatment of Copper/Zinc Concentrates by Pressure Hydrometallurgy." CIM Bulletin, 57 (1964), pp. 857-66.

EXXON. Energy Outlook 1977 - 1990. n.p.: EXXON, n.d.

Faith, W.L., D.B. Keyes, and R.L. Clark. Industrial Chemicals. 3rd ed. New York: John Wiley & Sons, 1965.

Federal Energy Administration. Economic Thickness for Industrial Insulation. Washington, D.C.: Government Printing Office, August 1976.

Federal Energy Administration. Energy Consumption in the Manufacturing Sector. Washington, D.C.: Federal Energy Administration, 1977.

Finniston, H.M. "Fewer Joules for Steelmaking." New Scientist, 63 (July 11, 1974), pp. 65-67.

Fletcher, J.B. "In-Place Leaching at Miami Mine, Miami, Arizona." Transactions of AIME. Vol. 250. New York: AIME, December 1971, p. 310.

Foard, J.E. and R.R. Beck. "Copper Smelting--Current Practices and
    Future Developments." Paper presented at AIME Annual Meeting,
    New York, February 1971.

Fort, M. "Some Practical Aspects of Electric Boosting." Glass Tech-
    nology, 5 (October 1964), pp. 199-201.

Franklin, William E., David Bendersky, William Park, and Robert Hunt.
    Potential Energy Conservation from Recycling Metals in Urban
    Wastes. Kansas City: Midwest Research Institute, May 1974.

Freeman, S. David, et al., A Time to Choose: America's Energy Future.
    Cambridge, Massachusetts: Ballinger Publishing Company, 1974.

Freeman, S. David. Energy: The New Era. New York: Walker and
    Company, 1974.

Garrett, H.M. and J.A. Murray. "Improving Kiln Thermal Efficiency--
    Dry Process Kilns." Rock Products, 77 (August 1974), pp. 58-71.

General Motors Corporation. Industrial Energy Conservation: 101
    Ideas at Work. Detroit: General Motors, 1977.

Glenn, R.D. "Energy Conservation Opportunity in Steam Use." Energy
    Conservation Through Effective Energy Utilization. Washington,
    D.C.: Government Printing Office, 1974.

Glickman, Leon and David White. "National Benefits of Energy Conser-
    vation." Energy Conservation Through Effective Utilization.
    Washington, D.C.: Government Printing Office, 1976, pp. 21-42.

Gordian Associates. An Energy Conservation Target for Industry SIC
    29. New York: Gordian, June 1976.

Gordian Associates. Historical Trends and Future Projections for
    Energy Consumption in the Production of Primary Refined Copper.
    New York: Gordian, January 1975.

Gordian Associates. International Technology Transfer. Washington,
    D.C.: Technical Information Center, ERDA, 1977.

Gordian Associates. The Potential for Energy Conservation in Nine
    Selected Industries. Vol. 1, Plastics. Vol. 2, Petroleum Re-
    fining. Vol. 3, Cement. Vol. 4, Copper. Vol. 5, Aluminum.
    Vol. 6, Steel. Vol. 7, Glass. Vol. 8, Paper. Vol. 9, Synthetic
    Rubber. Washington, D.C.: Government Printing Office, 1975.

Gott, E.H. "The Economic Importance of Continuous Casting of Steel
    Slabs." Proceedings of the Third Annual Conference of the Inter-
    national Iron and Steel Institute. London: International Iron
    and Steel Institute, 1969, pp. 102-15.

Gray, John E. Energy Policy: Industry Perspectives. Cambridge, Massachusetts. Ballinger Publishing Company, 1975.

"Great Lakes Paper Launches First Closed Cycle Kraft Pulp Mill." Paper Trade Journal, 161 (March 15, 1977), pp. 29-34.

Gyftopoulos, Elias P. Study of Effectiveness of Industrial Fuel Utilization. Waltham, Massachusetts: Thermo-Electron Corporation, January 1974.

Gyftopoulos, Elias, Lazaros Lazaridis, and Thomas Widmer. Potential Fuel Effectiveness in Industry. Cambridge, Massachusetts: Ballinger Publishing Company, 1974.

Hall, E.H. "Evaluation of the Potential for Energy Conservation in Industry." Energy Conservation: A National Forum. Ed. T. Veziroghu. Coral Gables, Florida: Clean Energy Research Institute, December 1975, pp. 55-76.

Hall, E.H. et al. Evaluation of the Theoretical Potential for Energy Conservation in Seven Basic Industries. Columbus, Ohio: Battelle Columbus Laboratory, July 1975.

Hall, F.K. "Wood Pulp." Scientific American. 230 (April 1974), pp. 52-62.

Hammond, A.L. "A Timetable for Expanded Energy Availability." Science, 184 (April 19, 1974), pp. 367-69.

Hanlan, J.F. Handbook of Package Engineering. New York: McGraw Hill, 1971.

Harkki, S.U. and J.T. Juusela. "New Developments in Flash Smelting." Paper presented at the Metallurgical Society of AIME, New York, 1974.

Hatsopoulos, George and J.H. Kernan. Principles of General Thermodynamics. New York: John Wiley & Sons, 1961.

Hayden, J.E. and W.H. Levers. "How to Conserve Energy While Building, Expanding a Refinery." Oil and Gas Journal, 71 (May 21, 1973), pp. 109-16.

Hayes, Denis. Energy: The Case for Conservation. Washington, D.C.: Worldwatch Institute, January 1976.

Haynes, Virgil. Energy Use in Petroleum Refineries. Oak Ridge, Tennessee: Oak Ridge National Laboratory, September 1976.

Helm, H.B. "Converting Batch Type Annealing Furnace from Radiant Tube to Direct Firing." Iron and Steel Engineer, 48 (August 1971), pp. 80-84.

Henderson, J.M. "Environmental Overkill--The Natural Resource Impact." Mining Congress Journal, 60 (December 1974), pp. 18-23.

Heu, J.P. "Heat Balance and Calculation of Fuel Consumption in Glass-making--Conclusions." The Glass Industry, 52 (February 1971), pp. 59-61.

Hobson, G.D. Modern Petroleum Technology. 4th ed. New York: John Wiley & Sons, 1973.

Hoffman, K.G., M. Beller, and A. Doernberg. Current BNL Reference Energy System Projections: Base Case Sept 19, 1975. Upton, New York: Brookhaven National Laboratory, September 1975.

Holgate, J.K. and P.H. Pinchbeck. "Use of Formed Coke: BSC Experience 1971/1972." Journal of Iron and Steel Institute, 211 (August 1973), pp. 547-66.

Hottel, H.C. and T.B. Howard. New Energy Technology--Some Facts and Assessments. Cambridge, Massachusetts: MIT Press, 1971.

Hovis, J.F. "Design and Operations of Reheat Furnaces--Energy Directed." Efficient Use of Fuels in the Metallurgical Industries. Chicago: Institute of Gas Technology, December 1974, pp. 479-98.

Hovis, James. "Energy Conservation--A Must." Iron and Steel Engineer, 51 (August 1974), 53-57.

How to Determine Economic Thickness of Insulation. New York: National Insulation Manufacturers Association, 1961.

Hunt, Robert, Frank Seabury, and Philip Valence. Energy Efficiency and Electric Motors. Cambridge, Massachusetts: Arthur D. Little, August 1976.

Hunter, W.L. "Continuous Charging and Preheating of Prereduced Iron Ore." Efficient Use of Fuels in the Metallurgical Industries. Chicago, Illinois: Institute of Gas Technology, December 1974, pp. 425-32.

Iammartino, N.R. "Cement's Changing Scene." Chemical Engineering, 81 (June 24, 1974), pp. 102-06.

Iannazzi, Fred D. "Comparison of Fiber and Energy Values of Waste Paper." Paper presented to TAPPI Pulping/Secondary Fibers Conference, Washington, D.C., November 6-9, 1977.

IGT Highlights. Supplement, April 11, 1977.

Institute of Gas Technology. _Efficient Use of Fuels in the Metal-_
  lurgical Industries. Chicago, Illinois: Institute of Gas Technology, December 1974.

International Nickel Company. "The Oxygen Flash Smelting Process of
  the International Nickel Company." _Trans. Can. Inst. Mining_
  Met., 58 (1955), p. 158.

Juusela, J., S. Harkki, and B. Anderson. "Outokumpu Flash Smelting
  and Its Energy Requirement." _Efficient Use of Fuels in the Metal-_
  lurgical Industries. Chicago, Illinois: Institute of Gas Technology, December 1974, pp. 555-76.

"Kaiser Acts to Effect 2% Energy Savings at Mead." _American Metal_
  Market, 82 (February 26, 1975), p. 8.

Kaiser, V.A. "Computer Control in the Cement Industry." _Proceedings_
  of the IEEE, 58 (January 1970), p. 76.

Kalmes, D.J. _A Review of the Lodding Paper Making Project_. Cambridge,
  Massachusetts: M/K Systems, January 1973.

Kaplan, S.I. _Energy Use and Distribution in the Pulp, Paper, and_
  Boardmaking Industries. Oak Ridge, Tennessee: Oak Ridge National Laboratory, August 1977.

Kaplan, S.I. _Energy Demand Patterns of Eleven Major Industries_. Oak
  Ridge, Tennessee: Oak Ridge National Laboratory, September 1974.

Kay, H. "Recuperators--Their Use and Abuse." _Iron and Steel Insti-_
  tute, 46 (June 1973), pp. 231-40.

Kellogg, H. "New Copper Extraction Processes." _Journal of Metals_,
  26 (August 1974), pp. 21-23.

Kellogg, H. "Prospects for the Pyrometallurgy of Copper." _Proceed-_
  ings of the Latin American Congress on Mining and Extractive
  Metallurgy. Santiago, Chile: n.p., August 1973.

Kemmetmueller, R. "Dry Coke Quenching--Proved, Profitable, Pollution-
  Free." _Iron and Steel Engineer_, 50 (October 1973), pp. 71-78.

Kirby, D.C., E.L. Singleton, and T.A. Sullivan. _Electrowinning_
  Aluminum from Aluminum Chlorides. Washington, D.C.: Government
  Printing Office, 1970.

Kirk, T.K. and J.M. Harkin. "Lignin Biodegradation and the Biocon-
  version of Wood." _AICHE Symposium Series_, 133 (1973).

Knight, Kenneth, George Kozmetsky, and Helen Baca. Industry Views of the Role of the Federal Government in Industrial Innovation. Austin, Texas: University of Texas, January 1976.

Kouvalis, A. Industrial Practice and Technology Review. The Pulp and Paper Industry. Argonne, Illinois: Argonne National Laboratory, April 1977.

Kreider, Kenneth and Michael McNeil, eds. Waste Heat Management Guidebook. Washington, D.C.: Government Printing Office, January 1977.

Kruesi, P.R. "Cymet Copper Reduction Process." Mining Congress Journal, 60 (September 1974), pp. 22-23.

Kuhn, M.C. "Anaconda's Arbiter Process for Copper." CIM Bulletin, 67 (February 1974), pp. 62-73.

Kunnecke, M. and B. Piscaer. "Choosing Insulation for Rotary Kilns." Rock Products, 76 (May 1973), pp. 138-148.

KVB, Inc. Industrial Boiler User's Manual, Vol. 2. Washington, D.C.: Federal Energy Administration, January 1977.

Larson, D.H. "A Projection of the Energy Demand by the Iron and Steel Industry." Paper presented at the Energy Conservation Through Effective Utilization Conference, Henniker, New Hampshire, August 19-24, 1973.

Larson, Dennis, Mark Fejer, and John Nesbitt. "Improving Energy Efficiency in Reheating, Forging, Annealing and Melting." Paper presented at Industrial Efficiency Seminar, AGA Marketing Conference, Atlanta, Georgia, March 5-7, 1975.

Larson, D.H. and M. Fejer. Improving Glass Melting Furnace Operations Using Improved Combustion Control Methods. Chicago: Institute of Gas Technology, May 1973.

Latest Energy Recovery Technology: New Coke Processing Method--CDQ." Tekko Kai Ho. Tokyo, July 11, 1974.

Laws, W.R. "Reducing Fuel Costs." Iron and Steel Institute, 47 (April 1974), pp. 105-13.

Lea, F.M. The Chemistry of Cement and Concrete. London: Edward Arnold, Ltd., 1956.

Leone, J.G. "New Melting Process Saves Fuel." The Glass Industry, 55 (February 1974), pp. 16-17.

Lewis, F.M. and R.B. Bhappu. "Economic Evaluation of Available Processes for Treating Oxide Copper Ores." International Journal
of Minerals Processing, 3 (1976), pp. 133-50.

Linsky, B. et al. "Dry Coke Quenching, Air Pollution and Energy: A
Status Report." Journal of Air Pollution Control Association,
25 (September 1975), 918-24.

Liddell, Donald. Handbook of Non Ferrous Metallurgy: Principles
and Processes. New York: McGraw Hill, 1945.

Listhuber, F.E. et al. "Steel and Plates of High Quality Made From
Continuously Cast Slabs." Iron and Steel Engineer, 51 (April
1974), pp. 92-98.

A.D. Little, Inc. Steel and the Environment: A Cost Impact Analysis.
Cambridge, Massachusetts: A.D. Little, May 1975.

Locklin, D.W. et al. Design Trends and Operating Problems in Combustion Modification of Industrial Boilers. Columbus, Ohio: Battelle Columbus Laboratory, April 1974.

Lorenzi, Otto De. Combustion Engineering. New York: Combustion
Engineering Company, 1947.

MacDonald, R.G. Pulp and Paper Manufacture. Vol. 1, The Pulping of
Wood. 2nd ed. New York: McGraw Hill, 1969.

MacLean, R.D. "Portland Cement, A Changing Industry." Rock Products,
76 (January 1973), pp. 92-95.

Mahan, W.M. and B.D. Daellenbach. "Thermal Energy Recovery by Basic
Oxygen Furnace Off Gas Preheating of Scrap." Efficient Use of
Fuels in the Metallurgical Industries. Chicago: Institute of
Gas Technology, December 1974, pp. 457-66.

Marcellini, R. and J. Geoffroy. "Development of a New Process to Preheat Coal Blends Used for Coking." AIME Proceedings of Ironmaking Conference. Vol. 31. New York: The Metallurgical Society
of AIME, 1972, pp. 166-73.

Margiloff, I.B. and R.F. Cascone. "The Scientific Design Fluid Bed
Cement Process." Paper presented at Rock Products Cement Industry Seminar, Chicago, December 8, 1975.

Marting, D. and R. Davis. "Coaltec System for Preheating and Pipeline
Charging of Coal to Coke Ovens." AIME Proceedings of Ironmaking
Conference. Vol. 31. New York: The Metallurgical Society of
AIME, 1972, pp. 174-84.

Matthys, William. "Retrofitting Heat Traps on Boilers." Power, 119
    (December 1975), pp. 21-22.

Maubon, A. "Technical and Economical Considerations of the IRSID/CAFL
    Oxygen Converter Gas Recovery System." Iron and Steel Engineer,
    50 (September 1973), pp. 87-97.

McChesney, H.R. "Recovery of Heat from Metal Processing Furnaces."
    Paper presented at Waste Heat Recovery Conference, Institute of
    Plant Engineers, London, September 25-26, 1974.

McChesney, H.R. "State of the Art--Regenerative and Recuperative Heat
    Recovery." Efficient Use of Fuels in the Metallurgical Indus-
    tries. Chicago: Institute of Gas Technology, December 1974,
    pp. 71-110.

McCord, George. Energy Conservation Trends in the Cement Industry.
    Society of Mining Engineers of AIME, Preprint No. 75-H-308, 1977.

McGowan, Terry. "Lighting Design Materials and Methods: All About
    Sources." Progressive Architecture, 9 (September 1973), pp.
    108-117.

Macrakis, Michael S., ed. Energy: Demand, Conservation, and Insti-
    tutional Problems. Cambridge, Massachusetts: MIT Press, 1974.

Meadows, Mark. "Speer Sees Growth Stymied." American Metal Market,
    81 (September 18, 1974), p. 1.

Mehta, P.K. "Trends in Technology of Cement Manufacture." Rock
    Products, 73 (March 1970), pp. 83-87.

Meyers, Peter. "The Potential for Energy Conservation in the Pulp and
    Paper Industry." Paper Trade Journal, 159 (February 17, 1975),
    pp. 68-71.

Miller, A. "Process for the Recovery of Copper from Oxide Copper-
    Bearing Ores by Leach, Liquid Ion Exchange, and Electrowinning
    at Ranchers Bluebird Mine, Miami, Arizona." The Design of Metal
    Producing Processes. Ed. R.M. Kibby. New York: American Insti-
    tute of Mining, Metallurgical, and Petroleum Engineers, 1969,
    pp. 337-67.

"MM Aluminum Upgrading Smelters in Northwest." American Metal Market,
    83 (March 30, 1976), p. 6.

Monroe, E.S. "Energy Conservation and Vacuum Pumps." Chemical Engi-
    neering Progress, 71 (October 1975), pp. 69-73.

Moore, C.A., et al. "Economic Potential of Kraft Product." TAPPI,
    59 (January 1976), pp. 117-120.

Myers, John. "Energy Conservation and Economic Growth: Are They Incompatible?" The Conference Board Record, 12 (February 1975), pp. 27-32.

Myers, John et al. Energy Consumption in Manufacturing. Cambridge, Massachusetts: Ballinger Publishing Company, 1974.

Nagano, T. and T. Suzuki. "Extractive Metallurgy of Copper." TMS-AIME, 1 (1976), pp. 439-57.

Nesbitt, J.D. "Increased Efficiency of Gas Utilization in Industrial Processing." Institute of Gas Technology, Chicago, Illinois. (Unpublished paper.)

"New Kawasaki Unit Uses Furnace Gas to Generate Power." American Metal Market, 81 (December 24, 1974), p. 3.

Niemela, T. and S.U. Harkki. "The Latest Development in Nickel Flash Smelting at the Harjavalta Smelter." Paper presented at the Joint Meeting MMIJ - AIME, Tokyo, Japan, May 1972.

"Non Ferrous Metal Data 1974." American Bureau of Metal Statistics, New York, 1975.

Nydick, S.E. et al. A Study of Inplant Electric Power Generation in the Chemical, Petroleum Refining and Pulp and Paper Industries. Washington, D.C.: Government Printing Office, June 1976.

Nydick, Sander and John Dunlay. Recommendations for Future Government Sponsored R & D in the Paper and Steel Industries. Waltham, Massachusetts: Thermo-Electron Corporation, August 1976.

Olson, Richard C. "Using Ceramic Fiber Refractories in Heat Treating Furnaces." Metal Progress, 103 (April 1973), pp. 85-91.

Osborne, M.J. "The APPA - TAPPI Whole Tree Utilization Committee." TAPPI, 57 (December 1974), pp. 5-7.

Pater, V. and J. Webster. "Methods of Charging Preheated Coal." Developments in Iron Making Practice. London: The Iron and Steel Institute, 1973, pp. 53-62.

Peacey, J. and W. Davenport. "Evaluation of Alternative Methods of Aluminum Production." Journal of Metals, 26 (July 1974), pp. 24-28.

Penberthy, Larry. "In Rebuttal: Factors Which Justify the Increased Cost of Electricity Over Natural Gas for Glass Making." The Glass Industry, 47 (June 1966), pp. 319-321.

Peray, K.E. and J.J. Waddell. The Rotary Cement Kiln. New York:
    Chemical Publishing Company, 1972.

Perry, George and James Monteaux. "Practical Analysis Can Cut Plant
    Energy Use, Trim Operating Costs." Oil and Gas Journal, 74
    (February 23, 1976), pp. 79-85.

Perry, John H. Chemical Engineer's Handbook. 14th ed. New York:
    McGraw Hill, 1969.

Persson, J.A. and D.G. Treilhard. "Electrothermic Smelting of Copper
    and Nickel Sulfides and Other Metal Bearing Constituents."
    Journal of Metals, 25 (January 1973), pp. 34-39.

Portland Cement Association. Concrete Information. Skokie, Illinois:
    Portland Cement Association, 1971.

Portland Cement Association. The Making of Portland Cement. Skokie,
    Illinois: Portland Cement Association, 1964.

Portland Cement Association. Energy Conservation in the Cement In-
    dustry. Skokie, Illinois: Portland Cement Association, January
    1975.

Preinvestment Data for the Aluminum Industry. Studies in Economics
    of Industry, No. 2. New York: United Nations, 1966.

Price, F.C. "Copper Technology on the Move." Chemical and Engineer-
    ing News, April 1973.

Proceedings of Symposium on Soviet Dry Quenching. Washington, D.C.:
    Sponsored by Patent Management, Inc. and VO Licensintorg, 1973.

Putnam, A.A. et al. Evaluation of National Boiler Inventory. Colum-
    bus, Ohio: Battelle Columbus Laboratory, October 1975.

Queneau, R.E. and R. Schuhmann. "The Q-S Oxygen Process." Journal
    of Metals, 26 (August 1974), pp. 14-16.

Rains, R.K. and R.H. Kadlec. "The Reduction of Al_2O_3 to Aluminum in
    a Plasma." Metallurgical Transactions, 1 (June 1970), pp. 1501-
    06.

"Rancher's Big Blast Shatters Copper Ore Body for In-Situ Leaching."
    Engineering and Mining Journal, 173 (April 1972), pp. 98-100.

Ray, Dixy Lee. The Nation's Energy Future. Washington, D.C.: Govern-
    ment Printing Office, December 1973.

R-ECON, A Method for Determining Economic Thickness of Add On Thermal
    Insulation. Mt. Kisco, New York: Thermal Insulation Manufac-
    turers Association, 1975.

Reding, J.T. and B.P. Shepard. Energy Consumption Fuel Utilization
    and Conservation in Industry. Washington, D.C.: Government
    Printing Office, August 1975.

Reding, J.T. and B.P. Shepard. Energy Consumption: Paper, Stone,
    Clay, Glass, Concrete, and Food Industries. Midland, Michigan:
    Dow Chemical Company, April 1975.

"Refining Process Handbook, 1974." Hydrocarbon Processing, 53 (Septem-
    ber 1974), pp. 106-214.

Resource Planning Associates. A Brief Analysis of the Impact of En-
    vironmental Laws on Energy Supply and Demand. Cambridge, Massa-
    chusetts: Resource Planning Associates, October 1974.

Resource Planning Associates. An Analysis of Constraints on Industrial
    Energy Conservation. Cambridge, Massachusetts: Resource Planning
    Associates, June 1974.

Resource Planning Associates. Co-Generation of Electricity and In-
    dustrial Steam. Cambridge, Massachusetts: Resource Planning
    Associates, January 1977.

Resource Planning Associates. Macro-Economic Impacts of Meeting Con-
    servation Targets in 10 Energy-Intensive Industries. Cambridge,
    Massachusetts: Resource Planning Associates, December 1976.

Reynolds, David P. "Remarks--41st Annual Meeting of the Aluminum
    Association," October 31, 1974. Release of the Reynolds Metal
    Company.

Rohrer, W.M. "Commercial Options in Waste Heat Recovery Equipment."
    Waste Heat Management Guidebook. Washington, D.C.: Government
    Printing Office, January 1977, pp. 141-54.

Rohrer, W.M. and K. Kreider. "Sources and Uses of Waste Heat." Waste
    Heat Management Guidebook. Washington, D.C.: Government Printing
    Office, January 1977, pp. 5-8.

Rosenberg, R.B. et al. "Energy Use Patterns in the Metallurgical
    Industries." Efficient Use of Fuels in the Metallurgical Indus-
    tries. Chicago, Illinois: Institute of Gas Technology, Decem-
    ber 1974, pp. 17-50.

Rosenkranz, R.D. Energy Consumption in Domestic Primary Copper Produc-
    tion. Washington, D.C.: Government Printing Office, 1976.

Ross, Marc and Robert Williams. Assessing the Potential for Energy
    Conservation. Albany, New York: The Institute for Public Policy
    Alternatives, July 1975.

Rydholm, S.A. Pulping Processes. New York: Interscience Publishers, 1965.

Sadler, A.M. Paper presented at AIChE Meeting, New York, November 30, 1967.

Sales Literature for Fuel Efficiency, Inc. Newark, New Jersey.

Sales Technotes from the Combustion Engineering Preheater Company. Wellsville, New York.

Salisbury, J.K. Kent's Mechanical Engineers Handbook. New York: John Wiley & Sons, 1950.

Schmitt, A.J. "Sheet Quality and Modern Press Section Arrangement." TAPPI, 56 (October 1973), pp. 56-59.

Schweiger, Bob. "Industrial Boilers--What's Happening Today." Power, 121 (February 1977), pp. s1-s24.

Shand, E.B. Glass Engineering Handbook. 2nd ed. New York: McGraw Hill, 1958.

Sheldon, Allen C. "Energy Use and Conservation in Aluminum Production." Energy Use and Conservation in the Metals Industry. New York: The Metallurgical Society of the American Institute of Mining, Metallurgical and Petroleum Engineers, 1975.

Shields, Carl. Boilers--Types, Characteristics and Functions. New York: McGraw Hill, 1961.

Shreve, R.N. Chemical Process Industries. 3rd ed. New York: McGraw Hill, 1967.

Silverthorne, P.N. "Power Factor Correction for Energy Conservation." ASHRAE Journal, DGW 13 (May 1975), pp. 28-32.

Sisson, William. "Nomogram Determines Loss of Steam and Fuel Due to Leaks." Power Engineering, 79 (September 1975), p. 55.

Sisson, William. "Nomogram Estimates Fuel Savings by Economizers or Air Preheaters." Power Engineering, 80 (April 1976), p. 97.

Slinn, R.J. Sources and Utilization of Energy in the U.S. Pulp and Paper Industry. New York: American Paper Institute, March 1973.

Snell, Foster D., Inc. Discussion Paper on Energy Conservation of the Copper Industry. Washington, D.C.: Federal Energy Administration, March 1975.

Snell, Foster D., Inc. Discussion Paper on Energy Conservation in the Glass Industry. Washington, D.C.: Federal Energy Administration, March 1975.

Sobotka and Company. Industrial Energy Study of the Petroleum Refining Industry. Washington, D.C.: Bureau of Mines, May 1974.

Sparrow, F.T. and T.F. Dougherty. Energy Conservation in the Iron and Steel Industry. Houston, Texas: University of Houston, April 1977.

Standen, A., ed. Kirk-Othmer Encyclopedia of Chemical Technology. 2nd ed. New York: John Wiley & Sons, 1963.

Stanford Research Institute. Patterns of Energy Consumption in the United States. Washington, D.C.: Government Printing Office, January 1972.

Stirling, R.D., J.C. Blessing and S.L. Fredericks. "Oxy-fuel Burners Streamline Cement Production." Rock Products, 76 (November 1973), pp. 44-47.

Stone, J.K. "Worldwide Roundup of Basic Oxygen Steelmaking." Iron and Steel Institute, 53 (April 1976), pp. 89-92.

Subramanian, K.N. and N.J. Themelis. "Copper Recovery by Flotation." Journal of Metals, 24 (April 1972), pp. 33-38.

Sugasawa, K. et al. Production of Formed Coke and Trials in Blast Furnace. Japan: Mitsubishi Research, October 1974.

Taylor, J.C. "Recent Trends in Copper Extraction. Efficient Use of Fuels in the Metallurgical Industries. Chicago: Institute of Gas Technology, December 1974, pp. 635-44.

Themelis, N.J. et al. "The Noranda Process." Journal of Metals, 24 (April 1972), pp. 25-32.

Thermo-Electron Corporation. Potential for Effective Use of Fuel in Industry. Cambridge, Massachusetts: Thermo-Electron Corporation, April 1974.

Thompson, R.E. et al. "A Study to Assess the Potential for Energy Conservation Through Improved Industrial Boiler Efficiency." Washington, D.C.: Government Printing Office, October 1976.

Tooley, Fay V., ed. Handbook of Glass Manufacture. New York: Books for Industry, 1974.

Toth, C. and A. Lippman. "The Quest for Aluminum." Mechanical Engineering, 95 (September 1973), pp. 24-28.

Treilhard, Donald. "Copper-State of the Art." Chemical Engineering, 80 (April 16, 1973), pp. p-z.

Trinks, W. and M.H. Mawhinney. Industrial Furnaces. 5th ed. New York: John Wiley, 1961.

Troxell, George, Harmer Davis, and Joe Kelly. Composition and Properties of Concrete. 2nd ed. New York: McGraw Hill, 1968.

Tsikarev, D.A. "Experimental Blast Furnace Smelting Using Molded Coke." Koks Khim, 4 (1974), pp. 58-59.

"Union Camp Adds Big Third Press to Board Machine." TAPPI, 57 (June 1974), p. 15.

United Nations Industrial Development Organization. Copper Production in Developing Countries. New York: United Nations, October 1970.

U.S. Bureau of Reclamation. Concrete Manual. 7th ed. Denver, Colorado: Bureau of Reclamation, 1963.

U.S. Department of Commerce. 1972 Census of Manufacturers: Papermills, SIC 2611. Washington, D.C.: Department of Commerce, February 1974.

U.S. Department of Commerce, Bureau of Domestic Commerce. 1975 Industrial Outlook. Washington, D.C.: Government Printing Office, 1974.

U.S. Department of Commerce. Energy Conservation Program Guide for Industry and Commerce. Washington, D.C.: Government Printing Office, 1974.

U.S. Department of Commerce. Census of Mineral Industries. Washington, D.C.: Government Printing Office,

U.S. Department of the Interior. Energy Perspectives 2. Washington, D.C.: Government Printing Office, June 1976.

U.S. Department of the Interior. Minerals Yearbooks. Washington, D.C.: Government Printing Office,

U.S. Department of the Interior. United States Energy Through the Year 2000. Washington, D.C.: Government Printing Office, 1974.

U.S. Senate. Industry Efforts in Energy Conservation. Washington, D.C.: Government Printing Office, 1974.

Veziroglu, T. Nejat, ed. Energy Conservation: A National Forum. Coral Gables, Florida: Clean Energy Research Institute, December 1975.

"Voluntary Industrial Energy Conservation." Progress Report 3. Washington, D.C.: Department of Commerce and Federal Energy Administration, April 1976.

Wagener, D. "Discussion." AIME Proceedings of 31st Ironmaking Conference. Vol. 31. New York: The Metallurgical Society of AIME, 1972, pp. 297-98.

Wahlgren, H. "Forest Residues--The Timely Bonanza." TAPPI, 57 (October 1974), p. 65.

Weddick, A.J. "The Noranda Continuous Smelting Process for Copper." Efficient Use of Fuels in the Metallurgical Industries. Chicago: Institute of Gas Technology, December 1974, pp. 645-60.

Williamson, Richard et al. Analysis of Energy Futures for the United States. Washington, D.C.: Energy Research and Development Administration, 1977.

Wilson, Carrol. Energy: Global Prospects 1985-2000. New York: McGraw Hill, 1977.

Wilson, W.G. "In Plant Generation: Profitable Today, More Profitable Tomorrow." TAPPI Manuscript, 1975.

Winger, John G. et al. Outlook for Energy in the United States to 1985. New York: The Chase Manhattan Bank, June 1972.

Woodcock, J.T. "Copper Waste Dump Leaching." Proceedings of Australian Institute of Mining and Metallurgy. Sydney: Australian Institute of Mining and Metallurgy, December 1967, pp. 47-66.

Woolf, P.L. "Improved Blast Furnace Operation." Efficient Use of Fuels in the Metallurgical Industries. Chicago: Institute of Gas Technology, December 1974, pp. 263-96.

Worner, H.K. and B.S. Andrews. "Integrated Smelting-Converting Slag Cleaning in a Single Furnace." Paper presented at TMS - AIME Annual Meeting, Dallas, Texas, February 1974.

Worthington, R.B. Autogenous Smelting of Copper Sulfide Concentrate. Washington, D.C.: Government Printing Office, 1973.

Yee, D.H. et al. "Chlorination Process for the Recovery of Copper from Chalcopyrite." SME Transactions, 254 (1973), 301-03.

INDEX